Accident Reconstruction 2002

SP-1666

All SAE papers, standards, and selected books are abstracted and indexed in the Global Mobility Database

Published by:
Society of Automotive Engineers, Inc.
400 Commonwealth Drive
Warrendale, PA 15096-0001
USA
Phone: (724) 776-4841
Fax: (724) 776-5760
March 2002

Permission to photocopy for internal or personal use of specific clients, is granted by SAE for libraries and other users registered with the Copyright Clearance Center (CCC), provided that the base fee per article is paid directly to CCC, 222 Rosewood Drive, Danvers, MA 01923. Special requests should be addressed to the SAE Publications Group. 0-7680-0934-0/02.

Any part of this publication authored solely by one or more U.S. Government employees in the course of their employment is considered to be in the public domain, and is not subject to this copyright.

No part of this publication may be reproduced in any form, in an electronic retrieval system or otherwise, without the prior written permission of the publisher.

ISBN 0-7680-0934-0
SAE/SP-02/1666
Library of Congress Catalog Card Number: N98-42939
Copyright © 2002 Society of Automotive Engineers, Inc.

Positions and opinions advanced in this paper are those of the author(s) and not necessarily those of SAE. The author is solely responsible for the content of the paper. A process is available by which the discussions will be printed with the paper if is is published in SAE Transactions. For permission to publish this paper in full or in part, contact the SAE Publications Group.

Persons wishing to submit papers to be considered for presentation or publication through SAE should send the manuscript or a 300 word abstract to: Secretary, Engineering Meetings Board, SAE.

Printed in USA

PREFACE

For the SAE 2002 World Congress, SAE has continued to provide a forum for the presentation of state-of-the-art, original research into advanced accident reconstruction methodologies. For the sessions on Accident Reconstruction, the organizers are pleased to present these original papers covering a wide range of topics relating to the field of vehicular crash analysis. Paper topics include: vehicle structural modeling, crush energy analysis, computer modeling, low speed collision analysis, pedestrian impact analysis, crash data recorded analysis, light filament analysis and the analysis of accident reconstruction uncertainty.

The selected papers form this SAE Special Publication, Accident Reconstruction 2002 (SP-1666).

A review committee made the paper selections, based on abstracts submitted by the authors. Reviewers included members of industry, academia, manufacturing and government. The invited papers were submitted to reviewers for comment on content and scope, organization and presentation, quality of data and technique validity, conclusion soundness, and editorial quality. Authors were encouraged to work closely with the reviewers in order to optimize the ultimate quality of the paper. At the conclusion of the review process, if desired, the reviewers have the opportunity to provide a written discussion that follows the paper in this published volume. The reader is encouraged to consult the written discussions in order to gain a deeper insight into the paper.

The organizers thank the authors for their contribution towards improving the state-of-the-art. As an understanding of the science of vehicle crashes improves, countermeasures can be developed in order to mitigate the significant damage done from injury-causing motor vehicle crashes.

The organizers also thank the members of the review committee for their valuable contribution of time and expertise. The members of the 2002 review committee are:

Greg Anderson
Scalia Safety Engineering

Christopher D. Armstrong
KEVA Engineering

Mark Arndt
Transportation Safety Technologies

Gary Bahling
General Motors Corporation

Mark Bailey
Macinnis Engineering Associates

Ken Baker
Ken Baker & Company

Cleve Bare
MP Holcomb Engineering

Wade Bartlett
Mechanical Forensics Engineering Services

Fawzi Bayan
FTI Consulting

Tom Bohan
Medical & Technical Consultants

Kris Bolte
NTSB

Matt Brach
Exponent Failure Analysis

Bill Cliff
Macinnis Engineering Associates

Jeff Croteau
Exponent

Terry Day
Engineering Dynamics Corporation

Eric Deyerl
Quan, Smith and Associates

C.R. Eddie
C.R. Eddie Engineering, Inc.

Jerry Eubanks
ACCA

Richard Fay
Fay Engineering

Al Fonda
Fonda Engineering

Thomas Fuggar
Accident Research & Biomechanics, Inc.

Geoff Germane
Germane Engineering

Wes Grimes
Collision Engineering Associates

Ron Heuser
Engineering Accident Analysis

George Hicks
Ingenium Engineering Services

Phil Hight
Accident Research and Analysis

Michael Holcomb
M.P. Holcomb Engineering

Stein E. Husher
KEVA Engineering

John Kerkhoff
KEVA Engineering

Dave King
Macinnis Engineering Associates

Rolly Kinney
Kinney Engineering

Ernest Klein
Collision Research and Analysis

Richard Lambourne
TRL

Mickey Marine
Thomas Engineering, Inc

Brian McHenry
McHenry Software

Jim Neptune
Neptune Engineering

Gus Nystrom
Amador Newtonian Engineering

Michael D. Pepe
Wolf Technical Services

Thomas Perl
Collision Safety Engineering

Mark Pryor
Collision Research and Analysis

Ric Robinette
FAY Engineering

Donald Rudny
Rudny & Sallmann Engineering

Karl Shuman
KEVA Engineering

Gunter Sigmund
Macinnis Engineering Associates

Anthony Stein
Safety Research Associates

John Steiner
Roger Clark Associates

Greg Stephens
Collision Research and Analysis

Amrit Toor
Intech Engineering Ltd.

Alan Thebert

Tom Vadnais
Vadnais & Wood

Michael S. Varat
KEVA Engineering

Matt Weber
Design Research Engineering

Jud Welcher
Biomechanical Research and Testing

Steve Werner
Exponent

Dennis Wood
Dennis Wood and Associates

Ron Wooley
Wolley Engineering Research Corporation

Finally, the organizers thank the SAE for providing a forum such as this, both by holding the annual Congress meeting and by placing the papers in a special publication so that others may benefit from the work presented by these authors and by the interaction with other sessions attendees.

Raymond M. Brach
University of Notre Dame

Michael S. Varat and Stein E. Husher
KEVA Engineering

Session Organizers

TABLE OF CONTENTS

2002-01-0535 **Using Event Data Recorders in Collision Reconstruction** 1
    Richard Fay, Ric Robinette, Darrell Deering and John Scott
        Fay Engineering Corp.

2002-01-0536 **Methods of Occupant Kinematics Analysis in Automobile Crashes** 15
    Jon E. Bready, Ronald P. Nordhagen,
    Thomas R. Perl and Michael B. James
        Collision Safety Engineering, L.C.

2002-01-0540 **Low Speed Collinear Impact Severity: A Comparison Between Full Scale Testing and Analytical Prediction Tools with Restitution Analysis** 23
    A.L. Cipriani, F. P. Bayan, M.L. Woodhouse, A.D. Cornetto,
    A.P. Dalton, C.B. Tanner, and and T.A. Timbario
        FTI/SEA Consulting
    E. S. Deyerl
        Quan, Smith & Associates

2002-01-0542 **Identification and Interpretation of Directional Indicators in Contact-Damaged Paint Films - Applications in Motor Vehicle Accident Reconstruction** 39
    Neil Clark and Roger E. Clark
        Roger Clark Associates

2002-01-0546 **Evaluating the Uncertainty in Various Measurement Tasks Common to Accident Reconstruction** 57
    Wade Bartlett
        Mechanical Forensics Engineering Services, LLC
    William Wright and Oren Masory
        Florida Atlantic Univ.
    Raymond Brach
        Notre Dame Univ.
    Al Baxter
        Suncoast Reconstruction
    Bruno Schmidt
        Southwest Missouri State Univ.
    Frank Navin
        University of British Columbia
    Terry Stanard
        Klein Associates

2002-01-0548 **Accelerations and Shock Load Characteristics of Tail Lamps From Full-Scale Automotive Rear Impact Collisions** 71
    Lindsay "Dutch" Johnson and Jeffrey Croteau
        Exponent Failure Analysis Associates
    Joseph Golliher
        Origin Engineering, LLC

2002-01-0549 **Possible Errors Occurring During Accident Reconstruction Based on Car "Black Box" Records** 81
    Marek Guzek and Zbigniew Lozia
        Warsaw University of Technology

2002-01-0550 **Revision and Validation of Vehicle/Pedestrian Collision Analysis Method** 101
    Amrit Toor, Michael Araszewski, Ravinder Johal, Robert Overgaard and Andrew Happer
        INTECH Engineering Ltd.

2002-01-0551 **Seventeen Motorcycle Crash Tests into Vehicles and a Barrier** 113
    Kelley S. Adamson
        Unified Building Sciences & Engineering, Inc.
    Peter Alexander
        Raymond P. Smith & Associates
    Ed L. Robinson and Gary M. Johnson
        Robinson & Associates, LLC
    Claude I. Burkhead, III
        Advanced Engineering Resources
    John McManus
        Consulting Engineering Services
    Gregory C. Anderson
        Scalia Safety Engineering
    Ralph Aronberg
        Aronberg & Associates Consulting Engineers, Inc.
    J. Rolly Kinney
        Kinney Engineering, Inc.
    David W. Sallmann
        Rudny & Sallmann Engineering Ltd.

2002-01-0552 **Narrow Object Impact Analysis and Comparison with Flat Barrier Impacts** 141
    Alan F. Asay, Dagmar B. Jewkes and Ronald L. Woolley
        Woolley Engineering Research Corp.

2002-01-0553 **Single Vehicle Wet Road Loss of Control; Effects of Tire Tread Depth and Placement** 161
    William Blythe
        Consulting Engineer
    Terry D. Day
        Engineering Dynamics Corp.

2002-01-0554	**Large School Bus Side Impact Stiffness Factors** **187**	

Kristin Bolte, Shane Lack and Larry Jackson
   National Transportation Safety Board

2002-01-0556	**Crush Energy Considerations in Override/Underride Impacts** **195**	

Micky C. Marine, Jeffrey L. Wirth and Terry M. Thomas
   Thomas Engineering, Inc.

2002-01-0557	**Curb Impacts – A Continuing Study In Energy Loss and Occupant Kinematics** **211**	

Steven E. Meyer, Joshua Hayden, Brian Herbst,
Davis Hock and Stephen Forrest
   Safety Analysis & Forensic Engineering (SAFE)

2002-01-0558	**The Use of Single Moving Vehicle Testing to Duplicate the Dynamic Vehicle Response from Impacts Between Two Moving Vehicles** **219**	

C. Brian Tanner, John F. Wiechel and Philip H. Cheng
   S.E.A., Inc.
Dennis A. Guenther
   Ohio State Univ.
Richard Fay
   Fay Engineering

2002-01-0559	**A Simulation Model for Vehicle Braking Systems Fitted with ABS** **235**	

Terry D. Day and Sydney G. Roberts
   Engineering Dynamics Corp.

2002-01-0560	**A Demographic Analysis and Reconstruction of Selected Cases from the Pedestrian Crash Data Study** **255**	

Jason A. Stammen
   Vehicle Research and Test Center, National Highway Traffic Safety
   Administration
Sung-won (Brian) Ko and Dennis A. Guenther
   The Ohio State Univ.
Gary Heydinger
   FTI/SEA, Inc.

2002-01-1305	**Re-Analysis of the RICSAC Car Crash Accelerometer Data** **271**	

Raymond M. Brach
   University of Notre Dame
Russell A. Smith
   U. S. Naval Academy

2002-01-0535

Using Event Data Recorders in Collision Reconstruction

Richard Fay, Ric Robinette, Darrell Deering and John Scott
Fay Engineering Corp.

Copyright © 2002 Society of Automotive Engineers, Inc.

ABSTRACT

This paper will give an overview of the information available from different types of vehicle data recorders, and ways the data can be used in the analysis of vehicle collisions. Reference will be made to current research relating to Event Data Recorders (EDR's). The methods used in downloading the data will be presented. Preservation of this evidence and admittance into court will be discussed. Traditional uses of this data in the determination of collision severity, seat belt use or non-use, and supplemental restraint system functionality will be discussed. Examples will be provided of additional, less traditional, uses involving the correlation of data recorded by EDR's during collision events with the results of reconstructions based on accident scene and vehicle crush data. Topics will include Delta-V, pre-crash vehicle motions and operation, driver perception/reaction times, driver actions, and vehicle acceleration/deceleration. Analysis of the pre-impact motions of vehicles with antilock brakes, impacts involving vehicles with prior damage, and chain-reaction accidents will be discussed.

INTRODUCTION

Historical Overview[1]

The history of research efforts relating to event data recorders can be traced back to the 1970's. At that time, NHTSA was using analog Disc Recorders to process and record crash parameters such as deceleration-time histories. Out of 1000 Disc Recorder-equipped vehicles, 23 crashes were analyzed with Delta-V's up to 20 mph.

Modern Event Data Recorders used in passenger cars and light trucks have existed since about 1990. Systems made today are still considered "early generation" systems. Significant enhancements have been added in the last decade or so, but widespread public awareness of these devices and accessibility to the data contained within has only started to come about in the last year or two, even in the accident reconstruction community.

In 1997, the National Transportation Safety Board (NTSB) recommended that the National Highway Traffic Safety Administration (NHTSA) should pursue vehicle crash information gathering using event data recorders. The National Aeronautics and Space Administration (NASA) Jet Propulsion Laboratory (JPL) also recommended that NHTSA study the feasibility of installing and obtaining crash data for safety analyses from crash recorders on vehicles. In 1998, pursuant to these recommendations, NHTSA created a Working Group of representatives from Government, Universities, the Original Equipment Manufacturer Industry, Aftermarket Industry, and the public. A formal report of their findings is expected soon.

The NHTSA Working Group, SAE, and ISO Working Group 7 "Accident Investigation", all appear to be studying currently available and near-term technology. The Japan Drive Recorder Committee, which includes representatives from the Ministry of Transportation, University of Tokyo Scholars, the Japan Automobile Research Institute, the Japan Automobile Manufacturer's Association, and Japan Auto Parts Industries Association is apparently researching two systems, the Accident Data Recorder (ADR) and the Driving Monitoring Recorder (DMR).

In 1999 and 2000 the NTSB held two international symposia related to transportation recorders. Topics covered technical, legal and privacy issues.

In addition to the upcoming report, NHTSA's Working Group has gathered a substantial amount of information that is publicly available through the Department of Transportation's Docket Management System on the Internet. It can be accessed at http://dms.dot.gov/. The Docket ID is 5218.

EVENT DATA RECORDER TYPES

There are almost as many varieties of Event Data Recorders as there are OEM's and after market suppliers. Not all manufacturers employ EDR technology and, those who do employ it do not use a common format for storing data.

There is a possibility that in the long term a common format will be developed, but regulatory activity may be required for this to occur. This could have both positive and negative effects. The current absence of regulatory activity appears to have encouraged many different approaches and new developments.

Today, most crash sensing is done via solid state accelerometers. Many current EDR systems use trigger logic associated with the airbag deployment system, but other triggers are also possible, and some devices record continuously to a looped memory or tape.

Some EDR's record signals from the vehicle's electrical systems and some incorporate additional video and audio signals. The authors have real world accident event experience with both of these types of Event Recorders and believe each has enormous potential for collecting pre-crash, crash, and post-crash data useful in understanding the circumstances of a crash.

Video systems, such as DriveCam, shown in Figure 1, MACBox, and BUS-WATCH, shown in Figures 2 and 3, can present a clear picture of the event from the camera view(s). BUS-WATCH incorporates features such as a notation bar documenting vehicle speed, brake application times, turn indicator actuation, temperature, and battery voltage on a time indexed scale. These need to be properly calibrated in order to provide accurate information. There are undoubtedly other systems like these in the marketplace.

Figure 1. DriveCam (Source: DriveCam Web Page)

Figure 2. BUS-WATCH camera (Source: BUS-WATCH Web Page)

Figure 3. BUS-WATCH VCR (Source: BUS-WATCH Web Page)

The video/audio systems can also record what was said in the time before, during, and after a crash event. The non-video/non-audio systems such as those used in passenger vehicles are ideal candidates for generating tabular and charted data relative to a collision event, which can be studied as snapshots in time. These have the advantage of not having to try to capture data from multiple images on a moving screen.

Passenger Vehicle EDR's

GM is in the forefront of the activity of installing EDR's in passenger vehicles and making the data publicly available. GM has installed EDR's in all their airbag-equipped vehicles since 1990. In May 1998, GM Added Pre-crash event capability. In 1999, GM installed their new generation of EDR, which captures pre-crash information including vehicle speed, brake application, engine RPM, and throttle for 5 seconds prior to the crash. Figure 4 shows a typical SDM.

Ford Motor Company started installing a crash recorder in one model in May of 1997. In May 1998, other models were equipped with recording devices. Ford equipped almost all 1999 model year vehicles with event data recorders.[1]

Figure 4. DELPHI SDM (Source: DELPHI Web Page)

Diesel Engine Control Modules

Another classification of data recorders is engine control modules. SAE 2000-01-0466, "Utilizing Electronic Control Module Data in Accident Reconstruction" by Goebelbecker and Ferrone, provides useful information on this topic.

Large diesel engines such as those made by Detroit Diesel, Cummins, Caterpillar, Mack, and others can be used in various over-the-road trucks such as Freightliner, Kenworth, Mack, Peterbilt, and Volvo. These trucks can be either Class 7 (26K to 33K lb.) or Class 8 (33K lb. and over). Engine data standardization is addressed in SAE J1587.

Handheld computers can be used to download and print out the data contained within the Electronic Control Modules (ECM's). Other manufacturer's engines can be read with a competitor's computer, but the data might not print out.

Typical data available for accident reconstruction and failure analysis is illustrated in the following table:

Vehicle Speed and Highest Speed Reached During a Trip	Engine Speed	Clutch Engagement
Gear Selection	Trip Distance	Cruise Control ON/OFF
Brake Activation	Monitoring of Hard Braking	Throttle Position
Governed Speed	Oil Pressure	Air Pressure
Speeding Triggers (Example 66 mph)	Engine Coolant and Transmission Temperature	Location - Optional GPS Feature

Figure 5. Typical Data Elements for Collision Reconstruction and Failure Analysis

Depending on the module, the duration of stored data can vary with changes to an adjustable time parameter. There can be up to two minutes of data surrounding the triggering event (a sudden deceleration). The event trigger sensitivity can also be adjusted (programmed) from 0 mph/sec up to 15 mph/sec. Multiple panic stop events may be recorded relative to a real-time clock. Some modules can also generate daily and monthly reports.

Custom hardware and software applications permit downloading and detailed viewing of the various reports. As an example, Detroit Diesel Electronic Control (DDEC) Reports software will allow the user to view the following reports:

Trip Activity	Vehicle Speed/RPM	Over speed/ Over rev
Engine Load/ RPM	Vehicle Configuration	Periodic Maintenance
Hard Brake Incidents	Last Stop Record	DDEC Diagnostics
Profile	Monthly Activity	Daily Engine Usage/Life to Date

Figure 6. Available On Highway Reports for Extracted DDEC Data

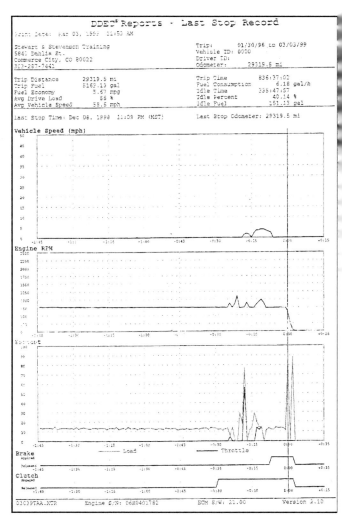

Figure 7. Sample Graphical Output for DDEC Reports Last Stop Record

Time	Vehicle Speed (mph)	Engine Speed (rpm)	Brake	Clutch	Engine Load (%)	Throttle (%)	Cruise	Diagnostic Code
-1:00	74.0	1613	No	No	66.50	70.00	No	No
-0:59	74.5	1612	No	No	66.00	70.00	No	No
-0:58	74.5	1616	No	No	65.50	69.60	No	No
-0:57	74.5	1612	No	No	71.50	73.20	No	No
-0:56	74.5	1619	No	No	73.00	74.00	No	No
-0:55	74.5	1615	No	No	72.00	72.80	No	No
-0:54	74.5	1615	No	No	74.00	74.40	No	No
-0:53	74.5	1618	No	No	75.50	75.60	No	No
-0:52	74.5	1617	No	No	81.00	79.60	No	No

Figure 8. Sample Graphical Output for DDEC Reports Hard Brake (Partial Output)

Automated Collision Notification

Non-OEM's generally consider Automated Collision Notification systems and Event Data Recorders as related technologies. OEM's appear to generally distinguish them as separate technologies.[1] After market recording devices are available which can analyze the vehicle's deceleration to determine if the vehicle has been in a crash and summon help via cell phone technology depending on severity. The General Motors "On Star" and Ford "Rescue" systems are two examples of OEM systems.

Figure 9. On-Star Automated Collision Notification (Source: General Motors Corporation Web Page)

METHODS FOR DOWNLOADING

There are several processes by which the crash metrics can be extracted from an event data recorder. Currently, special tools are needed to retrieve the data. One such example is the GM Tech1 Tool shown in Figure 10. In the future, a self-read feature may be incorporated in vehicles. This would allow anyone access to the data.

Event data recorders may contain information that is considered proprietary, so the capability to download data from some devices is only realistically available to the manufacturer.

Figure 10. General Motors Corporation Tech1 Tool

Figure 11. Printout from Tech-1 Tool

As an example, Ford has a current engineering tool not available to the public. Configuration of the tool is sensitive to the manufacturer of the airbag sensor but Ford envisions a common tool for the future.[1]

A common tool can be provided by the OEM or it can be outsourced. An example of an outsourced common tool is the Crash Data Retrieval Tool manufactured by Vetronix Corporation. This tool was first offered for public sale in 2000 for about $2,500. This does not include the cost of the required laptop computer. Vetronix Corporation has indicated they have reached an agreement with Ford so that their tool will be able to retrieve the data from Ford airbag modules.[2] It should be noted that the CDR does not translate all of the Hexidecimal data contained in the EDR.[3]

Figure 12. Vetronix Corporation's Crash Data Retrieval Tool

Vetronix also offers training for its tool and has also presented the tool at various technical sessions.

The special tools referred to above consist of hardware and software. The person downloading the data needs an external reader. The raw data is encoded, so an interface is needed to decode the data as well as to deduce its validity. This is currently done with an enabled PC, 486 or better, running Windows 95 or higher. The Crash Data Retrieval Tool is sold as a kit with an interface box, cables, an associated power supply and software.

On General Motors' vehicles, event data can be accessed through the Diagnostic Link Connector (DLC) or read directly from the Sensing and Diagnostic Module (SDM) by way of this interface using the computer's serial data port.

TYPES OF DATA AVAILABLE (ELEMENTS)

The data available vary by manufacturers and models. The different manufacturer's systems are mutually incompatible. Each type of recorder stores different data elements.

What is recorded is influenced by technology challenges, supplier availability, legal issues, consumer acceptance, and cost implications among other issues. Currently, OEM's determine what data is collected, how much data is needed, the data quality, data rates, and bandwidths. The manufacturer also sets prioritization of included elements. Usefulness, ease of implementation and reliability are additional considerations.[1]

The following tables show data typical of that recorded by Ford modules.

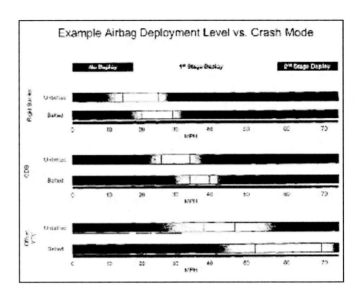

Figure 13. Mock Ford Data

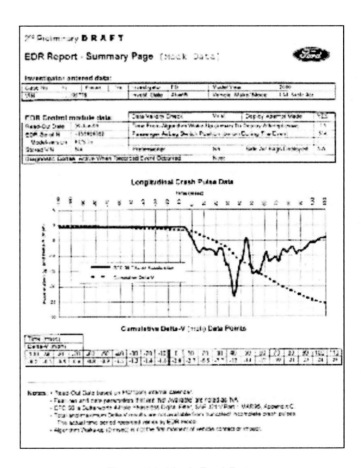

Figure 14. Mock Ford Data

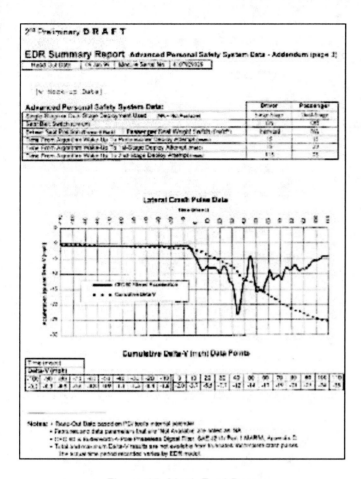

Figure 15. Mock Ford Data

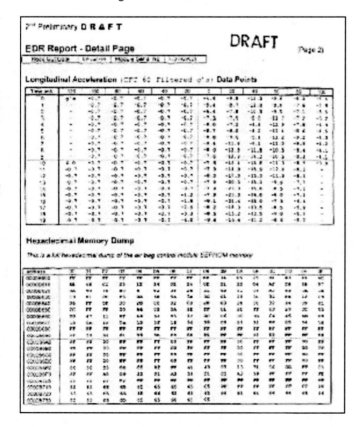

Figure 16. Mock Ford Data

Figure 17 details a general list of data elements, which may be currently available to manufacturers or available in the future. These elements include pre-crash, crash, and post-crash data useful to determination of causation, evaluation of injury potential and reconstruction. The development of a minimum recommended data set would be helpful to the reconstruction community.

Lateral Impact Delta-V	SIR Warning Lamp Status
Longitudinal Impact Delta-V	Seatbelt Status by Seating Location and Number of Occupants by Location
PDOF	Seatbelt Pretensioner Status
Longitudinal and Lateral Acceleration Vs. Time	Airbag Inflation Time
Yaw, Pitch and Roll Measures	Load Distribution
Cell Phone Use Measurements	Rear Impact Crash Severity Descriptor
Pre-crash Driver Inputs Including Braking, Steering and Throttle.	Suspension and Suspension Pulse History, Active Suspension
ABS, Traction Control, Stability Control Information	Crash Time
Wheel Rotation	Voltage/ Lamp Illumination Status
Pavement Friction	Crash Location
Rollover Sensor to Determine Tripped and Un-tripped Rollovers.	Vehicle ID and Equipment Status

Figure 17. Data Elements

COLLISION ANALYSIS

Event Data Recorders improve the accident analysis process. "Event Recorders not only simplify the reconstruction process, but they also increase the accuracy of the reconstruction resulting in more detailed conclusions..."[4] The data obtained can facilitate more detailed reconstruction and analysis than previously possible, and can also reduce the time required. The data has also been used to reconcile conflicting reports from witnesses to a collision event.

The "black box" data is similar to eyewitness testimony except that it is generally more reliable. Some uses of information in crash reconstruction include occupant seatbelt use, g-force and Delta-V comparison to injury risk, vehicle crashworthiness studies, crash pulse shape, and analysis of vehicle and driver performance.

EDR's give the reconstructionist an opportunity to look at data that was not previously available. They do not replace collision reconstruction. However, they are an invaluable tool in most cases, provided that the data is properly interpreted.

The Haddon Matrix[5] has been used to show information available with and without EDR's. The following matrix shows the type of data that can be collected from a crash without on-board data recording.

INFORMATION AVAILABLE WITHOUT EDR's

	Human	Vehicle	Environment
Pre-Crash		Skid Marks	
Crash		Calculate Delta-V	
Post-Crash	Injury	Collision Damage	Environment After Collision

Figure 18a. Haddon Matrix

Without EDRs, an engineer is left to reconstruct a motor vehicle collision based on traditional techniques, physical evidence, and the implementation of mathematical models based on the laws of physics. The presence of EDR data adds more physical evidence that increases the credibility of a reconstruction. The following matrix shows the type of data that can be collected from a crash with on-board data recording.

ADDITIONAL INFORMATION AVAILABLE WITH EDR's

	Human	Vehicle	Environment
Pre-Crash	Seatbelt Use, Steering, Braking	Speed, ABS, Other Controls	Conditions During Crash
Crash	Airbag data, Pretensioners	Crash Pulse, Measured Delta-V, Yaw, Inflation Time	Location
Post-Crash	ACN (Automatic Collision Notification)	ACN	ACN

Figure 18b. Haddon Matrix

The following graphs (Figures 19, 20, and 21) are representative of the type of data that can be obtained from the Crash Data Retrieval Tool manufactured by Vetronix Corporation.[2] The data can show vehicle speed, throttle position, brake application, engine RPM, the collision pulse, seat belt use and other information. Figure 21 is a tabular listing of numerical values for these parameters.

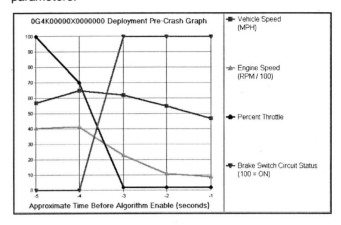

Figure 19. Deployment Pre-Crash Graph from Vetronix Corporation's Crash Data Retrieval Tool

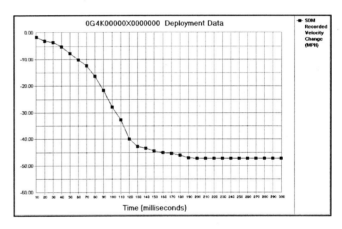

Figure 20. Sensing and Diagnostic Module (SDM) Recorded Velocity Change Graph from Vetronix Corporation's Crash Data Retrieval Tool.

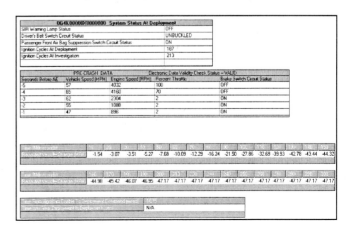

Figure 21. Tabular Output from Vetronix Corporation's Crash Data Retrieval Tool

Seatbelt Use or Nonuse

Traditionally, seat belt usage has been determined by the presence or absence of loading indicators on the webbing, or other indicators of strain or abrasion on the attachment and anchorage hardware, such as shown in Figure 22.

Figure 22. Loading Indicator on Seatbelt Webbing

Seatbelt usage information from EDR's can be helpful when the studied collision involves an airbag equipped vehicle and traditional indicators of seatbelt usage may be absent because the occupant may not travel far enough to significantly load the belt, as depicted in Figure 23.

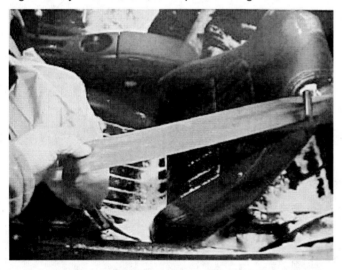

Figure 23. Absence of Load Indicator on Seatbelt

Use or non-use of the driver's seatbelt is reported via the Vetronix Crash Data Retrieval Tool as shown in Figure 24.

Figure 24. Driver's Belt Switch Circuit Status "Unbuckled"

5-Car Chain Reaction Collision

Another area where this data has proven useful is in the analysis of multi-car impacts. The following graph was used in the analysis of a 5-car chain reaction collision.

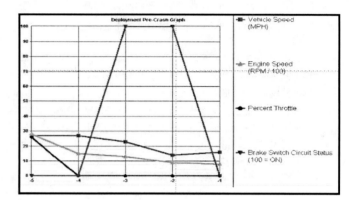

Figure 25. Deployment Pre-Crash Graph for Vehicle Number 4 in a 5-Car Chain Reaction Collision

In this impact sequence, the first three vehicles were reportedly stopped and there were conflicting accounts on whether the fourth or fifth vehicle initiated the collision event.

The above data comes from the fourth vehicle. About 3 to 4 seconds prior to airbag deployment, the driver of this vehicle got off of the throttle and applied the brakes. The engine RPM dropped off during that time as well, and continued to decrease. About 2 seconds prior to deployment, the vehicle forward speed increased even though the engine RPM was continuing to reduce and the vehicle was traveling uphill. Coincidentally, the driver's foot came off the brake at about that time.

The airbag ultimately deployed in the frontal collision with the rear of vehicle 3. Consistent with eyewitness testimony, and the damage to the vehicles, it was evident that the collision initiating impact was between the rear of the fourth vehicle and the front of the fifth vehicle. The impact did not occur as described by the driver of the fifth vehicle, who indicated that the driver of the fourth vehicle initially impacted the three stopped vehicles.

Braking Onset

EDR's can tell a crash reconstructionist something about vehicle deceleration and vehicle speed prior to a deployment or near-deployment event. This is especially helpful in cases where no marks can be found on the road, such as in cases involving antilock braking, or braking without skidding. It is also helpful in cases involving a lack of documentation of skid marks, or scenes where skid marks are destroyed by the elements or other traffic. In some instances the pre-crash data shows a Speed-Time relationship for several seconds. The slope of the Speed-Time curve can provide an indicator of the deceleration rate the vehicle experienced during that time.

This could be useful in quantifying the frictional characteristics of the roadway or the braking efficiency of the vehicle.

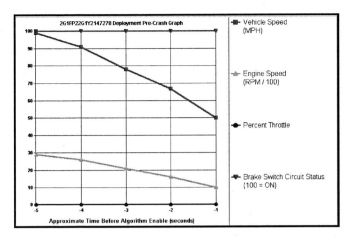

Figure 26. Speed-Time Curve Shown in Deployment Pre-Crash Graph

The average vehicle deceleration documented in the above graph appears to have been about 0.56 g during the time period shown.

Data Limitations

The proprietary design of event recorders is something ultimately chosen by the manufacturers. As a result, different acquisition rates, filtering, and output intervals are used. Only part of the Hexidecimal data contained in the EDRs is publicly obtainable for translation.

The collection of data for the previous graph starts at algorithm enable, or t = 0 seconds (t = 0 seconds is not shown). Algorithm enable occurs in a very short time after first contact, but prior to airbag deployment. Additionally, the data in the previous graph only goes back 5 seconds before algorithm enable. There is no record of the vehicle status before this. There is no positive time value for any of the data elements, namely, speed, engine rpm, throttle position, or brake switch circuit status.

Also, for the Vetronix Corporation Crash Data Retrieval Tool, there are only 5 discrete data points shown for each of the data elements. These points have been interconnected for ease of visualization, but the straight lines connecting the points are only an approximation of the actual curves.

Using the graph from the previous 5 - Car Chain Reaction example (Figure 25), one should note that the brake switch was in the off position 4 seconds prior to deployment and in the on position 3 seconds prior to deployment. The actual brake switch activation could have occurred anywhere in between -3.9 seconds to -3.0 seconds. In some cases, a more precise determination could be made from analysis of roadway evidence. It should also be noted that the data points for each data element could be out of phase with one another in time. Another limitation is that the recorder can store at most two event files.

In many instances, the EDR speed data is only documenting the longitudinal components of speed and Delta-V. For events such as a side impact, the EDR data must be viewed accordingly.

Pedestrian Impacts

Depending on vehicle and pedestrian weights and the collision geometry, these types of impacts might be sufficient to generate a near-deployment file. However, the crash sensing algorithm used in GM vehicles is not enabled below about 2g. When this criteria has been met, the recorded data has been helpful in the analysis of vehicle speed and driver reaction times, in addition to traditional methods of analysis.

Delta-V Determination

A vehicle equipped with an EDR is an instrumented vehicle capable of recording crash data. EDR data has been correlated with Federal Motor Vehicle Safety Standard (FMVSS) 208 and New Car Assessment Program (NCAP) tests and has been shown to be a reliable indicator of crash Delta-V.[1,5]

The availability of longitudinal Delta-V is extremely useful to the reconstructionist. Since this data is uniaxial and unidirectional, it needs to be used in conjunction with other data and analysis to reconstruct an accident. For instance, it has proven invaluable in the validation of accident reconstructions using the CRASH3[6] algorithm and derivatives, and conservation of momentum based analyses employing automated calculations. Conversely, a reconstruction that does not correlate with the electronically recorded Delta-V will be suspect.

Many programs rely on stiffness parameters derived from barrier impact tests and/or scene data in order to compute Delta-V. If appropriate care is not taken, the programs can yield inaccurate results for real-world impacts that may not resemble the barrier tests. On-board data recorders can provide accurate measurements, which can be correlated to Delta-V.

Comparing Reconstructions to Recorded Data

In the example below, EDR data for the 1999 Pontiac Grand AM was downloaded via a Tech1 Tool, see Figure 27, and interpreted by General Motors Corporation personnel. The interpretation of the recovered data was sent in a letter from General Motors. The letter contained External History Fault Codes (airbag deployment and removal of the SDM from the vehicle), Internal History Fault Codes (none), Ignition cycle count, Fault Codes and Cleared Data (normal part of new vehicle inspection process). The letter also provided information on the Near

Deployment Data File (none created), Warning Lamp Statistics Buffer (no warning lamp on), ADS (ADS not present), and One Deployment Event Data File.

The letter read, "This data indicated that there was no warning lamp on when the algorithm enabled. The warning lamp had been off for 65 ignition cycles (since the last time the codes were cleared). Time from Algorithm Enable to Deployment 32.5 milliseconds. Delta-V Data 6.14 mph. (This is the longitudinal component of the side impact.) All of the remaining data normally stored by the SDM was lost when power was interrupted. The near deployment event data files were also lost, but this is insignificant since there was no near deployment event recorded."

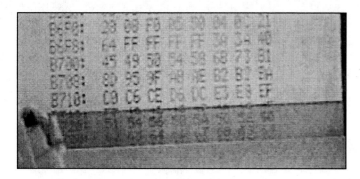

Figure 27. General Motors' Download Using Tech1 Tool

Figures 28 and 29 show vehicle crash analysis corresponding to the GM Tech1 crash data. Note that the recorded 6.14 mph longitudinal Delta-V is significantly different than the 35 to 39 mph total Delta-V.

ACCIDENT HISTORY

IMPACT VELOCITY (Basis: Scene Data and Linear Momentum)

	Total (mph)	Forward (mph)	Lateral (mph)	Sideslip (deg)
99GrAm	18.5	18.5	0.0	0.0
ChevrPU	49.8	49.3	6.9	8.0

DELTA-V (Basis: Damage)

	Total (mph)	Forward (mph)	Lateral (mph)	PDOF (deg)
99GrAm	38.8	-6.2	38.3	-80.8
ChevrPU	17.1	-16.7	-3.9	13.0

DELTA-V (Basis: Linear Momentum)

	Total (mph)	Forward (mph)	Lateral (mph)	PDOF (deg)
99GrAm	36.0	-5.8	35.6	-80.6
ChevrPU	15.9	-15.5	-3.6	13.2

Figure 28. EDCRASH[7] Computed Results Showing Damage Basis Longitudinal -6.2 mph Delta-V and Linear Momentum Basis Longitudinal -5.8 mph Delta-V

CASE NUMBER: None

** 360° MOMENTUM **

Page 2 of 4

INPUTS:	
The Wt of Veh 1 in Pounds is:	3300.00
The Min. Spd After Impact, Veh 1 is:	34.00
The Depart Angle, Veh 1 is:	280.00
The Wt of Veh 2 in Pounds is:	7285.00
The Min. Spd After Impact, Veh 2 is:	34.00
The Depart Angle, Veh 2 is:	270.00
The Angle of Approach is:	264.00

RESULTS:	
The Spd of Veh 1 in MPH is:	17.30
The Vel of Veh 1 in FPS is:	25.36
The Spd of Veh 2 in MPH is:	49.43
The Vel of Veh 2 in FPS is:	72.46

VECTOR SUM ANALYSIS DATA:

CALCULATED INFO		DIAGRAM DRAWN AT	
ITEM	MPH-LB	INCHES	DEGREES
W1V1	57090.00	0.48"	0.00°
W2V2	360097.55	3.00"	264.00°
W1V1+W2V2*	358682.94	2.99"	273.10°
V1 PDOF/ P	116700.44	0.97"	251.25°
W1V3	112200.00	0.93"	280.00°
W2V4	247690.00	2.06"	270.00°
W1V3+W2V4*	358729.82	2.99"	273.10°
V2 PDOF/ P	117071.63	0.98"	71.23°

Energy Absorbed = 218804.03 ft-lbs

Coefficient of Restitution (e) = 0.00

PDOF1 = 251.25° (E to G = 71.25°)
PDOF2 = 71.23° (F to H = 12.77°)

ΔV1 = 35.36 MPH Long = 6.14 Lat = -34.82
ΔV2 = 16.07 MPH Long = 0.00 Lat = -16.07

Scale of Diagram: 1 inch = 120032 mph-lb.

* *Note that these are VECTORAL additions.*

AR 98 Professional R5: © 1994-2001, Maine Computer Group

Figure 29. Momentum Analysis Showing Longitudinal 6.14 mph Delta-V Computed in AR98[8]

Another Example

Figure 30 shows a longitudinal Delta-V curve generated with EDSMAC4 in HVE[9] compared to a longitudinal Delta-V curve recorded by the Sensing and Diagnostic Module (SDM) of a 1999 Chevrolet Suburban involved in a frontal impact. This data was downloaded during examination of the Suburban using the Vetronix Corporation Crash Data Retrieval Tool.

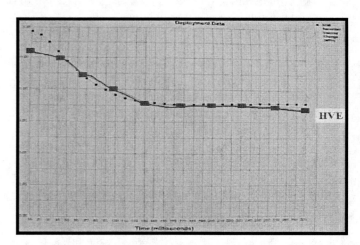

Figure 30. Comparison of SDM Recorded Velocity Change Graph to Longitudinal Velocity Change Graph Generated by EDSMAC4 in HVE[9]

Looking Ahead

In the future, it is possible that these modules will contain more data elements, shorter sample rates, and longer sample durations before and after the crash event. Refer to the DATA ELEMENTS section above for a list of elements that vehicle manufacturers are evaluating for inclusion on their recorded data.

PRESERVATION OF EVIDENCE

As with other types of evidence, the EDR data used to support accident analyses should be preserved. Sufficient records of this information should be maintained by the vehicle owner, persons downloading the data, and persons or entities having an interest in the data. NHTSA keeps EDR data from its investigations in various databases such as CIREN, FARS, NASS, and SCI. The requirements for preservation and protection of the data should also follow local, state, and federal laws where applicable.

ADMITTANCE INTO COURT

In the legal arena, the authors have observed a trend towards a greater demand for providing more detailed data to define crash conditions including pre-impact, impact, and post-impact phases. EDR's can be useful in meeting these demands, provided that they accurately reflect the crash metrics. The availability of EDR data has opened the door to a new generation of understanding and modeling automotive collisions.

The courts should value EDR technology. It can prevent injustices to victims of traffic collisions by its ability to impartially and accurately report data, which is then analyzed to assess liability. The data reported are generally more accurate than other data collected by investigators, which are accepted in courts all the time.

The reported speed and velocity changes are generally more accurate than other methodologies based on observation of post-crash vehicle deformation and scene data, which are readily accepted in courts.

Several considerations should be made regarding the admittance of EDR data into court. As with other evidence, the chain of custody should be documented. The authors' experience has been that this type of evidence has been more contested in criminal cases versus the somewhat less restrictive civil litigation cases.

In order to ensure validity and accuracy of translation of the data, the person downloading the data should always provide or have available raw data from the download, known as a "hex dump." This "hex dump" can be reported using Vetronix Corporation's Crash Data Retrieval tool.

EDR's should probably not ordinarily be used as a stand-alone tool for a reconstruction in a legal arena. Much of the reported EDR data can and should also be confirmed by alternative means where possible.

CONCLUSION

The application of EDR's to vehicles has provided useful data for accident investigation and reconstruction. EDR data previously available only to OEM's is becoming available to the general public through tools such as the Vetronix CDR kit. This data does not take the place of a collision reconstruction. However, it provides complimentary information useful for the reconstruction. In commercial vehicles, EDR data from diesel truck engines extends the knowledge of pre-collision driver and vehicle factors and can be important to the understanding of a collision. This data may include brake and clutch applications and engine speed during full brake application events. For bus applications, devices such as BUSWATCH and DriveCam provide video images and audio in addition to electronic data.

The electronic pre-collision data can be analyzed to determine such things as brake application and deceleration rates, which may be difficult to determine from tire marks by ABS brakes. Seat belt use is one of the outputs from EDR's. This is sometimes otherwise difficult to determine when there is an associated airbag deployment. The data from an EDR can be extremely useful in analyzing a chain reaction or pedestrian collision.

Future developments will hopefully provide more data accessibility, such as self-reading devices, and an expanded database for use in understanding the pre-impact phase of collisions.

REFERENCES

1. Minutes of the National Highway Transportation Safety Administration Event Data Recorder Working Group.

2. Vetronix Corporation, Santa Barbara CA, Crash Data Retrieval system downloads pre- and post-crash data from a vehicle's airbag module to a laptop computer. The Windows® based CDR software presents this data in an easy to read graphical and tabular form.

3. Rosenbluth, W. "Investigation and Interpretation of Black Box Data in Automobiles: A Guide to the Concepts and Formats of Computer Data in Vehicle Safety and Control Systems." ASTM Monograph Series, ASTM Stock No. MONO4. Copublished with SAE, SAE Order No. R - 313. Mayfield, PA, June, 2001.

4. Bolte, K., Jackson, L., Roberts, V., and McComb, S., National Transportation Safety Board, "Accident Reconstruction/Simulation with Event Recorders." International Symposium on Transportation Recorders, May 3 - 5, 1999, Arlington, VA.

5. Chidester, A. "Chip" and Hinch, J., National Highway Traffic Safety Administration and Mercer, T. and Schultz, K., General Motors Corporation, "Recording Automotive Crash Event Data." International Symposium on Transportation Recorders, May 3 - 5, 1999, Arlington, VA.

6. NHTSA, "CRASH3 User's Guide and Technical Manual." U.S. D.O.T., NHTSA, National Center for Statistics and Analysis, Accident Investigation Division, Washington D.C. 29590.

7. EDCRASH is a product of Engineering Dynamics Corporation, which has been presented in previous SAE papers.

8. Maine Computer Group, "Accident Reconstruction Professional Users Guide." 1994-2001, Sarasota Florida 34237.

9. EDSMAC4 and HVE are products of Engineering Dynamics Corporation, which have been presented in previous SAE papers.

ADDITIONAL SOURCES

Triodyne, Inc., of Niles, IL, authored an article titled, "Crash Data Retrieval Kit Recovers Reconstruction Data from GM Black Boxes," which describes the Crash Data Retrieval Kit, and the data recorded by GM vehicles. This was published in their "SafetyBrief," August, 2000, Volume 16, No. 5. This document references No. 5 above by Chidester, et al.

Reviewer Discussion

Reviewers Name: John C. Steiner, B.S.M.E.
Paper Title: Using Event Data Recorders in Collision Reconstruction
Paper Number: 2002-01-0535

Using Event Data Recorders in Collision Reconstruction is a paper that does a good job of presenting the overall availability and current technological trends of Event Data Recorders both from vehicle manufacturers such as General Motors and Ford Motor Company as well as after-market manufacturers such as DriveCam and Bus-Watch. Important facts and issues with the data limitations of the Event Data Recorder is also discussed in this paper which is important, in that the Engineer must be aware of. These limitations can be identified or explained by a trained Engineer who can analyze this data.

John C. Steiner
Roger Clark Associates

Methods of Occupant Kinematics Analysis in Automobile Crashes

Jon E. Bready, Ronald P. Nordhagen, Thomas R. Perl and Michael B. James

Collision Safety Engineering, L.C.

Copyright © 2002 Society of Automotive Engineers, Inc.

ABSTRACT

Understanding occupant kinematics is an important part of accident reconstruction, particularly with respect to injury causation. Injuries are generally sustained as the occupant interacts with the vehicle interior surfaces and is rapidly accelerated to the struck component's post-impact velocity.

This paper describes some methods for assessing occupant kinematics in a collision, and discusses their limitations. A useful technique is presented which is based on free-body analysis and can be used to establish an occupant's path of motion relative to the vehicle, locate the point of occupant contact, and determine the occupant's velocity relative to that contact location.

INTRODUCTION

During initial inspection of the accident vehicle, detailed vehicle motions have not yet been determined, so detailed occupant kinematics typically are not known. Therefore, some general understanding of occupant kinematics is necessary in order to direct attention to the vehicle interior surfaces where occupant contact marks might be expected. Methods used for determining the occupant's motion range from simple "rules of thumb" to more sophisticated analyses that estimate the vehicle's motion. Several rules of thumb, sometimes used to predict the location of occupant contact, can be useful, but are not always accurate for complex crash configurations that result in vehicle motion that is not immediately obvious.

This paper describes several methods for determining occupant kinematics, and discusses their limitations. The methods discussed treat the occupant as a point mass. A simple analysis based on free-body motion is presented, which can be used for planar accident configurations. This simplification is useful for evaluating general occupant travel path, but obviously does not describe the effects of body articulations.

OCCUPANT CONTACT MARKS

In a crash, all free bodies, including occupants, will continue to move with their pre-impact velocities (speed and direction relative to the ground coordinate system), while the crash forces change the vehicle's velocity state. A relative velocity, therefore, develops between the occupant and the vehicle. This relative velocity results in what has been termed the "second-impact", as the occupant strikes some part of the vehicle interior and his velocity is brought into conformity with that part of the vehicle. When a large relative velocity develops, large impact forces will be required to accelerate the occupant to match the vehicle component's changed velocity. These forces often leave evidence of the second-impact either on the occupant, the vehicle, or both. Information from medical records matched with markings on the vehicle interior surfaces can sometimes confirm occupant kinematics. Methods for analyzing occupant contact marks have been presented elsewhere [3,11], as well as methods for distinguishing occupant loading marks from normal usage marks in restraint systems [7,8].

When matching known or suspected occupant contact marks with the occupant kinematics, some variations can be expected due to the following factors:

1. The occupant may not have been in the normal seated position (NSP) at the time of impact. There is a variety of normal driving and riding postures among the population, and occupants often put themselves out of the NSP while performing some task or making some adjustment. Effects such as pre-impact braking and/or lateral deceleration associated with swerving or loss of vehicle control can also cause occupants to be displaced from their NSP.

2. Forces from restraints and/or contact with interior components may alter the occupant motion before his primary contact with the vehicle's interior. Forces associated with gripping the steering wheel, bracing, and seat friction are generally low in magnitude and will have a significant effect only in low severity collisions.

3. Vehicle pitch, yaw and roll motions may significantly alter the location of potential occupant contact. Since this paper is limited to planar motion, only the effects of yaw motion are discussed. However, the same principles apply to pitch and roll.

4. In collisions with substantial intrusion, the dynamic and residual displacement of interior components should be considered.

RULES OF THUMB

Several rules of thumb are often used for predicting occupant motion. These methods are useful for estimating the general direction the occupant moves with respect to the vehicle, and help locate occupant contact marks on interior surfaces. They are typically used in preliminary occupant kinematics analysis, for example, during initial vehicle inspections when the vehicle motion has not yet been established. Some of these methods are discussed below.

OCCUPANT MOVES TOWARD VEHICLE CRUSH - This method of predicting occupant motion simply assumes that all occupants move directly toward the area of vehicle crush. For example, if the front end of the vehicle were damaged, this rule would suggest that the occupant moved forward. If the rear of the vehicle were damaged, it would suggest that the occupant moved rearward. This method for predicting occupant motion is not always accurate; Figure 1 presents several accident scenarios in which the occupant will actually move away from the area of vehicle crush.

OCCUPANT MOVES PARALLEL TO THE PDOF - A second method of predicting occupant motion assumes that the occupant moves in a direction parallel to, but opposite of, the Principal Direction of Force (PDOF) [2,6,9]. This method of predicting occupant motion works well for accidents with minimal yaw rotation. However, with increased vehicle yaw rotation, predictions using this method become less accurate.

In a collision, forces act through various load paths while the impacting vehicles are in contact. The PDOF represents the direction of a single equivalent force, which is the resultant of all the actual forces. It acts exactly parallel to the change of the vehicle's c.g. velocity (ΔV). The PDOF (and thus the vehicle ΔV) can sometimes be estimated from the shape of the vehicle's crush profile and/or from the direction of vehicle crush in the area of direct contact. This methodology is useful for making estimations of the PDOF during vehicle inspections to preliminarily assess occupant motion. However, this method of estimating the PDOF is not appropriate for detailed evaluations of occupant kinematics. The PDOF is the direction of the ΔV and should be determined by subtracting the pre-impact velocity vector from the post-impact velocity vector of the vehicle.

OCCUPANT IMPACTS ROTATED INTERIOR - A third method of predicting occupant motion takes into account the effects of vehicle yaw rotations. In many accident scenarios, collision forces applied in a location and direction not passing through the vehicle c.g. can cause the vehicle to rotate (yaw) significantly before the second-impact. A procedure to account for vehicle yaw rotation is to first use either the location of vehicle crush or the PDOF method to estimate the direction the occupant would be expected to move relative to the vehicle absent vehicle yaw rotation. Then, visually rotate the vehicle about its c.g. an estimated amount the vehicle would yaw between initial vehicle collision and the occupant's contact with the vehicle interior. The location where the occupant's straight-line motion intersects the rotated vehicle interior is where the second collision might be expected (see Figure 2). This method is a quick and easy way to estimate the effect of yaw rotation on the location of the second-impact, however, it does not take into account horizontal movement of the vehicle's c.g. and will therefore not necessarily yield precise results.

EFFECTS OF VEHICLE YAW ROTATION

Vehicle yaw rotation affects occupant kinematics in two ways: it can cause an increase or decrease of the relative velocity between the occupant and vehicle interior; and the unrestrained occupant's path of travel relative to the vehicle will be curved. The three methods described above are helpful for predicting the general location of the second-impact from an assumed starting position, however, they do not establish the actual path or velocity of the occupant relative to the vehicle.

Methods for determining the velocity of the occupant relative to the vehicle have been evaluated in the literature [1,2,4,5]. One of the difficulties in this type of analysis is that the magnitude and direction of the relative velocity changes with time. Cheng, et. al. [1] suggest using a fixed time of 0.3 seconds from first impact to occupant contact with the vehicle interior. Marine, et. al. [5] show velocity vectors for several, specified times at a single location. A more useful method is to determine the path of the occupant, so that the actual time, velocity and location of the second-impact can be determined.

FREE-BODY ANALYSIS

A simple technique for evaluating occupant kinematics is to: 1) determine the vehicle's motion, 2) treat the occupant as a free-body and apply Newton's first law of motion, and 3) transform the occupant's motion into the vehicle coordinate system. This approach: 1) establishes the occupant's path of motion relative to the vehicle, 2) locates the second-impact and aids in identifying contact marks, 3) determines the occupant's velocity relative to that impact location, 4) determines the time to second-impact, 5) identifies possible ejection portals, and 6) aids in determining injury causation.

The laws of motion apply in the inertial (fixed) reference frame (i.e. ground coordinates), which is like visualizing the vehicle and occupant motion through a ground-based camera. Once the motion has been established in ground coordinates, it can then be transformed to permit examination of occupant motion relative to the vehicle, which is like visualizing the occupant's motion through a vehicle-mounted camera. This transformation can be done graphically or analytically. Figure 3 illustrates how this process can be done graphically. In the ground coordinate system, the occupant initially travels with the same direction and speed as the vehicle (see Figure 3a). The impact-induced motion of the vehicle relative to the ground can be drawn (based on a computer simulation or hand

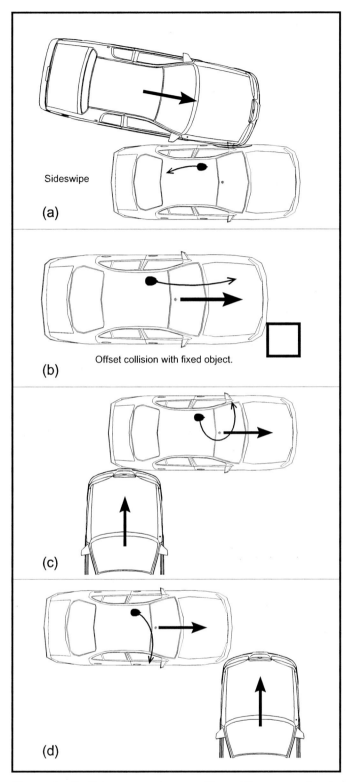

Figure 1 - Collision examples resulting with occupant not moving directly toward vehicle crush (Resulting occupant head motion drawn in vehicle coordinates).

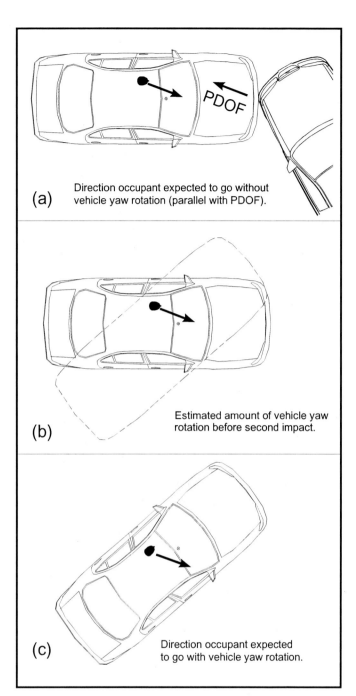

Figure 2 - Effect of vehicle yaw rotation on occupant straight-line motion.

calculations), and the free-body occupant positions for the same time intervals can also be drawn (see Figure 3b). It is then a simple process to draw the location of the occupant relative to the vehicle for each time increment (see Figure 3c).

Vehicle motion generated by SMAC or PC-CRASH can also be analytically transformed from the ground coordinate system to the vehicle coordinate system for use in visualizing free-body motion. Figure 4 is an example showing occupant motion in vehicle coordinates, which is derived from a PC-CRASH simulation.

Free-body motion can be determined from any initial seated location with the path plotted relative to vehicle geometry for reference. The vehicle c.g., overall dimensions, and documented occupant contact marks can also be specified as shown in Figure 4. The path of the free-body extends beyond the vehicle boundary because no actual impact is modeled. This analytical method allows documented occupant contact marks to be specified as "targets", useful for calculation of the relative speed between a free-body and the target(s). Using these targets, the vehicle motion can also be refined, iteratively, if the pre-impact location of the occupant is known. The analysis illustrated in Figure 4 gives the second-impact occurring between 240 and 250ms from initial impact, with the corresponding second-impact velocity of 23.7 kph [14.7 mph] relative to the documented contact marks.

Free-body motion based on SMAC or PC-CRASH can also be superimposed on a more detailed, scaled drawing of the accident vehicle for further examination, as depicted in Figure 5. In this example, high vehicle yaw rotation causes the occupant to travel in a curved path (relative to the vehicle) around the steering wheel, explaining why the occupant did not significantly load the steering wheel before striking the windshield at the known location.

In Figure 6, free-body motions starting from three seating positions are compared. This example illustrates how vehicle yaw rotation can affect the motion of the occupants starting from various seated positions. The analysis shows the occupants striking the right side interior panels from unique paths, and at different speeds and times.

The consequences of high yaw rotation on occupant motion are illustrated in Figure 7. In this example, a locomotive struck the right front of the vehicle at a railroad crossing. The driver was subsequently ejected through the rear window opening after striking the right front seat in a front-to-rear direction.

CONCLUSIONS

1. Several "rules of thumb" are often used for predicting the general location of occupant contacts in automobile crashes. These simple methods may not be accurate in complex collision scenarios resulting with high vehicle raw rotation, or when vehicle motion is not immediately obvious.
2. A technique for establishing the occupant's actual path of motion relative to the vehicle when a high rotational velocity is present is to: 1) determine the vehicle's motion, 2) treat the occupant as a free-body, and 3) transform the occupant's motion into the vehicle coordinate system.
3. Establishing the occupant's path of motion relative to the vehicle aids in determining: 1) the location of the second-impact and identifying contact marks, 2) the time from initial collision to the second-impact, 3) the occupant's velocity relative to the impact location, 5) possible ejection portals, and 6) occupant injury causation.
4. Occupant motion can be determined either analytically or graphically.
5. Use of known occupant contacts as "targets" can be applied to iteratively refine vehicle motion if the pre-impact location of the occupant is known.

CONTACT

Collision Safety Engineering, L.C.
150 South Mountainway Drive
Orem, Utah 84058
Office 801-229-6200
Fax 801-229-6202
Internet cselc.com

BIBLIOGRAPHY

1. Cheng, Philip H., Guenther, Dennis A., "Effects of Change in Angular Velocity of a Vehicle on the Change in Velocity Experienced by an Occupant during a Crash Environment and the Localized Delta V Concept." SAE 890636.
2. Fay, Richard J., Raney, Anne U., Robinette, Ric D., "The Effect of Vehicle Rotation on the Occupant's Delta-V." SAE 960649.
3. Nyquist, Gerald W., Kennedy, Everett P., "Accident Victim Interaction with Vehicle Interior: Reconstruction Fundamentals." SAE 870500.
4. Backaitis, Stanley H., DeLarm, Leon, Robbins, D. Hurley, "Occupant Kinematics in Motor Vehicle Crashes." SAE 820247.
5. Marine, Micky C., Werner, Stephen M., "Delta-V Analysis from Crash Test Data for Vehicles with Post-Impact Yaw Motion." SAE 820247.
6. Robinette, Ric D., Fay, Richard J., Paulsen, Rex E., "Delta-V: Basic Concepts, Computational Methods, and Misunderstandings." SAE 940915.
7. Bready, Jon E., Ronald P. Nordhagen, Richard W. Kent, "Seat Belt Survey: Identification and Assessment of Noncollision Markings." SAE 1999-01-0441.
8. Bready, Jon E., Ronald P. Nordhagen, Richard W. Kent, Mark W. Jakstis, "Characteristics of Seat Belt Restraint System Markings." SAE 2000-01-1317.
9. Baker, J.S., Fricke, L., The Traffic Accident Investigation Manual, Northwestern Traffic Institute.
10. Grimes, W., "The Effect of Crash Pulse Shape on Occupant Simulations." SAE 2000-01-460.
11. Severy et. al., "Motorist Head and Body Impact Analysis, Methodologies and Reconstruction." SAE 850097.

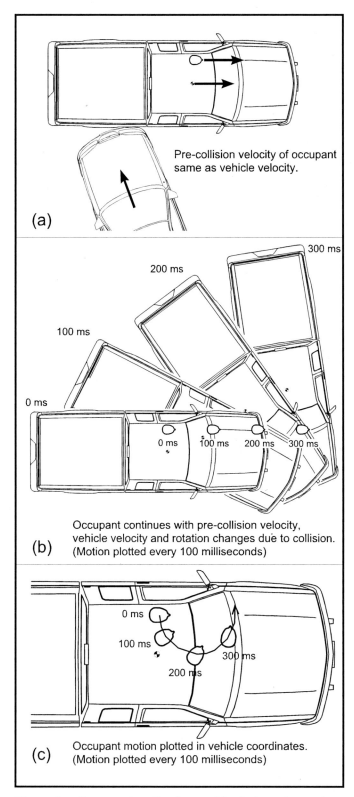

Figure 3 – Occupant motion is graphically transformed and drawn in the vehicle reference system in order to visualize the motion of the occupant relative to the vehicle.

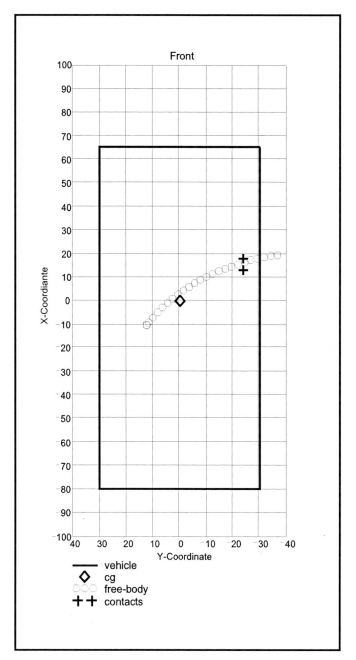

Figure 4 – Free-body motion plotted relative to vehicle geometry and occupant contacts, based on transformation of PC-CRASH vehicle motion (Free-body drawn 10 ms increments).

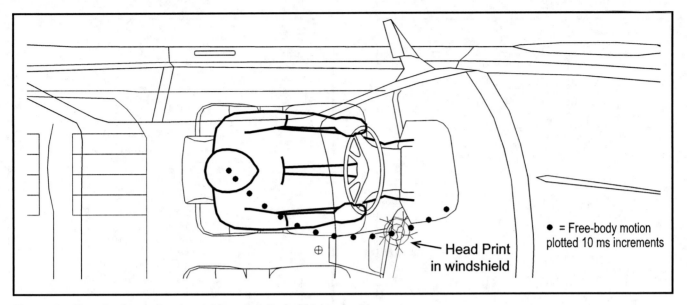

Figure 5 – Free body motion overlaid on detailed vehicle graphics.

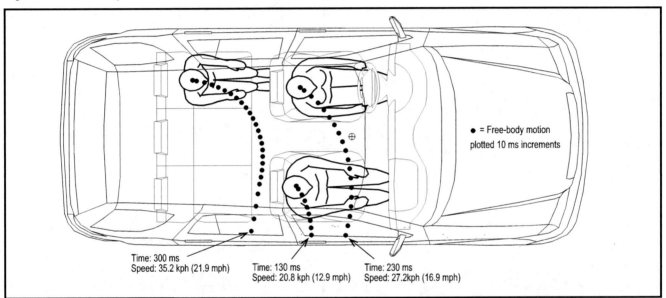

Figure 6 – Free body motion from several seating positions compared.

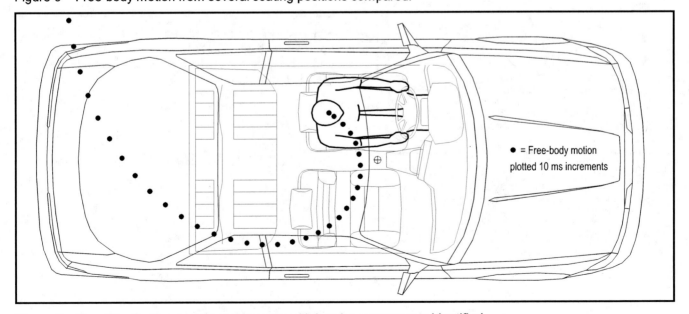

Figure 7 – Possible ejection portals and impacts with interior components identified.

Reviewer Discussion

Reviewers Name: Michael S. Varat
Paper Title: Methods of Occupant Kinematics Analysis in Automobile Crashes
Paper Number: 2002-01-0536

The determination of occupant kinematics is an analysis that is often required in accident reconstruction. Often, the level of detail required does not justify the expenditure of the engineering resources required of a model such as MADYMO or ATB. Therefore, the familiarity with planar occupant kinematics techniques is a valuable skill set required of the accident reconstructionist. The authors have presented an excellent technical overview of the basics of planar occupant kinematics analysis using the application of free body motion using Newton's laws of physics. One essential caution with any occupant kinematics analysis is that the engineer must take great care in the reconstruction of vehicle motions before attempting to use those motions to reconstruct the occupant dynamics. The accurate reconstruction of occupant dynamics first requires the accurate reconstruction of the vehicle dynamics.

2002-01-0540

Low Speed Collinear Impact Severity: A Comparison Between Full Scale Testing and Analytical Prediction Tools with Restitution Analysis

A. L. Cipriani, F. P. Bayan, M. L. Woodhouse, A. D. Cornetto, A. P. Dalton, C. B. Tanner and T. A. Timbario
FTI/SEA Consulting

E. S. Deyerl
Quan, Smith & Associates

Copyright © 2002 Society of Automotive Engineers, Inc.

ABSTRACT

Low speed collinear collisions have received some attention in the past in published technical literature. Underrepresented are full-scale instrumented tests utilizing vehicles equipped with foam core bumpers and closing speeds greater than 2.2 meters per second (m/s). Systematic testing was designed to obtain data in collisions between vehicles with similar and mixed bumper structures. Testing was performed at closing speeds ranging from 0.8 to 5.4 m/s. Following each test, vehicle bumper and other damage was documented. Data from the 30 tests for each category of bumper and mixed categories were analyzed to identify the test speed, load magnitude, velocity change, duration of impact and coefficient of restitution. In addition, the energy absorption characteristics and damage thresholds of the various types of bumper systems were obtained. The coefficient of restitution was analyzed as a function of closing speed, velocity change, impact duration, bumper type, and other parameters and compared to previously published data and equations. A modified momentum-energy-restitution (MER) method was also used to calculate predicted vehicle response for each test and compare it to the actual test results.

INTRODUCTION

Much has been published and conjectured about the difference in behavior of foam core and piston absorber bumper systems, but there is relatively little full scale, low-speed testing data available for vehicles equipped with foam core bumpers. One of the goals of this series of tests was to document the low-speed impact behavior of foam core bumper systems and their interaction with other foam core and piston absorber systems. These results can then be correlated with previously published tests and analysis.

One difference in analyzing low-speed collisions versus high-speed impacts is the inclusion of the effects of restitution. The coefficient of restitution is a measure of the rebound that occurs when a vehicle strikes another object. Mathematically, it is the ratio of separation velocity and closing velocity. A restitution value of one reflects a purely elastic impact, which cannot occur in real world vehicle-to-vehicle collisions. Zero restitution indicates a purely plastic impact. A small or zero restitution typically occurs at higher speeds or when the vehicles remain interlocked post-impact as in some under-ride type collisions. Several existing publications provided restitution values for various vehicle-to-vehicle and vehicle-to-barrier impacts, primarily with piston absorber bumper systems. The goals of this paper are to expand the restitution data for vehicle-to-vehicle impacts, to compare these data to the correlation presented by Antonetti [1], and to incorporate and analyze data from various tests as documented in various references [2-6]. A new correlation was developed using the data from the testing performed by the authors combined with data previously presented by Antonetti [1] and others [2-6]. In addition, correlations are presented for piston absorber to piston absorber impacts and foam core to foam core impacts separately.

One way to predict a vehicle's response to a low-speed collision is by way of the MER method. This method uses restitution data and energy from barrier data to determine velocity change. Bailey, et al's [7] previous work compared velocity change calculated by the MER method to velocity change from staged low-speed impacts. Bailey, et al also measured average isolator compression in specific vehicle-to-vehicle impacts. In Bailey, et al's [7] paper, the average isolator compression was compared to actual vehicle-to-barrier impacts to determine energy and restitution.

In accident reconstruction, the question often posed is whether collision severity is sufficient to cause certain injuries to the occupants of the vehicles. The first step in such an analysis is to determine the severity of the impact in terms of velocity change, average acceleration, and peak acceleration. The ultimate question about injuries is the domain of the biomechanical engineer and will not be addressed in this paper. In many low-speed collisions a specific level of severity cannot be precisely determined. This is particularly true when no permanent crush can be identified or when only photographs and repair estimates are available. However, an upper limit modified MER analysis can be undertaken. In such an analysis, a maximum velocity change is calculated by using conservative estimates of all parameters lacking precise values. One method of performing such an analysis is detailed in the following steps:

- Determine the appropriate vehicle weights including occupants and loads. A conservative approach to incomplete weight data will minimize the weight of the vehicle of interest and maximize the other.

- Calculate vehicle stiffness coefficients, "A" and "B", for the year, make, and model group of vehicles to determine the crush energy involved. Assume an extent of crush that is conservative, and include some minimal amount of permanent crush in cases where none appears in photographs or repair estimates. In addition, using the maximum possible width for the contact area will also maximize velocity change.

- Compare the force of impact computed for each vehicle. The force at impact must be equal for each vehicle, but cannot exceed the smaller of the two.

- Use the appropriate restitution for the relative approach or closing speed at impact. This is an iterative process, which will converge on the calculated upper limit for velocity change.

- The average acceleration can then be obtained by dividing the velocity change by a minimum impact duration of 100 milliseconds (msec). Assuming a half triangle wave, the peak acceleration is simply 2 times the average acceleration.

The presented method varies from Bailey; et al in that energy is calculated by way of bumper stiffness' ("A" and "B" coefficients) from crash tests performed by the Department of Transportation (DOT) and other published sources. Data sources such as Neptune's compilation [8] and crash plots using the Campbell [9] or Prasad [10] methods may also be used. The presented method offers an upper limit to velocity change, average acceleration, and peak acceleration. This paper also compares calculated values using the above method with the actual test values of velocity change, average acceleration, and peak acceleration. Correlation between the results will validate the use of this modified MER method as a means to determine an upper limit vehicle response in low-speed collisions.

FULL SCALE TESTING

TEST PROCEDURE

A matrix of 30 low-speed collinear collisions were staged using 4 different test vehicles: 2 with foam core bumper systems and 2 with piston absorber systems. Tests were conducted with the front of the bullet vehicle impacting the rear of the target vehicle. The matrix was assembled as follows: piston absorber into piston absorber, foam core into foam core, piston absorber into foam core, and foam core into piston absorber. The vehicles consisted of a 1986 Honda Accord, a 1988 Mazda 929, a 1989 Chevrolet Cavalier, and a 1985 Chevrolet Celebrity. The bullet and target vehicles were interchanged so that each vehicle's front and rear bumpers were tested up to the 5.4 m/s closing speed. The bullet vehicle speed was increased by propelling it to the desired speed measured via radar gun and 5th wheel until impact with the target. The target vehicle brakes were released and this vehicle was allowed to roll freely. In some of the higher speed tests, the vehicle was brought to a controlled rest. The bullet vehicle was released just prior to impact. The test matrix was designed to test vehicles at speed increments of 0.8 m/s to approximately 5.4 m/s. Table 1 presents the actual tested speeds.

Table 1. Testing Matrix.

Test #	Bullet	Target	Actual Speed (m/s)
1-7	Honda	Mazda	0.9, 0.8, 1.6, 2.9, 4.1, 4.6, 5.0
8-10	Mazda	Honda	0.9, 2.1, 2.5
11-15	Mazda	Cavalier	1.1, 1.7, 3.1, 4.0, 4.9
16-21	Cavalier	Celebrity	0.8, 1.7, 2.9, 3.8, 4.2, 5.8
22-24	Celebrity	Cavalier	0.8, 1.7, 2.7
25-27	Mazda	Honda	3.7, 4.6, 5.2
28-30	Celebrity	Cavalier	3.7, 4.6, 5.8

TEST VEHICLES

The 4 test vehicles were inspected, weighed, and documented photographically and geometrically prior to testing. The requirements for selecting the vehicles were that they be fairly common and generally representative of the average passenger car, be capable of rolling freely, have all major components such as engines, interiors, and transmissions in place, and that they have undamaged front and rear bumper structures. The vehicles were unaltered prior to testing and all fluids were drained. Table 2 shows each vehicle's specifications. Appendix A provides a more complete listing of test vehicle data including an illustrated parts diagram of each bumper system. The "A" and "B" values presented in Table 2 were calculated using the Campbell [9] and Prasad [10] methods. The method that produced the best correlation result was used.

Table 2. Vehicle Specifications.

	Honda	Mazda	Cavalier	Celebrity
Type	Sedan	Sedan	Coupe	Sedan
Bumper	Foam	Foam	Piston	Piston
OAL (m)	4.57	4.90	4.55	4.78
OAW (m)	1.70	1.70	1.68	1.75
Weight (kg)	1146	1495	1052	1249
Front "A" (lb/in)	232.1	278.1	213.8	203.0
Front "B" (lb/in^2)	75.6	79.1	58.0	52.9
Rear "A" (lb/in)	156.2	352.7	227.3	197.4
Rear "B" (lb/in^2)	32.1	133.5	65.5	60.2

INSTRUMENTATION

Tri-Axial Accelerometer

The primary means of data collection for the tests was two IST EDR-3 tri-axial accelerometers. These units are self-contained "black boxes" which require no external connections during the testing sequences. One EDR-3 was mounted in each vehicle during testing. They were rigidly mounted to the transmission tunnel along the centerline, near the center of gravity of each vehicle. Each was mounted horizontally and as close to level as possible. A sampling frequency of 3.2 kHz was set for each channel. Acceleration data in the X, Y, and Z directions were recorded for analysis. A full series of tests was run with a vehicle acting as a bullet and another as the target before the data was downloaded from the EDR's to a laptop computer. The units were then moved to new vehicles for the next series of tests. This data storage capability greatly simplified and expedited the testing process. Each EDR was calibrated prior to crash testing.

The EDR units were triggered to record 500 msec of data before and after detecting a peak acceleration of 0.5 g's in any of the three axes. The units recorded a total of 1 second of data for each test collision.

Figure 1 shows overlaid sample data from the two EDR's from the full scale testing. The figure shows data from only the X direction and indicates that one vehicle had a positive acceleration (the target) while the other vehicle had a negative acceleration (the bullet).

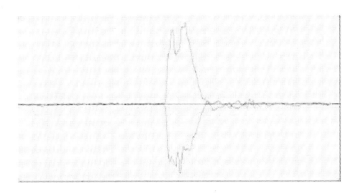

Figure 1. Sample IST Accelerometer Data.

5th Wheel

A 5th wheel was constructed using the frame and tire from a 0.50 meter unicycle with the pedals and seat removed. The shaft of the unicycle wheel was coupled to the input shaft of a 60 pulse per revolution optical encoder that served as the input to a Shimpo model DT-5TG digital tachometer. In addition to providing a digital readout, this tachometer has the ability to generate a rapidly updated analog signal proportional to its input. In addition, the unit is programmable to allow scale factors to be input for both the digital readout and the analogue output.

The 5th wheel was calibrated by measuring the distance traveled by the tire when it was rolled through a single revolution. This distance represented the travel distance of the 5th wheel for one revolution, or 60 pulses, of the optical encoder.

The unicycle frame was attached by lightweight tubing to a separate frame that mounted to the passenger's side door of each bullet vehicle through suction cups. The frames were adjustable so that the axis of the unicycle wheel could be kept parallel to the plane of the ground for all of the vehicles. In addition, there were joints in the frame that allowed the unicycle frame to rotate about pitch and yaw axes so that good wheel contact could be maintained during vehicle maneuvering.

The 5th wheel directly measured the instantaneous velocity of the bullet vehicle as it was being pushed up to impact speed, and displayed the speed to the nearest tenth of a mile per hour. This allowed the impact speed to be generally controlled in each of the tests. In addition the analog output of the device gave a record of the bullet vehicle speed going into and coming out of the impact. Data were collected by way of LabTech Notebook data acquisition software. Figure 2 is an example of 5th wheel data acquired during testing.

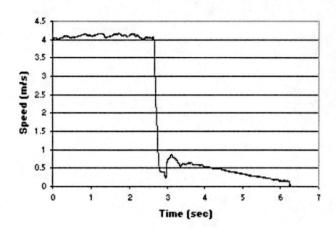

Figure 2. Sample 5th Wheel Data.

Radar Gun

A Stalker ATS Profession Sports Radar gun was used to also track the closing speed of the bullet. The Radar gun was calibrated prior to crash testing using the manufacturers recommended procedures.

Photographs and Measurements

After each test the front of the bullet vehicle and the rear of the target vehicle were photographically documented. Appendix B provides a set of representative photographs from these tests. In addition, the vehicle's overall length was measured after each test to determine if any permanent crush resulted from testing. No repairs were made to the vehicles between tests.

DATA REDUCTION

Two different methods were used to reduce the data recorded by the IST accelerometers. The first method utilized the DynaMax software that accompanied the accelerometers. A 4th order phaseless Butterworth filter was applied to the raw data in the direction of travel to remove noise and vibration.

The second method used DADiSP 4.0 numerical method software to integrate the acceleration data. The raw data were filtered by way of a moving average to remove noise and vibration. Acceleration data were also offset to remove any error from non-level mounting of the accelerometers. Impact duration and velocity change were determined for each test and compiled. Figure 3 shows an example of the DADiSP 4.0 output.

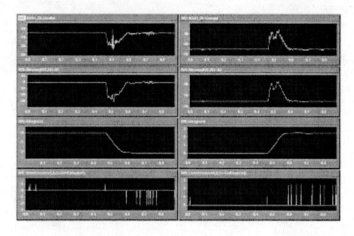

Figure 3. Sample DADiSP 4.0 Graphs.

The various recording devices provided a clear indication of the initial impact time reference. The separation time of the vehicles was distinguished using filtering or averaging of the data. The test data were viewed in the unfiltered or un-averaged format and the target and bullet vehicle data were compared to obtain the separation time. Determination of Impact duration and velocity change for each collision were then compiled and are shown in Appendix C.

Test 1 was not used for restitution or MER analysis as the target vehicle's brakes were inadvertently engaged, thus affecting the results.

ANALYSIS

RESTITUTION ANALYSIS

In reviewing Antonetti's paper, a marginally better fit to the data was discovered by using a 3rd order logarithmic fit. This fit provided a better match to the data at low approach speeds above 0.25 m/s. Below 0.25 m/s, the logarithmic fit goes to infinity. A comparison of these fits is shown in Figure 4 of Appendix D. The correlation coefficient for the logarithmic fit is 0.906 whereas the exponential fit coefficient is 0.904. The equation for Antonetti's original fit and the new logarithmic fit are as follows:

$$e = 0.5992 \exp^{-0.2508v + 0.01934v^2 - 0.001279v^3}$$

$$e = 0.47949 - 0.274781 \log(v) - 0.07673 \log(v)^2 - 0.01895 \log(v)^3$$

Each of the 30 collisions provided additional data that were added to the data presented by Antonetti [1]. Third order exponential and logarithmic fits were explored. A graph of the comparison between the 3rd order exponential and logarithmic best-fit curves can be seen in Figure 5. The data were also examined individually for each bumper system. Figure 6 shows 3rd order logarithmic curve fits for foam core bumper systems, piston absorber bumper system, and all data. The piston type absorber data was over-represented and tends to pull the all data curve close to the piston type absorber curve. As expected, the trend illustrates restitution increases as closing speed decreases. The foam core type bumper systems tend to produce a higher restitution than the piston type absorbers. The correlation coefficient, r, for the all data curve fit is 0.806, for the form core data is 0.911, and for the piston type absorber is 0.866.

In general, the data confirmed the overall curve fit, whether in exponential or logarithmic form, produced by Antonetti [1]. Additionally, the logarithmic fit for the foam core data produces a good correlation to the available data. The following three equations were obtained from the best-fit curves:

Piston Composite (Piston to Piston)

$$e = 0.44871 - 0.24963 \log(v) + 0.04988 \log(v)^2 - 0.12660 \log(v)^3$$

Foam Core Composite (Foam Core to Foam Core)

$$e = 0.53391 - 0.31054 \log(v) + 0.22723 \log(v)^2 - 0.28015 \log(v)^3$$

All Data Composite

$$e = 0.47477 - 0.26139 \log(v) + 0.03382 \log(v)^2 - 0.11639 \log(v)^3$$

MER METHOD COMPARISON

The MER method is a method used in reconstructing low-speed collinear collisions. The data from the 30 staged tests were compared to an analysis of each of these collisions using photographically estimated vehicle damage and the MER equations shown below:

Conservation of Momentum

$$m_T v_T + m_B v_B = m_T v_T + m_B v_B$$

Conservation of Energy

$$\frac{1}{2} m_T v_T^2 + \frac{1}{2} m_B v_B^2 = \frac{1}{2} m_T v_T^2 + \frac{1}{2} m_B v_B^2 + E_T + E_B$$

Coefficient of Restitution

$$v_T - v_B = e(v_B - v_T)$$

Crush Energy

$$E = L(AC + \frac{BC^2}{2} + \frac{A^2}{2B})$$

Target Vehicle Delta-V

$$\Delta v_T = \frac{(1+e)}{1 + \frac{m_T}{m_B}} \sqrt{\frac{2(E_T + E_B)(m_T + m_B)}{(1-e^2) m_T m_B}}$$

Crush Force

$$F = AL + BLC$$

Relative Approach Velocity

$$v_{RA} = \frac{1 + \frac{m_T}{m_B}}{(1+e)} \Delta v_T$$

This modified MER procedure was used for each crash test and compared to the data collected during the staged collisions. The 30 staged crash tests showed an impact duration, delta-t, of 0.098 to 0.215 seconds. No correlation was found between approach speed, velocity change, or damage and Impact duration.

Several methods can be used to calculate the peak acceleration experienced by the vehicles in a low-speed collision. The first method simply utilizes the peak force and divides by the weight of the vehicle in question, in accordance with Newton's Second Law. The second and third methods assume the shape of the impact pulse as either a half sine wave or half triangle wave. The average of both waves are 0.636 and 0.50, respectively. The peak acceleration is then calculated by dividing the average acceleration by either value. The authors determined that the only method that consistently over-predicts the peak acceleration relative to the tests was to assume a half triangle wave. The peak acceleration is twice the average acceleration.

The results of the comparison between actual crash data and the predicted modified MER values indicate an over-prediction in velocity change, average acceleration, and peak acceleration by the modified MER method as can be seen in Appendix C. Accordingly, this data verifies that the modified MER method can be used as a tool in determining the upper limit of velocity change, average acceleration, and peak acceleration in low-speed collisions under these conditions.

Predicting the precise velocity change, average acceleration, and peak acceleration in low-speed collisions with accuracy is typically difficult, if not impossible, given the limitations of data available from the real world crash. Even in an ideal case, a staged crash test with identical vehicles may not yield identical results to those of an actual collision. According, the modified MER method described herein provides an effective technique to analyze low-speed collisions. The results of such an analysis provide an upper limit value for velocity change or acceleration.

CONCLUSION

Low speed crash testing was performed to evaluate restitution of different bumper constructions, especially foam core type bumpers. The overall correlation of restitution to closing speed put forward by Antonetti [1] is considered valid with minor modifications. A 3^d order logarithmic fit appears to provide a minor improvement in correlation overall, and appears to fit much better at closing speeds of 0.25 to 1.25 m/sec. In addition, this paper provides a separate 3^d order logarithmic curve for the foam core to foam core bumper collision. The correlation of this curve to the data is considered to be good and usable for future analysis needs. More test data would further refine these correlations.

A modified MER method of analysis is presented. This method allows the analyst to determine an upper limit value for velocity change, average acceleration, and peak acceleration. When compared to the 30 crash tests performed during this study, the modified MER method always over-predicted these parameters. This method accordingly provides an effective technique to analyze low-speed collisions.

ACKNOWLEDGMENTS

The authors would like to thank Vincent Antonetti for all his help and providing data and software for data analysis and Gregory Hoshal at IST.

REFERENCES

1. Antonetti, V.W., "Estimating the Coefficient of Restitution of Vehicle-to-Vehicle Bumper Impacts", SAE 980552.
2. Howard, R.P., Bomar, J., and Bare, C., "Vehicle Restitution Response in Low Velocity Collision", SAE 931842.
3. Malmsbury, R.N. and Eubanks, J.J., "Damage and/or Impact Absorber (Isolator) Movements Observed in Low Speed Crash Test Involving Ford Escorts", SAE 940912.
4. Siegmund, G.P., King, D.P., and Montgomery, D.T., "Using Barrier Impact Data to Determine Speed Change in Aligned, Low Speed Vehicle-to-Vehicle Collisions", SAE 960887.
5. Siegmund, G.P., Bailey, M.N., and King, D.J., "Characteristics of Specific Automobile Bumpers in Low-Velocity Impacts", SAE 940916.
6. Szabo, T.J. and Welcher, J., "Dynamics of Low Speed Crash tests with Energy Absorbing Bumpers", SAE 921573.
7. Siegmund, G.P., Bailey, M.N., and King, D.J., "Automobile Bumper Behavior in Low-Speed Impacts", SAE 930211.
8. http://www.neptuneeng.com
9. Campbell, K.L., "Energy Basis for Collision Severity", SAE 740565.
10. Prasad, A.K., "Energy Dissipated in Vehicle Crush-A Study Using the Repeated Test Technique", SAE 900412.

CONTACT

Alfred L. Cipriani, P.E. or Fawzi P. Bayan, P.E.
Principal Mechanical Engineer
Vehicle Accident Reconstructionist
FTI/SEA Consulting
1110 Benfield Boulevard
Millersville, MD 21108
(800) 635-9507

SYMBOLS

"T" and "B" subscripts denote target and bullet vehicles respectively.

m mass of vehicle

v pre-impact velocity of vehicle

v post-impact velocity of vehicle

e intervehicular restitution

E energy

A crush coefficient

B crush coefficient

C depth of crush

L length of crush

Δv change in velocity of vehicle

F crush impact force

Appendix A

1986 Honda Accord 4 Door Sedan
VIN: JHMBA7430GC044106

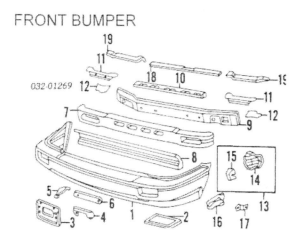
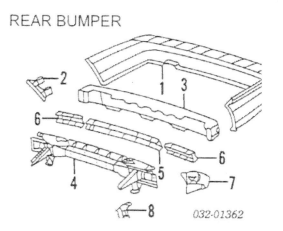

1988 Mazda 929 4 Door Sedan
VIN: JM1HC221XJ0101575

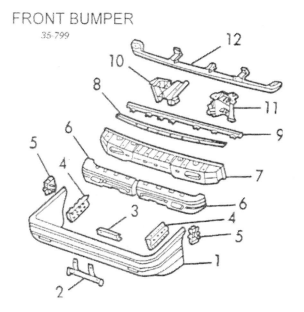
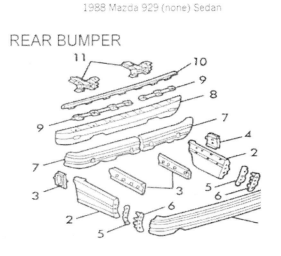

1989 Chevrolet Cavalier 2 Door Coupe
VIN: 1G1JC1114KJ299443

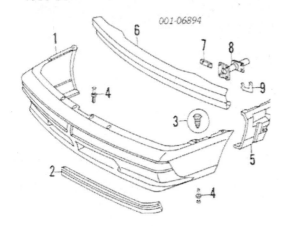

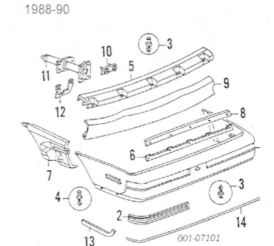

1985 Chevrolet Celebrity 4 Door Sedan
VIN: 2G1AW19R4F1209175

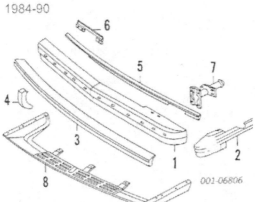

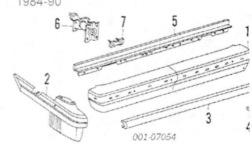

Appendix B

Photo 1. 0.8 m/s. Rear of Mazda.
Minor blemishes to plastic bumper cover from Honda license plate mounting bolts. No permanent crush damage.

Photo 2. 0.8 m/s. Front of Honda.
Note license plate mounting bolts. No permanent crush damage.

Photo 3. 1.6 m/s. Rear of Mazda.
No permanent crush damage.

Photo 4. 1.6 m/s. Front of Honda.
No permanent crush damage.

Photo 5. 2.9 m/s. Rear of Mazda.
No permanent crush damage.

Photo 6. 2.9 m/s. Front of Honda.
No permanent crush damage.

Photo 7. 4.1 m/s. Rear of Mazda. More pronounced scuffing. No permanent crush damage.

Photo 8. 4.1 m/s. Front of Honda. More pronounced scuffing. No permanent crush damage.

Photo 9. 4.6 m/s. Rear of Mazda. Bumper cover detached at right rear wheel well and bumper rotated downward slightly.

Photo 10. 4.6 m/s. Front of Honda. Permanent crush damage to left front corner of bumper.

Photo 11. 5.0 m/s. Rear of Mazda. No additional damage noted.

Photo 12. 5.0 m/s. Front of Honda. More severe damage, including buckling of engine compartment hood.

Photo 13. 0.8 m/s. Rear of Celebrity. No permanent crush damage.

Photo 14. 0.8 m/s. Front of Cavalier. No permanent crush damage.

Photo 15. 1.7 m/s. Rear of Celebrity. No permanent crush damage.

Photo 16. 1.7 m/s. Front of Cavalier. No permanent crush damage.

Photo 17. 2.9 m/s. Rear of Celebrity. No permanent crush damage.

Photo 18. 2.9 m/s. Front of Cavalier. No permanent crush damage.

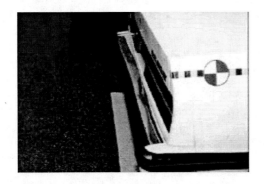

Photo 19. 3.8 m/s. Rear of Celebrity. Right side of bumper permanently crushed inward approximately one inch.

Photo 20. 3.8 m/s. Front of Cavalier. No permanent crush damage.

Photo 21. 4.2 m/s. Rear of Celebrity. Right side of bumper permanently crushed inward approximately three inches.

Photo 22. 4.2 m/s. Front of Cavalier. Some permanent crush damage.

Photo 23. 5.8 m/s. Rear of Celebrity. Additional permanent crush damage.

Photo 24. 5.8 m/s. Front of Cavalier. Additional permanent crush damage.

Appendix C

Test #	Vehicle	V (init) (m/s)	Test DELTA Vx (m/s)	MER DELTA Vx (m/s)	Test Gx (ave) (g)	MER Gx (ave) (g)	Test Gx (peak) (g)	MER Gx (peak) (g)	Restitution	DELTA t (sec)
1	HONDA	0.9	(0.6)	N/A	(0.4)	N/A	(1.0)	N/A	0.22	0.104
1	MAZDA	0.0	0.4	N/A	0.3	N/A	0.6	N/A		0.104
2	HONDA	0.8	(0.8)	(3.4)	(0.4)	(3.4)	(1.3)	(6.8)	0.62	0.137
2	MAZDA	0.0	0.6	2.6	0.3	2.6	0.9	5.2		0.137
3	HONDA	1.6	(1.4)	(3.4)	(0.8)	(3.5)	(2.6)	(7.0)	0.54	0.124
3	MAZDA	0.0	1.0	2.6	0.6	2.7	1.8	5.4		0.124
4	HONDA	2.9	(2.2)	(3.4)	(1.5)	(3.5)	(4.5)	(7.0)	0.34	0.101
4	MAZDA	0.0	1.7	2.6	1.1	2.7	2.9	5.4		0.101
5	HONDA	4.1	(3.2)	(3.4)	(1.7)	(3.5)	(6.0)	(7.0)	0.38	0.130
5	MAZDA	0.0	2.4	2.6	1.3	2.7	4.2	5.4		0.130
6	HONDA	4.6	(3.6)	(3.6)	(2.5)	(3.6)	(7.2)	(7.3)	0.37	0.101
6	MAZDA	0.0	2.7	2.7	1.9	2.8	4.9	5.6		0.101
7	HONDA	5.0	(3.9)	(4.2)	(2.7)	(4.2)	(8.2)	(8.5)	0.35	0.098
7	MAZDA	0.0	2.9	3.2	2.0	3.3	5.4	6.5		0.098
8	MAZDA	0.9	(0.6)	(2.5)	(0.3)	(2.6)	(1.0)	(5.2)	0.50	0.119
8	HONDA	0.0	0.7	3.4	0.4	3.4	1.4	6.8		0.119
9	MAZDA	2.1	(1.4)	(2.5)	(0.8)	(2.6)	(2.5)	(5.2)	0.46	0.113
9	HONDA	0.0	1.7	3.4	1.1	3.4	3.3	6.8		0.113
10	MAZDA	2.5	(1.7)	(2.5)	(1.1)	(2.6)	(3.4)	(5.2)	0.55	0.108
10	HONDA	0.0	2.1	3.4	1.4	3.4	4.1	6.8		0.108
11	MAZDA	1.1	(0.6)	(2.4)	(0.3)	(2.5)	(0.9)	(5.0)	0.31	0.155
11	CAVALIER	0.0	0.8	3.4	0.3	3.5	1.0	7.0		0.155
12	MAZDA	1.7	(1.1)	(2.4)	(0.5)	(2.5)	(1.6)	(5.0)	0.52	0.155
12	CAVALIER	0.0	1.5	3.4	0.6	3.5	1.9	7.0		0.155
13	MAZDA	3.1	(1.8)	(2.4)	(0.8)	(2.5)	(3.1)	(5.0)	0.38	0.150
13	CAVALIER	0.0	2.5	3.4	1.2	3.5	4.0	7.0		0.150
14	MAZDA	4.0	(2.2)	(2.4)	(1.3)	(2.5)	(4.1)	(5.0)	0.34	0.119
14	CAVALIER	0.0	3.1	3.4	1.8	3.5	6.3	7.0		0.119
15	MAZDA	4.9	(2.8)	(2.9)	(1.6)	(2.9)	(5.4)	(5.8)	0.34	0.121
15	CAVALIER	0.0	3.8	4.1	2.2	4.2	6.6	8.3		0.121
16	CAVALIER	0.8	(0.8)	(3.1)	(0.3)	(3.2)	(1.0)	(6.4)	0.73	0.167
16	CELEBRITY	0.0	0.6	2.6	0.2	2.7	0.9	5.4		0.167
17	CAVALIER	1.7	(1.2)	(3.1)	(0.5)	(3.2)	(1.3)	(6.4)	0.27	0.176
17	CELEBRITY	0.0	0.9	2.6	0.4	2.7	1.3	5.4		0.176
18	CAVALIER	2.9	(1.9)	(3.1)	(0.7)	(3.2)	(2.2)	(6.4)	0.20	0.190
18	CELEBRITY	0.0	1.6	2.6	0.6	2.7	2.2	5.4		0.190
19	CAVALIER	3.8	(2.6)	(3.3)	(1.4)	(3.4)	(3.3)	(6.7)	0.25	0.128
19	CELEBRITY	0.0	2.2	2.8	1.2	2.8	3.3	5.7		0.128
20	CAVALIER	4.2	(2.9)	(3.6)	(1.7)	(3.7)	(4.2)	(7.4)	0.26	0.115
20	CELEBRITY	0.0	2.5	3.0	1.5	3.1	4.3	6.2		0.115
21	CAVALIER	5.8	(3.8)	(4.0)	(2.6)	(4.1)	(7.7)	(8.2)	0.24	0.102
21	CELEBRITY	0.0	3.3	3.4	2.3	3.5	6.9	6.9		0.102
22	CELEBRITY	0.8	(0.6)	(2.7)	(0.2)	(2.8)	(0.8)	(5.6)	0.31	0.171
22	CAVALIER	0.0	0.5	3.2	0.2	3.3	1.5	6.6		0.171
23	CELEBRITY	1.7	(1.1)	(2.7)	(0.4)	(2.8)	(1.6)	(5.6)	0.36	0.215
23	CAVALIER	0.0	1.2	3.2	0.4	3.3	1.5	6.6		0.215
24	CELEBRITY	2.7	(1.6)	(2.7)	(0.7)	(2.8)	(2.1)	(5.6)	0.26	0.166
24	CAVALIER	0.0	1.8	3.2	0.8	3.3	2.8	6.6		0.166
25	MAZDA	3.7	(2.3)	(2.8)	(1.5)	(2.9)	(4.3)	(5.7)	0.45	0.104
25	HONDA	0.0	3.1	3.7	2.0	3.7	6.7	7.5		0.104
26	MAZDA	4.6	(2.7)	(4.0)	(1.8)	(4.1)	(5.8)	(8.2)	0.40	0.107
26	HONDA	0.0	3.7	5.3	2.4	5.4	7.1	10.7		0.107
27	MAZDA	5.2	(2.9)	(4.0)	(2.0)	(4.1)	(6.4)	(8.2)	0.28	0.100
27	HONDA	0.0	3.8	5.3	2.6	5.4	8.8	10.7		0.100
28	CELEBRITY	3.7	(2.1)	(2.8)	(1.1)	(2.9)	(3.1)	(5.7)	0.26	0.128
28	CAVALIER	0.0	2.5	3.4	1.4	3.4	3.7	6.8		0.128
29	CELEBRITY	4.6	(2.8)	(3.0)	(1.2)	(3.0)	(3.7)	(6.1)	0.34	0.167
29	CAVALIER	0.0	3.4	3.5	1.4	3.6	5.2	7.2		0.167
30	CELEBRITY	5.8	(3.3)	(3.7)	(1.7)	(3.8)	(5.2)	(7.5)	0.28	0.135
30	CAVALIER	0.0	4.1	4.4	2.1	4.5	6.5	8.9		0.135

Appendix D

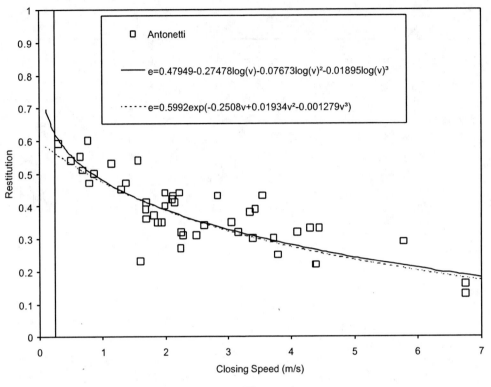

Figure 4.

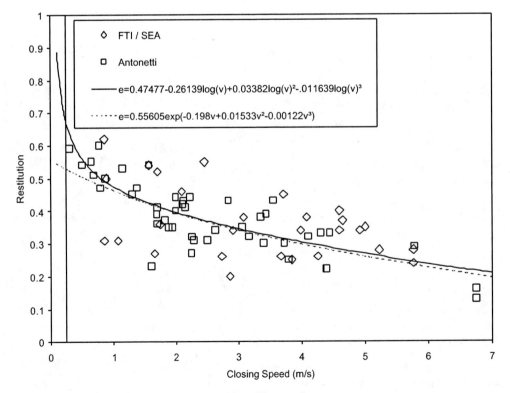

Figure 5.

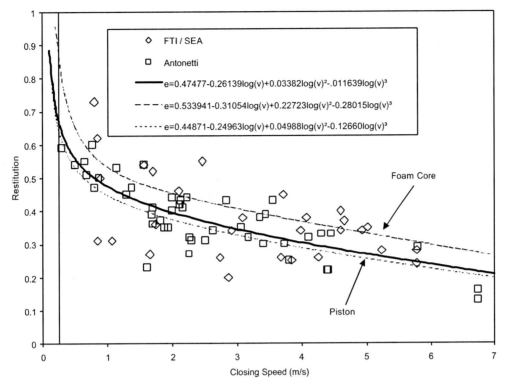

Figure 6.

Identification and Interpretation of Directional Indicators in Contact-Damaged Paint Films - Applications in Motor Vehicle Accident Reconstruction

2002-01-0542

Neil Clark and Roger E. Clark
Roger Clark Associates

Copyright © 2002 Society of Automotive Engineers, Inc.

ABSTRACT

Virtually every motor vehicle on the road today has large portions of its exterior coated with some type of protective, decorative finish—materials that are collectively known as "paint". What many observers dismiss as a vanity feature are actually complex, laminated structures that respond to damage events in a limited- and predictable number of ways. In doing so, these laminates of non-crystalline solid layers develop a finite range of recognizable failure patterns.

The majority of motor vehicle accidents do create damage to at least one of the painted exterior surfaces of the involved vehicle. The informed accident investigation and reconstruction professional who can identify the patterns within these damage features, and interpret them correctly, has an added tool for determining the relative motions between the vehicle, and the event which produced the damage.

INTRODUCTION

One of the core objectives of Vehicular Accident Reconstruction is the complete- and accurate understanding of the motions of a vehicle throughout an accident event. This goal is both particularly important and especially difficult when the vehicle has sustained multiple contacts—whether as the result of interaction with multiple objects, or from multiple contacts with the ground during a rollover event.

Quite often, the only objective evidence of "what happened" to a vehicle in a crash is recorded in the physical damage sustained by the vehicle itself. The first step in Accident Reconstruction is to select the tools that are most appropriate to the valid interpretation of evidence, and then to apply those tools in the appropriate manner. In the last three decades, governmental agencies and private Accident Reconstruction professionals have devised- and refined methods for analyzing the "macro" damage features associated with high-energy, linear collisions associated with serious occupant injury. But a large percentage of claims- and lawsuits develop from accidents that may not produce substantial damage to the involved vehicles.

In 2000 as in previous years, fully two-thirds of all accidents reported to law enforcement agencies were classified as "Property Damage Only" events [1]. But even if most of those non-injury crashes caused less-than-major damage to the involved vehicle(s), It is likely that many of these vehicles would have sustained minor or even "micro" damage to a painted body panel—in the form of dents, scrapes and scratches. These often-overlooked damage features can record details of the damaging event, which the reconstructionist can interpret to understand the relative motions between the "struck" vehicle and the "striking" object.

Paint damage analysis can be used to confirm other directional indicators, or to establish contact motion when other indicators are inconclusive- or contradictory. The directional features in the paint film of a damaged panel often makes it possible to characterize the relative speeds- and motions of contacts between vehicles in disputed collisions, and can validate- or challenge alleged Hit-&-Run accidents, or even establish the direction-of-rotation in roll-over events.

Far from being random or rare artifacts, we estimate that directional indicators can be found in roughly 70 percent of painted panels that have been damaged by direct-contact events. This paper will describe a number of features that may be found in damaged paint that can provide reliable indicators of the motion of the damaging event, relative to the vehicle.

HISTORY OF THIS PROJECT

This project is based upon paint damage analyses begun by Roger E. Clark in the mid- to late 1970's, in an effort to evaluate the validity of motor vehicle accident claims where only minor contact damage was found. Having worked for a number of years in the field of materials testing and experimental stress analysis, Mr. Clark was familiar with the uses of brittle coatings to study the patterns of surface stresses in components, and he realized that the presence of similar features in damaged automotive paints could indicate the orientation, motion and magnitude of the damaging event.

Close examination of minor-damage sites found that equivalent "micro" paint damage patterns often developed in superficial contacts that mainly involved the paint film itself, while causing little if any damage to the underlying panel. In other words, paint films could be damaged by forces acting within the paint film itself, not merely in response to damage to the underlying panel.

Mr. Clark first described these features in an unpublished research project for the School of Safety and Systems Management at the University of Southern California [2]. These findings were put to use to interpret the subtle features of glancing contacts common in minor "Hit-&-Run" accident claims.

Since that initial work, his observations have been tested through the inspection of damaged automotive finishes of dozens of vehicles from numerous foreign- and domestic marques, as well as in the "repair" paint finishes on cars and trucks, and the paints of after-market accessories. Most of the same paint failure patterns have been documented in the damaged finishes of many non-automotive surfaces in the driving environment: these include the spray-applied finishes on items such as motorcycle helmets, bicycles and mailboxes, and the brush- or roller-applied "house paints" used on fixed objects such as buildings and parking poles.

REPEATABLE DAMAGE-INDUCING TESTS

In the early field work, uncertainty decreased as the number of subjects increased, but a nagging element of uncertainty was always present. Prior to 1994, all of our experience with- and knowledge of directional indicators in damaged paint had come from "after-the-fact" examinations of the exterior panels of production motor vehicles involved in real-world accidents. The subject vehicles were found in salvage yards, and we rarely had a complete understanding of the event(s) that had produced the damage. And while care was taken to only select examples where direction-of-motion could be determined with certainty by other means (usually the bending- or displacement of adjacent structures), we desired a high degree of certainty—of repeatability—and of observability while the tests were being performed.

Any testing needed to satisfy a number of requirements:
- The testing process had to be able to subject a wide range of vehicle panels to realistic contacts by a broad cross-section of real-world objects;
- The testing apparatus needed to do more than simply scratch a painted panel; it needed to strike the panel with sufficient force to deform the underlying panel, as often occurs in real-world accidents.
- The test panels needed to be sized for ease of handling during testing, and easy storage afterward;
- The tests needed to be repeatable and safely observable;

All of these requirements combined to rule out vehicle-to-vehicle collisions.

The most straightforward, cost-effective and space-efficient means of bringing impactors into repeatable, observable contact with test panels was to mount an impactor (the "bullet") to a pendulum, and the "to-be-struck" sample (the "target") to a rigid support. When the pendulum was released, its arcing path produced the added benefit of generating distinct points of engagement- and disengagement, as well as rising-, peak- and falling engagement forces during the contact.

A pendulum with a nominal length of 3.05 meters (10 feet) was assembled from sections of 3/4-inch plumbing pipe. A low-friction pivot was achieved by assembling a "T"-bar at the top of the pendulum, which fit into eye-bolts threaded into a ceiling beam. A plywood platform fastened to the bottom of the pendulum provided the mounting surface for a variety of impactors. Weights could be added to the platform to ensure that the contact had the momentum needed to carry the impactor through its contact with the test sample.

The pendulum was held in a raised position by a metal gate latch on a pivoting arm that engaged a striker on the edge of the pendulum platform. This provided a repeatable fall height of 61 cm. (24 inches). A steel laboratory table 86.4 cm (34 inches) high was positioned beneath the pendulum, and fitted with a plywood platform, which allowed a variety of "target" panels to be secured to the table with screws to approximate the dimensional stability of the part as installed on a vehicle.

To gauge the speed of the pendulum before- and during contact, free- and loaded tests were videotaped against a background of a measuring tape. Frame counting of these taped tests determined that the pendulum achieved at-impact apex speeds of approximately 13 Kph (8.0 mph). While this speed is low in absolute terms, it would represent a significant closing speed in real-world accidents.

TEST PANELS

Slightly damaged panels from a variety of automobiles were collected from local body repair shops; the best candidates were fenders and door skins, which had been rendered irreparable by cuts or punctures - leaving large undamaged areas on which our tests were conducted. Only panels with distinguishable "factory" paint finishes were accepted.

All of the panels were struck with a cinderblock to replicate contacts with masonry surfaces. Selected panels were then struck with samples of component from other vehicles, to simulate car-to-car contacts; these impactors included:
- The sturdy tips cut from three automobile fenders;
- The corner cut from the rigid composite front-panel of a U.S. compact sedan;
- The corner of a resilient-core bumper, and
- The impact strip from the rear bumper of a compact pickup truck.

References to these pendulum tests will appear throughout this paper.

Samples of each panel were also subjected to a secondary (and admittedly unscientific) series of hand-delivered blows from a ball-peen hammer. The hammer tests delivered both perpendicular- and glancing blows to painted metal- and non-metallic panels to observe the damage response of the different materials and their paints:

Perpendicular impacts to the painted panels tested the paint's response to two different types of contacts: the concave distortion of the panel, which occurs when an area is dented "inward" by a localized contact to the exterior, and the bulging distortion of a panel dented "outward" by an impact, as occurs when a painted panel is formed against some interior component.

A separate series of glancing hammer swipe tests were used to bruise the paint layers while causing minimal deformation to the foundation material, in order to learn whether damage induced directly within the paint layer(s) was similar to paint damage caused by distortion of the underlying panel. Mechanized, repeatable tests assuring greater repeatability could be performed in the future.

Together, these experiments not only validated the conclusions based on field work by replicating the patterns found in real-world accidents, but also identified several previously unrecognized patterns of paint failure that have subsequently been verified on real-world accident vehicles.

THE STRUCTURE OF PAINT FILM

In order to understand how a car's paint film can be damaged, it is necessary to understand how a modern automotive finish is "constructed." Even though the thicknesses of paint films are measured in the thousandths-of-an-inch, their responses to damage are best understood when the paint film is viewed as a three-dimensional material having depth and structure, in addition to length and breadth.

At the beginning of the automotive century, the main reason for painting an automobile body was to protect it from the environment; the aesthetics so important to the luxury car buyers was (at least initially) insignificant to the mass-market consumer; recall the anecdote that Henry Ford offered his Model "T" in any color the buyer wanted - so long as it was black [3]. Efficiency was everything in mass-producing affordable automobiles: Hand-finishing each body could take weeks to complete, so the paints for vehicles like the Model "T" were chosen for ease-of-use, rapid curing time, and low cost. Of all of the paints then available, the Model "T" was painted with black enamel because it covered- and dried quickly [3].

As cars and trucks transcended mere utility, so did the automotive paint job. Then as now, customers craved the unique - and color allowed a vehicle to stand out from its drab competitors. Using a fast-drying, spray-applied lacquer developed by Dupont for General Motors, Oakland Motors stole the 1924 auto shows with their "True Blue" Oakland Six [3]. Automakers were soon turning out models in an array of colors from the same assembly line. And we, the public, loved it.

For decades, the typical automotive body was constructed of stamped steel panels that were etched with mild acids in preparation for the typical "paint job": a layer of adhesion-promoting primer, covered by one- or two layers of pigmented paint. This simplistic "Panel Preparation / Primer / Topcoat" finish [5,6] is rarely seen on modern cars and trucks. The modern paint film is often composed of a half-dozen separate material layers, deposited individually in carefully controlled environments [5,6,7,8]. Each individual layer is formulated to provide some specific attribute that contributes to the durability of the paint film, adjacent layers must be chemically compatible with one another, and the entire paint film must respond uniformly to stresses produced by flexing, impacts and temperature extremes—and the inevitable door-ding.

And where the term "panel" once almost exclusively meant "steel", a finish-bearing panel can now be composed of steel, aluminum, molded plastic or some form of composite material (i.e.; fiberglass). Most of the vehicles on the road today have bodies containing both metallic and nonmetallic panels.

PAINT FILMS ON METALLIC PANELS

The most quality-critical paint films are those applied to sheet metal panels, because the "paint" provides the primary environmental protection for the panel. The typical paint film applied to steel panels consists of at least four separate layers, each applied in discrete steps to produce a laminated paint film that can range in thickness from 0.13- to 0.23 mm. (0.005- 0.009 inch) [8,9,10].

The foundational layers of the paint film include panel treatments to resist corrosion, and primers to enhance paint-to-panel adhesion, while the topcoat layer(s) of the paint film finish provide its appearance characteristics: the pigmented "color coat" often contains metallic "flake" materials or "pearl" additives to increase visual interest. Finally, the color-coat is typically sealed with a transparent "clear-coat", which excludes the environment from the paint film, and resists ultra-violet degradation [7,8,9,10].

Figure 1 represents typical paint film on a sheet steel panel, with the discrete layers listed in the order of application:

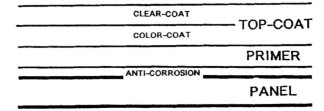

Figure 1 – Paint Film on Metallic Panel

- The PANEL is typically stamped from sheet steel, although aluminum panels are increasingly used.
- The ANTI-CORROSION treatments include galvanizing and electro-deposited materials.
- The PRIMER improves paint adhesion to the treated panel.

- The TOPCOAT provides the panel's visible color and smooth finish. Modern topcoats usually consist of:
  - The base COLOR-COAT to provide the color, and
  - The CLEAR-COAT to seal the color-coat against weathering, while providing added shine and visual depth.

Body areas subject to gravel-strikes may receive a resilient "anti-chip" layer between the PRIMER and the COLOR COAT layers to absorb the energy that could otherwise damage the paint film layers and disrupt the adhesion between layers [9]. Such damage can lead to the loss of a "chip" (to be defined later, along with other paint-damage-specific terms).

PAINT FILMS ON NON-METALLIC PANELS

Because non-metallic components are not prone to corrosion, the paint film applied to those surfaces can be relatively simple [9,10]. A typical paint film structure for a non-metallic panel is shown in Figure 2, described in the order of application to the panel:

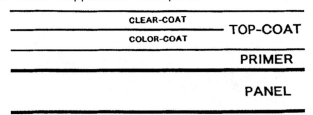

Figure 2 – Paint Film on Non-Metallic Panel

- The PANEL may be made of rigid composite (fiberglass), semi-rigid plastics (Ram Flex) or resilient plastics (urethane or polypropylene)
- The PRIMER on fiberglass and polyurethane panels furnishes a uniform foundation color for the topcoat; polypropylene panels need acrylic resin primer to assure paint adhesion.
- The TOPCOAT provides the visible color and smooth finish. Modern topcoats usually consist of:
  - The COLOR-COAT, which provides the color, and;
  - The CLEAR-COAT, which seals the color-coat, and provides added shine and visual depth.

MOST VEHICLES USE BOTH PANEL TYPES

Current-model vehicle bodies are usually a collection of metal panels, fitted with rigid- and resilient non-metallic components. Each panel type benefits from a different combination of panel treatments finish materials and application- and curing techniques:

The paints applied at the factory to metallic panels are typically formulated around "thermo-setting" materials, which are applied by sprayers, and then "baked" with ovens or heat lamps to cause the top-coat to flow to a smooth layer. As the top-coat cools, it sets to a hard, glossy finish that resists scuffs- and scratches [6,8].

A hard finish is also desirable on non-metallic rigid components, but baking would damage the foundation material. These items receive paints that are formulated to give acceptable finish and hardness when cured at lower (room) temperatures [6,8].

And finally, resilient components need paints that are able to flex- and recover with their underlying panel, in order to survive minor bumps without failing. The paints destined for these applications are blended with "plasticizers" (also called "flex agents"), which allow them to remain intact- and bonded to their resilient panels after minor impacts [6,8].

No vehicle body using all three types of components could be painted properly using any single kind of paint, yet a vehicle assembled from parts coated with paints from three different sources would practically guarantee mismatched tones and textures. This problem is addressed by painting each type of panel on a separate spur of the same assembly line, drawing from a common vat of "base" paint. Prior to application, the base paint is modified with specific additives to provide the characteristics appropriate to each type of panel. The sub-assembled metal body is "baked" to cure its thermo-setting paint before being fitted with heat-sensitive rigid- and flexible components [9,10].

Due to the specialized chemistry of the individual material layers, and the need to apply- and cure each specific layer individually, the finished paint film possesses distinct "inter-coat" or "interface" boundaries [9,11]. This project confirmed both basic materials concepts and in-the-field experience that failures of a layered paint film are somewhat more likely to occur at the layer boundaries [12]. In other words, the adhesion between dissimilar material layers within a paint film tends to be inferior to the "intra-coat" cohesion within the individual layers themselves. [11]. A notable exception to this rule was first identified in our testing, and later recognized in real-world damage. The over-all durability of the paint film depends both upon the interaction- and "inter-coat adhesion" between adjacent layers of the paint film, and the "intra-coat cohesion" characteristics within a specific individual layer.

THE NATURE(S) OF PAINT FILMS

Paint finishes can respond to damage situations in a variety of ways, and while they rarely behave in purely one manner, their responses to damage can be broadly understood as somewhere between truly brittle and truly ductile materials. The paint films used on rigid panels tend to reside at the brittle end of the spectrum, and to respond to damage as a non-crystalline solid (such as glass), while the paint films applied to resilient panels tend to be ductile.

One of the first- and most significant finding of our field observations was that paints often develop fractures in response to stresses in their underlying panels - and that these fractures can indicate the orientation, motion and magnitude of these stresses.

These ideas are far from new, and have been utilized by materials testing professionals in the form of brittle-film stress diagnostic coatings. Brittle-film products applied to test components can produce a two-dimensional "contour map" of the distribution and magnitude of tensile stresses in the component surface.

Perhaps the best known of the brittle-film diagnostic products goes by the **Stresscoat**® trade name. **Stresscoat**®, and similar brittle-film materials, are formulated to generate visible fractures in response to tensile- and shearing forces; the patterns of the resulting fractures reveal the patterns- and magnitude of surface stresses in components and assemblies [12]. By nature- and purpose, brittle-film diagnostic products differ significantly from protective paint finishes in at least two respects: First, in order to respond rapidly and predictably to component stresses, brittle-films must be applied to form a single, uniform and extremely thin material layer (0.01-mm. to 0.015-mm [0.004-in to 0.006-in.]) [13]; and Second, because they are expected to fail in response to minute forces, brittle-films need a useful life only as long as the duration of the test cycle.

Those characteristics are almost completely opposite of an owners' expectations of the paint film that covers their new car or truck; vehicle buyers demand that the paint on their vehicle be durable enough to withstand years of environmental assault with their "good looks" intact, in addition to surviving minor insults such as gravel-strikes and "door dings". These requirements necessitate a multi-layered "laminate" that is far thicker than any brittle-film product.

But even for these distinct differences, automotive paints and brittle-film products do have one thing in common: they tend to be less ductile than the surfaces they cover. As a result, the finish layers have the capacity to develop fractures that record the nature- and distribution of the stresses acting across- and within their underlying panel.

Paint films do not merely "crack" and "chip" when damaged: depending on the nature- and motion of the contact, and the characteristics of the paint film itself, the finish can respond with various combinations of fractures in tension- and shear, by buckling, by delaminating, and by tearing. All of the paint damage features identified in this work are a variation on two elemental fracture types: straight fracture lines typical of tensile- and shearing forces that develop ahead of a moving impactor, and the circular- and semi-circular fractures that appear to be linked segments of straight fractures form at right angles to radiating tensile lines of force produced by a localized contact.

Individually, each damage element conveys some information about the contact that caused the damage; when the two elemental fractures combine, they produce highly reliable indicators of contact direction. The development of these patterns is dependent upon the direction of the contact relative to the painted panel, and whether the contact involves protruding elements that can penetrate the paint film. It is this very dependency which make the fracture patterns very reliable indicators for direction-of-motion.

The precise mechanisms responsible for each type of visible paint failure are by no means fully understood—and are subject enough for a separate paper. This paper will deal with the visible, physical characteristics of stress-induced paint failures, and will make only limited observations on the force(s) and mechanism(s) responsible for a particular failure pattern.

COLLISION CHARACTERISTICS

Damage-producing contacts span the spectrum from a simple scratch at parking speed, to a violent collision that can render a vehicle unrecognizable. For this study, the angular nature of the contact largely determines the type- and appearance of damage that are likely to appear in the paint film.

RIGHT-ANGLE CONTACTS

The most symmetrical damage patterns are those produced by perpendicular blows to the painted panel. Contacts in which the forces to the panel are applied at large angles (approaching perpendicular) are expected to dent- or penetrate the panel, with relatively little lateral motion across the panel. These sorts of contacts produce distinctly different paint failures, according to whether the force is applied from the painted surface of the panel (as in an external dent), or from the underside of the painted panel (as when a panel is forced against an underlying structure).

CONCENTRIC RING ("BULLS-EYE") FRACTURES

Field observations had identified patterns that were later confirmed by controlled testing: when paint had been strained by right-angle external impacts, the paint surrounding the impact site tended to develop concentric sets of ring fractures, which resemble a "bulls-eye" target.

The centroid of the ring fractures typically indicate the center of the contact, so the presence of ring fractures indicate the existence of a localized, inward-moving contact. Figure 3 depicts an idealized set of concentric rings of the sort that can be found surrounding the center of a localized impact:

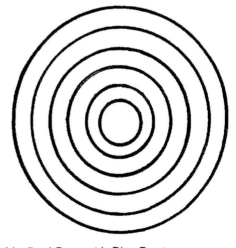

Figure 3 – Idealized Concentric Ring Fractures.

The inward application of force against the panel produces a bowl-shaped distortion of the panel—which should cause the material across the "bowl" to elongate in all directions. If the panel is made of a ductile material such as sheet metal, plastic distortion should remain after the force is released, in the form of a "dent". If the panel is an elastic material, it may regain its pre-contact shape, after the overlying paint has sustained damage, and exhibits concentric fractures. The fact that the paint fractures develop in rings around the impact center demonstrates that the paint fractures perpendicular to radial tension. Circumferential tension must also develop as the panel undergoes concave distortion, although the absence of such fractures implies that these forces to not exceed the paint's strength.

Something other than pure tensile stress is responsible for the formation of ring fractures. Figure 4 is an idealized depiction of the concentric ring fractures generated by a hammer strike to the exterior of a painted bumper cover molded of PVC plastic:

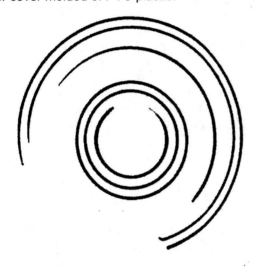

Figure 4 – Locations of Test-Generated Ring Fractures.

The test result differs markedly from the idealized concept; whereas the "idealized" concept depicts ring fractures developing at regular intervals from the center, the test produced two groups of ring fractures— separated by a broad band of non-fractured paint. We theorize that the fractures developed in response to a combination of bending-and tensile stresses in the panel; the inner fractures formed where the panel had seen concave bending around the impactor, while the outer group developed where the panel bent in a convex manner—apparently where constrained by the top- and bottom corners of the bumper cover. The fact that equivalent fractures occur in paint over zones subjected to opposite bending motions indicates that fractures can originate- and propagate both upward and downward through the paint film.

Figure 5 shows the actual set of concentric ring fractures produced by a ball-peen hammer strike to the painted bumper cover:

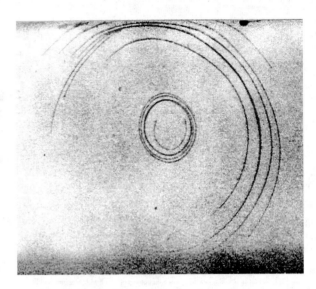

Figure 5 – Concentric Ring Fractures.

As shown in both of the preceding Figures, the fractures do not always link to form complete rings: the formation of any portion of a ring fracture presumably indicates that the surface tension across that particular point on the panel has reached the tensile limit of the paint material; the absence of a fracture presumably means that the stresses across that particular area were sub-critical.

NOTE: The mottled appearance of the panel, and the fuzzy appearance of the fractures are the product of a dye penetrant process needed to make the fractures visible. The elasticity of the panel and its modified finish had caused the ring fractures to re-close, becoming almost invisible to the naked eye. The impact center was located by noting the centroid of the ring fractures. So the presence of ring fractures not only indicate the occurrence of a localized, inward-moving contact, but can be used to locate the center of the contact site.

Examples of concentric-ring fractures are often found in accident damage. Figure 6 is a photograph of the ring-fractures found in the paint of a flexible bumper cover, where the tail-pipe end of the struck vehicle had penetrated the cover:

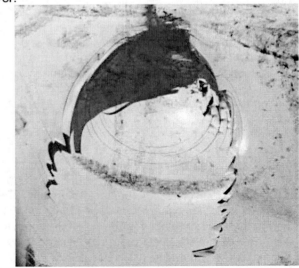

Figure 6 – Ring Fractures from Penetrating Contact.

One particular variant of concentric ring fractures occurs when the indenting contact is applied to a panel with outward-bowing curvature. When a localized contact presses inward against a curved panel, the panel material surrounding the contact undergoes non-uniform stresses--which cause relatively flat fractures to appear only on opposite sides, across the curvature, rather than forming a circle. Those areas on opposing sides of the curvature undergo exaggerated flexing—which can disrupt the paint film down to the panel.

Figure 7 is a photograph of such damage to the left quarter panel of a 1969 Chevrolet Nova, which had sustained a minor rear-end contact by a smaller- and much lighter vehicle:

Figure 7 – Parallel Fractures Above and Below a Dent in a Curved Panel.

The owner of the Nova had pointed to this diamond-shaped damage as "compression buckling" to prove the severity of the rear-end collision. Examination found creases in the panel skin below the side moulding, from an object that had pressed inward against the panel below its curved upper surface.

As the indention progressed inward- and spread upward, the concentric propagation of force was apparently altered by the cylindrical contour of the panel—and the indentation took on a "diamond" shape. The panel material at the "top" of the diamond actually bent—causing the paint film to fracture in series of slightly-concave lines—causing the entire paint film to debond from the panel itself; the panel at the "bottom" of the diamond did not distort to the point of paint damage—perhaps because of the reinforcing contours of the near-by wheel arch.

The mechanism of this diamond-shaped indentation can be reproduced by pressing inward against the side of an empty aluminum beverage can.

While the fracturing of the paint associated with this feature is produced by bending of the panel--rather than expansion of the panel beneath the paint film, it appears often enough in case work to be considered useful.

STELLATE ("STAR-BURST") FRACTURES

The complement to damage produced by a localized external contact would be one in which an object strikes the opposite surface of the panel—or more commonly, when the painted panel is forced downward against an underlying structure. For comparability, the same ball-peen hammer was used to deliver a perpendicular blow to a separate section of the same bumper cover used in the "ring fracture" test.

When such a force is applied to the underside of a painted panel, the panel "balloons" outward beneath the paint. As with the external impact, the panel is presumed to stretch in all directions. But unlike the external strike, circumferential distortion appears to predominate, because the paint on the exterior of the bulge fails along straight lines that radiate outward from the impact site—forming the "stellate" pattern of a classic "burst" fracture [11,12].

Figure 8 is a photograph of an actual burst-type paint fracture produced by a ball-peen hammer strike to the underside of the same painted PVC plastic bumper cover. As before, dye penetrant was needed to make the fractures distinct after the PVC panel had regained its original contour:

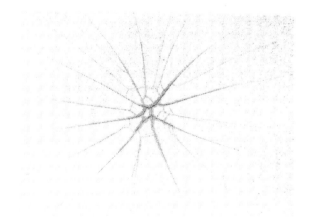

Figure 8 – Stellate Fracture in Resilient Paint.

Stellate fractures are far more obvious on sheet metal panels, due both to the relative brittleness of their paint films—and the tendency for the panel to retain distortion. Figure 9 is a tracing of the burst-type fractures produced by a ball-peen hammer blow to the underside of a steel fender:

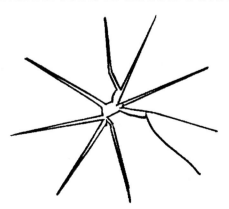

Figure 9 – Tracing of Stellate Fractures.

Figure 10 is a photograph of that same stellate fracture—whose fissures radiate outward from the impact site:

Figure 10 – Stellate Fracture in Rigid Paint.

The expansion of the panel beneath the inelastic paint of the rigid panel causes the fractured paint to debond from the panel surface into separate, triangular "teeth"; because the sheet metal plastically retains much of its distortion, these thin "teeth" are flexed out- and apart, forming obvious gaps, and making the teeth prone to flaking.

The same 1976 Chevy Nova that sustained the diamond-shaped dent also bore a small stellate fracture high on the left quarter-panel. A small object had shifted forcefully within the unlined trunk, generating a thumb-sized bulge in the panel:

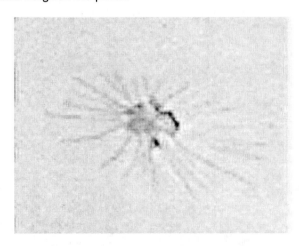

Figure 11 – Stellate Fracture from Internal Contact.

While dents and displacement are more obvious indicators of motion in rigid panels, "ring" and "stellate" patterns can be useful in analyzing damage to resilient panels - which regain their shape once force has been released. This can be useful, since resilient body components are becoming ever more common.

In general, the relatively hard, multi-layered paint films applied to metallic panels are the most susceptible to stress fracturing and delamination in damaging contacts. The top-coat paint applied to rigid non-metallic panel tends to be less likely to fracture and chip in damaging contacts—perhaps because the finish is not "baked", or because the lack of need for anti-corrosion treatments results in a thinner paint film with fewer material layers, and fewer inter-coat boundaries where failures can occur.

The plasticized paints applied to resilient panels are the least likely to fracture and chip in damage situations, due mainly to the inherent flexibility of the finish material, and possibly enhanced by the minimal layering of dissimilar materials. When plasticized paints are stressed to the point of failure, they tend to slip, buckle and tear.

SUPERFICIAL VS. PENETRATING CONTACTS

Most real-world accidents involve compound motions between the struck vehicle and the striking object. We will divide such moving contact events into two broad categories: In *Superficial* contacts, the contacting object for the most part slides across the surface of the paint film, and induces its damage through the paint film by shifting the uppermost layer(s) against underlying layers in the finish. In *Penetrating* contacts, the contacting object mechanically enters the paint film to induce force- and damage directly within the layers.

SUPERFICIAL CONTACTS

Moving contacts can merely scuff the paint surface, or cause material transfer between surfaces; more forceful non-penetrating contact can displace the upper paint film layers (described below), or generate chains of static-type fractures (described after that). And each contact may form patterns with usable directional indicators.

Following the hammer-blow tests which produced the ring- and stellate fractures, we used the same hammer to direct glancing hammer blows across the surfaces of painted panels at extremely shallow angles—causing no apparent distortion to the panel. In most cases, the contact merely scuffed the paint surface. But on a door skin from an early-`90's Mazda MPV mini-van, the hammer strikes generated a string of "ripples" in the paint film. These ripples only developed on the section of the panel that had received the "second" color of a factory-applied "two-tone" paint scheme.

Figure 12 shows one of these lines of evenly spaced "ripples"; the hammer stroke was from left-to-right:

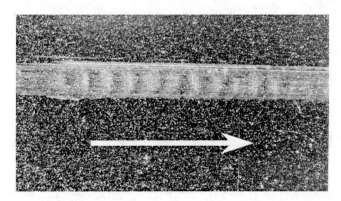

Figure 12 – Ripples in Rigid Paint Film.

The ripples stood slightly above the surrounding paint surface, and resembled the compression pleats that form in a rug that has been pushed from one end across a smooth floor. The rug analogy is not perfect, since the sliding rug forms pleats because its opposite ends have moved closer together; in this case, the strip of rippled paint had not obviously detached from the surrounding material to experience a shortening in length, so the formation of ripples suggested that the paint had stretched beyond its original dimensions.

Whatever the mechanism, the ripples stood slightly above the surrounding undamaged paint film. The curvature of individual ripples bulged in the direction of the impactor motion, consistent with the involved paint film having been shifted by the impactor.

Figure 13 is a tracing of the photograph, highlighting the curvature of the ripples, as well as the locations of the light-colored lines that predominate on the flanks of the ripples that faced the passing hammerhead:

Figure 13 – Tracing of the Ripples

These light-colored lines on the "approach" flanks of the ripples are believed to be shallow contact scuffs in the paint's clear-coat layer.

Three features of these ripples indicate that they had developed <u>ahead</u> of the impactor:
1. The surface scuffs within the rippled finish are most dense on the ripple surface facing the advancing impactor, with the ramp of each developing ripple minimizing contact to the opposite surface.
2. This, in turn, suggests that the ripples formed sequentially during contact, and were not the product of elastic recovery by the paint film after the hammer had passed.
3. The centers of the ripples have a distinct curvature, which bulges in the direction of motion. A tension mechanism would probably tend to pull material flat, rather than generate ripples that rise above the surrounding paint surface.

DELAMINATION OF DAMAGED PAINT

Probing the contact area broke away small, irregular chips—indicating that the ripples had both fractured from the color-coat, and also caused it to debond from the under-layers.

Clear adhesive tape was pressed across the ripple and removed, and it pulled away a chain of three triangular chips to expose more gray primer--which indicated that the paint film had failed at the top-coat/primer boundary. The resulting "hole" is shown in Figure 14:

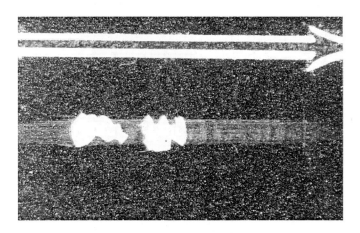

Figure 14 – Chips from the Rippled Paint.

The upper edge of the right "hole" has a distinct saw-toothed profile, with straight sides that angle outward from the centerline of the damage at 45-degree angles. Each side crack leads to the interior of a convex face that curves back toward the center of the damage—and the root of the next side crack.

The straight fractures on the opposite sides of these chips define a nearly perfect 90-degree angle characteristic of a shear failure in a brittle material, with fractures propagating forward- and outward at roughly 45 degrees to either side of the line of contact. [12]. The straight side fractures also match the linking cracks that generate between ring fractures in moving, indenting contacts, while the bulging face duplicates the curved ring fracture that develops ahead of an advancing impactor—as will be discussed in the next section.

As a note, the undersides of the chips revealed a silvery metal-flake—even though the topcoat color was a deep blue. Examination determined that the deep blue color had been achieved by applying a transparent blue topcoat over a metallic silver "mid-coat".

MOTION DURING INDENTING CONTACT

When an impactor simultaneously moves across a painted surface while pressing inward to deform the underlying panel, the paint around the moving impactor is subjected to shifting (shearing) forces as well as tensile stresses. This combination of forces is able to generate a series of over-lapping ring fractures in brittle paints as a result of the distortion of the underlying panel.

The ring fractures produced by a stationary impact have a counterpart in moving contacts, in the form of the "chained" ring fracture. While a stationary indentation generates a fixed circular zone of stress in the paint film, the stresses surrounding a moving contact shift in the direction of motion. In addition to producing concentric fractures in the paint film, these stresses can produce a linear chain of partial ring fractures, with each ring centered along the line of contact. Figure 15 is a line drawing of such a "chain" of partial ring fractures:

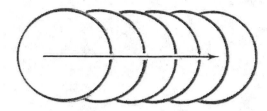

Figure 15 – Idealized Chain of Ring Fractures.

The incomplete nature of these partial ring fractures holds evidence that they were produced in succession round the advancing impactor, with bulges of the successive crescent-shaped fractures indicating the direction of moving impactor. It appears that as each successive ring fracture develops, its limbs can only propagate rearward until they are halted by a pre-existing fracture.

Figure 16 shows the damage found in the brittle paint of a mid-1970's domestic car, whose repainted door had sustained a shallow impact from a darker-colored vehicle. A smooth, dark-colored component of the striking vehicle initially broke through the paint of the struck vehicle, then rode onto the paint surface--leaving scuffs and dark transfers as it moved forward- and upward (left-to-the right in the image): As the object moved across the door, it generated an overlapping series of partial ring fractures in the paint:

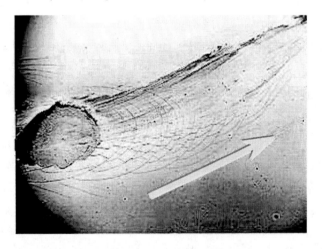

Figure 16 – Chain of Ring Fractures.

As the moving impactor produced the long indentation in the door skin, it subjected the overlaying paint to a moving source of radial tension; after the initial contact generated complete ring fractures around the initial contact site, the moving contact produced a succession of crescent-shaped partial-ring fractures, which all center upon the mid-line of the contact path. But these subsequent fractures form incomplete circles, and the limbs of each successive fracture terminate where they meet the fracture line of the immediately preceding ring.

As each fracture ring develops, it isolates the band of material between itself- and the preceding ring fracture. The central part of this isolated material lays in the path of the advancing contact; as the contact moves across each band, the impactor could crush the paint layer(s) that lay directly beneath it - and shift those layers in its direction of motion.

The movement of that contacted zone of finish material would generate stresses around the contacted zone: the material directly ahead of the advancing impactor would undergo compressive loading, while the material directly behind the moving impactor would undergo tensile forces; the movement of the contacted material between the undisturbed areas to either side of the contact would undergo shearing forces. In areas where such shear loading exceeded the strength of the paint film layers, it should generate fractures that propagate forward- and outward to each side of the line of contact. Such fractures are clearly visible in Figure 16, mainly below the contact path. As seen in the grazing hammer strikes, the cracks appear to be the product of shear within the paint film, because they form at right-angles to the ring fractures, and generally radiate forward- and outward from the contact zone at roughly 45 degrees to the direction of force. It is also possible that some of these fractures develop in response to tensile forces that form in the paint film behind the advancing impactor.

We called these features "linking cracks", because they link the bulging (convex) face of preceding ring fractures to the inward-curved (concave) surface of the successive ring fracture. Because the linking cracks are almost exclusively limited to the material bands between adjacent ring fractures, they appear to represent the final stress failure in the paint film that begins with the formation of the chained ring fractures. These "linking cracks" are remarkably similar to the side-cracks that defined the fan-shaped chips lifted from the rippled paint.

As can be seen in Figure 16, many of the linking cracks propagated at angles other than 45 degrees; some angle away from the principal direction of force by as much as 60 degrees—although none of the linking cracks appear to propagate at less than 30 degrees to either side of the direct-contact zone. The reasons for these variations are unclear, but the fractures may be developing at 45 degrees to lines-of-force that are diverging from the direction-of-motion (with their complimentary fracture obscured- or obliterated by the moving impactor). Alternately, they may simply be following the path of least resistance--perhaps through irregularities in the paint film, or around stress concentrations due to distortion of the underlying panel [12]. It is also possible that a rapidly-moving impactor may generate linking cracks that propagate rapidly enough to intercept the expanding ring fracture limbs and halt their progression. In doing so, the fractures isolate a fan-shaped zone of paint, quite similar to those lifted from the ripples in rigid paint.

RIPPLES IN RESILIENT PAINT

The modified paints applied to resilient panels also develop ripples in shallow contacts, but the inherent flexibility of the paint film allows a higher degree of distortion prior to the development of failures in the topcoat, or between the topcoat and the under-layers. As often as not, the ductile paint does not fully recover to its pre-stretched condition—leaving excess material that forms exaggerated ripple "pleats". When failures do occur, the paint buckles and tears instead of fracturing—forming distinctive "ruffled" edges.

Figure 17 shows several of the features that can be found in damage-induced failure of modified paint films. This paint was on the resilient bumper cover of a 1994 Ford Probe. A dark object with a sharp element had gouged the paint film in the compound direction indicated by the arrows:

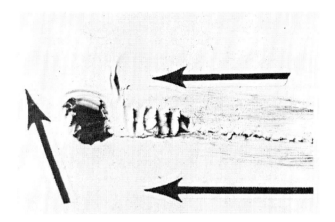

Figure 17 – Ripples in Resilient Paint.

The contact began when a narrow, dark object with a thin, sharp element moved across the paint surface; some of the dark surface material of the object deposited onto the white bumper paint, but the thin element cut through the paint film--loosening small flaps of paint that then twisted to gave the cut a jagged appearance.

The thin cut broadens abruptly at the center of the picture, stretching and buckling the white paint into a series of parallel pleats, separated by valleys of relatively flat paint; the paint film ultimately split across the tops of four of the ripples to expose the black cover material. The sharp element of the impactor moved past the vertical ripples before it stopped, and then moved sharply to the right--shifting a broad patch of paint that folded into multiple ripples that provide an unmistakable indicator of motion.

Close examination of the boundaries of the shifted paint patch found that the material had stretched and folded before tearing, leaving twisted ripples at the edges of the scrape that turned in the direction of motion. These features would indicate direction even if the displaced material had been lost.

A separate element of the same object made brief contact ahead of the bottom arrow, cutting- or tearing the paint film to expose the black bumper, and push up two sharp ripples which bend across the center in the direction of motion. Ripples formed in flexible paints are often able to "heal" if the contact does not break the paint film, because the paint film can recover much of its pre-distorted shape. If the paint film breaks, the finish will buckle, shift and tear, leaving discernible patterns that record the direction of motion.

DAMAGE FROM PENETRATING CONTACT

As described in our examples of moving superficial- and indenting contacts, essentially brittle paints can respond to moving damage by generating a combination of curved- and straight fractures. This tendency is amplified when the moving object actually penetrates- and travels within the paint film.

After the inherent brittleness- or ductility of the paint film, the greatest determinant of paint damage in a contact event appears to be the degree to which the paint film is penetrated by a moving impactor. When a moving impactor penetrates the paint film, it is able to introduce forces to several layers of the paint film simultaneously.

THE SINGLE FAN CHIP

As noted with the chip lifted from the rippled brittle paint, two types of fractures develop to isolate the individual paint chips: straight side fractures originate in pairs that diverge at approximately 45 degrees to each side of the line-of-contact, to enclose an angle of approximately 90 degrees. At about the same time, a curved fracture—bulging in the direction of motion—develops ahead of the moving impact, with limbs that scribe a regular arc.

Through some mechanism not yet understood, the straight "side" fractures and the bowing "face" fractures intercept- and halt one another's development to isolate an area of paint that is typically lost—leaving a quarter-round pit in the paint film. We dubbed these features "fan-chips" because of their resemblance to an unfolded paper fan. Figure 18 illustrates the shape- and presumed propagation of the "side" and "face" fractures in the formation of a "fan chip":

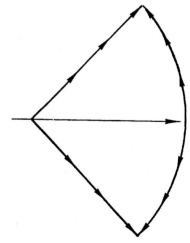

Figure 18 – Idealized Single Fan Chip.

Occasionally, the side cracks have been found to extend slightly beyond the bulging "face" fracture; since either fracture should halt the development of the other, this suggests that the straight side fractures may develop first- or more rapidly than the curved "face" fracture.

We note that the shape of the "fan chip" resembles the cross-section of the cone of glass that is punched out of plate glass when struck by a small projectile, such as a "B.B." or air gun pellet. Both failures are initiated by a small object that penetrates the material, generating concentric shear fractures that expand simultaneously forward and outward from the point of origin--classically at 45-degree angles to the line of force. Such a ballistic hole is shown in Figure 19:

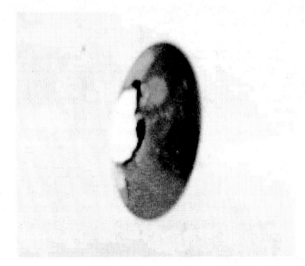

Figure 19 – Conical Ballistic Hole in Glass.

It is this resemblance that prompted us to think of paint films as laminated non-crystalline solids, similar to glass. The analogy is not a perfect one, because we have recently examined one ballistic hole in plate glass with distinctly concave sides, instead of the straight side fractures found in most ballistic holes—and in our fan chip holes.

The depth of the cone broken from the window is limited by thickness of the glass; this raised the question of what mechanism(s) governed the size of fan chips? When a penetrating impactor generates more than one "fan chip", they all tend to be of roughly the same size. The factors controlling- and limiting the dimensions of a fan chip are presently unknown.

MULTIPLE FAN CHIPS

The presence of even a single fan chip can establish the direction of contact, but our controlled testing and field experience alike have found that sustained scratching- or scraping contact usually generates multiple fan chips that may be set angle-to-face or even over-lap, so the face of one fan chip will obliterate the origination point of the succeeding chip.

When a number of fan chips are chained or "nested" together in this manner, the edges of the scrape takes on a saw-tooth appearance. An idealized chain of nested fan chips is illustrated in Figure 20:

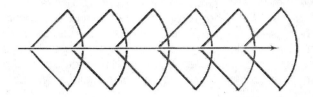

Figure 20 – Idealized Chain of Nested Fan Chips.

As with the solitary fan-chip, the direction-of-motion is indicated by the divergence of the straight side fractures, and the direction of the bulge of the face factures.

The factors that determine the "periodicity" or frequency of fan-chip formation along a contact furrow are still under investigation. In our experience, the side fractures of each new fan-chip tends to originate from the furrow edges at the point where the preceding curved "face" crosses the furrow. In general, narrow furrows tend to produce lines of slightly overlapping fan-chips, while broad furrows can produce fan-chips that nest deeply together—sometimes to the point that it is difficult to distinguish the straight side fractures from the curved face-fractures. It can also be difficult to determine the straightness- or curvature of extremely short fractures. In these cases, the best source of information sometimes comes from the first- or last fan-chip in the chain, from an isolated fan-chip elsewhere on that scratch, or from features in an adjacent scratch.

Our first impact tests with the pendulum involved striking masonry cinderblock across the various painted panels. The rough, granular surface of the cinder-block had numerous asperities, which generated multiple parallel lines of damage in the paint film, from mild scuffs that merely dulled the paint finish, to gouges which cut deeply into the paint film—displacing material as they progressed, and introducing stresses directly within the layers of the paint film.

One such contact across the door skin from the Mazda MPV mini-van produced a broad scrape mark which penetrated the color-coat and dislodging a series of chips—leaving behind a nested chain of fan-shaped chip-holes (Figures 21A and 21B). Most of these chip-holes exhibit the straight sides defined by fractures that radiated outward from both edges of the scrape in the direction of motion, and terminating in face fractures that bulged in the direction of motion.

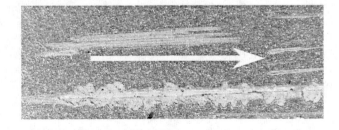

Figure 21A – Scrape with Multiple Fan Chips.

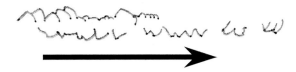

Figure 21B – Tracing of Side- and Face Fractures.

As noted about the linking cracks in the chained ring fractures of Figure 16, the side fractures of many fan chips diverge from the contact furrows at angles other than the 45-degree "rule" proposed by the shear mechanism model [12]. The fact that the mid-lines of individual fan chips frequently pointed away from the line-of-motion may be due to localized strong- or weak points within the paint film, or the interaction of forces within the paint as the fan chips develop. That said, the majority of matchable sets of side fractures enclose 90 degrees of arc, and leave distinct "quarter-round" pits in the paint film. With that variability in mind, the side fractures always radiate forward- and outward from their scratch- or scrape in the direction of the damaging motion, while the curved face fractures always "bulged" in the direction of motion.

In Figures 21 A/B, several of the face fractures are also associated with faint, curved remnants within the contact furrow. This feature recurred repeatedly during our work, and is described in a following section.

Occasionally, the fan chips found along the edges of a broad furrow are in fact side-by-side chains of fan-chips that originate from separate but parallel contact lines within the furrow. A sketch depicting the development of overlapping side-by-side fan chips is shown in Figure 22:

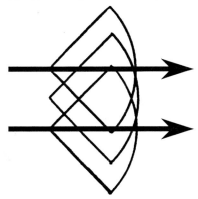

Figure 22 – Overlapping Fan Chips from Parallel Furrows.

Figure 23 is a macro photograph of a complex scrape inadvertently produced by the head of a mounting screw during pendulum testing of the door skin from a compact G.M. pickup truck. The broad, shallow contact caused the color coat to break away over a broad area, even though the contact caused only minor rippling of the light-colored primer, and did not distort the door skin itself. Close examination of this furrow found multiple, overlapping fan chips originating along at least three parallel furrows:

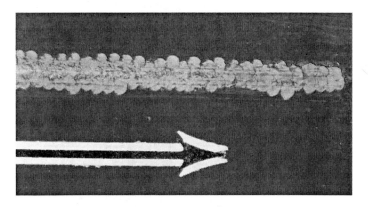

Figure 23 – Overlapping Fan Chips in Parallel Scrapes.

The edges of the scrape are composed of multiple curved fractures, representing the faces of densely packed fan chips. Because the side-fractures of each chip trace back to their origin along a furrow, the numerous small-radius chip holes had originated along several separate furrow lines.

Multiple lines of origin had not been recognized in the field prior to this test, but have since been found in real-world contacts against coarse surfaces. The curved chip faces are so densely packed that few straight side cracks are visible. This fact would normally complicate direction-of-motion identification, were it not for a number of "outlines" left behind by the departing fan chips.

ADHESIONS WITHIN DAMAGE

Fan-chips usually shear away cleanly from their underlying layer, leaving only the curved fracture edges in the material bordering the furrow. But occasionally, the departing fan chips leave behind remnants of the detached material. Close examination of the overlapping fan-chip holes in the scrapes shown in Figures 21A and 23 reveal multiple faint lines of color-coat that remain within the scrape. These remnant lines exhibit curvature corresponding to radii of adjacent fan chip holes—and many of these remnants are clearly extensions of partial fan chips at the edges of the scrape.

We have dubbed these remnants of departed fan-chips "adhesions"; the mechanism of their formation is beyond the scope of this project, but we believe they are left behind during the final detachment of the chip, as the curved "face" rises from the panel. As such, the shape of an "adhesion" should indicate direction as reliably as a fan chip.

Figure 24 is a sketch of an idealized chain of fan-chips, with dashed lines representing the locations of adhesions:

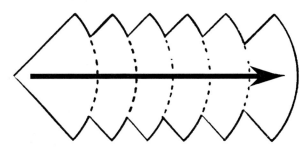

Figure 24 – Adhesions Within Chain of Fan Chips.

Adhesions were first recognized in the test-created damage shown in Figure 23, but have since been found on a number of accident-damaged vehicles. A remarkable example is shown in Figure 25, a photograph of damage on the door of a 1987 Toyota Corolla:

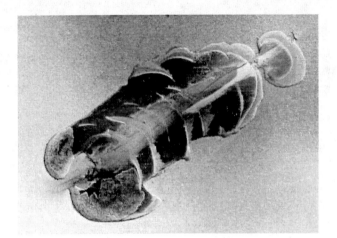

Fig. 25 - Adhesions Within Gouging Damage.

Shown approximately twice life-size, this gouge was produced by a thin object with a smooth surface moving across the panel (left-to-right and upward in this view) while indenting a door panel. The contact had stretched the metal and fractured the paint film approximately 0.5 cm (7/16-inch) to each side of the "furrow".

As the impactor moved through the paint, it generated deep fractures that propagated through the paint- and primer layers, down to the dark "electro-coat" corrosion treatment layer of the door skin. At least 12 separate chips were broken from the paint film, overlapping to such a degree that no straight side fractures are evident. Seven of these departing chips left behind curved adhesions of primer on the Electro-coat anti-corrosion layer, with the face fractures bulging the direction of motion.

The terminal end of this deep gouge displays a single, oval chip, which exhibits another directional indicator—the "stepped" delamination produced by an impactor rising out of a multi-layered paint film.

CONCHOIDAL DELAMINATION

Similar eccentric damage features were found at the terminal ends of several scratches produced in out tests. In each case, the impactor had risen out of the paint film at some angle, either because the impactor had risen from the panel, or because the curvature of the panel caused the impactor to depart from the surface. As the impactor ascended through the layers of the paint film, fractures apparently propagated ahead of the immediate contact area, preferentially along the intercoat boundaries to split the film into its separate layers. We believe that the rising impactor is then able to apply an upward-angled force to the edges of the debonded layers, which allows it to "pry" a stepped chip from sequentially higher layers of the paint film. The result was an eccentric chip hole, in which the "steps" are broadest in the direction of motion.

We chose to call the process "conchoidal delamination", because the stepped chip hole that result from the failure resembles the conchoidal fractures of non-crystalline solids encountered in geology [14]; such features are also known as "clam-shell" fractures, for their resemblance to the eccentric growth rings in the shells of clams and other bivalve mollusks. [12,14]

An idealized conchoidal chip hole is illustrated in Figure 26:

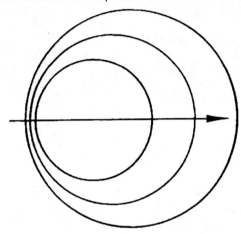

Figure 26 – Eccentric Conchoidal Chip Hole.

An excellent real-world example of a conchoidal fracture produced by a disengaging impactor pattern was found at the terminal end of the gouging contact of Figure 25 in the adhesions section; a magnified view of the end of that feature is presented here as Figure 27:

Figure 27 – Conchoidal Fracture from Rising Impactor.

As the impactor gouged downward- and rearward across the door skin, its inward component indented the door skin. The impactor then lifted out of the paint film—perhaps deflected away by an outward bend in the door skin. The rising impactor broke away a final, oval chip, in which the smaller hole in the primer is rimmed by a larger hole in the topcoat that is offset in the direction of motion.

The paint chip holes formed during conchoidal delamination differ from fan chip holes in at least two ways: for unknown reasons, the conchoidal chips tend to be larger than the fan chips liberated by furrowing contacts, and; the perimeters of conchoidal chip holes are entirely ovoid- or circular—lacking the straight side fractures characteristic of fan chips. This latter difference is probably of greater importance, since it indicates some dissimilarity in the mechanism of formation. We currently visualize the eccentric shape of ovoid chips as equivalent to a cone of material departing the paint film at an oblique angle, as shown in Figure 28:

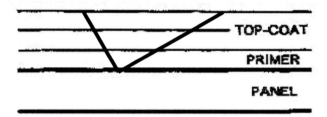

Figure 28 – Oblique Cone in Paint Film.

NOTE: As the impactor departed across these layer steps, it smeared a small quantity of the gray primer into the color-coat. Material displacements and transfers provide additional reliable means of determining the direction of a moving contact, and are subject enough for their own project.

Conchoidal chips can develop wherever the line of force applied by the impactor rises from the plane of the paint film at a moderate angle, estimated to be at least 15 degrees. This type of disengagement can occur when the impactor lifts away from the panel, or when the panel curves away from below the path of the impactor. This fact was demonstrated by our own testing, as well as conchoidal chipping damage found at the edges of body panels where the contact had stopped short of the panel edge. The exiting impactor apparently exerts a lifting force on the final ring of disturbed paint that had formed ahead of the advancing contact.

In fact in our field work, conchoidal delamination features are most often found where linear contact passes across a distinct downward curve- or fold in the panel surface, such as the gap between panels, or a recessed bodyline, as illustrated in Figure 29:

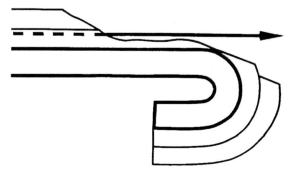

Figure 29 – Straight Contact Departing Edge of Panel.

The photograph in Figure 30 shows three inter-connected conchoidal chips produced in one of our tests, when a cinder block passed across a recessed contour line in the door from a Mazda MPV. In this view, the direction of motion was left-to-right, with disengagement occurring when the impactor crossed a recessed contour line:

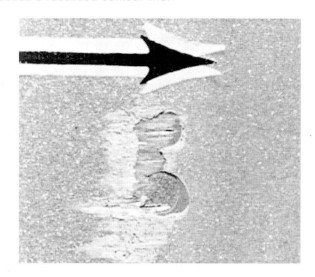

Figure 30 – Conchoidal Chip Holes at Recessed Contour.

Elements of the departing impactor broke away three circular chips grouped in the form of a wing nut. In this example, the upper- and middle chips had lost the topcoat to expose light gray primer that was then scraped by rough grains on the impactor's surface. In the large bottom chip, the silver topcoat had flaked away to expose a crescent of electro-coat panel treatment; behind that, a circular chip of the e-coat layer had detached to expose the metal panel. As before, the stepped appearance developed in the direction-of-motion.

A variation of conchoidal chipping can also occur where an object contacts the leading edge of a painted panel. Figure 31 is a photograph of a conchoidal chip hole found at the rear edge of a fender, which had been forced into the edge of the adjacent fender:

Figure 31 – Conchoidal Chip from Edge Contact.

Instead of a rising impactor, the contact to the rounded door edge had generated lines of force that rose through the paint film at an angle that shifted the paint up- and over the surrounding material at as many as three points. In doing so, the forces apparently generated three chips, which joined to form somewhat scalloped edges in two layers of the paint film. As with other conchoidal chips, the outermost layer is broken away over a larger area than the layer below—which also happened to be white.

As a note, the presence of the second color-coat indicates that the fender had been repainted before the accident that led to the fender being replaced. The quality of both color coats, combined with the absence of a primer layer between the color-coats suggested that the fender had been damaged- and refinished at the factory, rather than being repainted at a body shop.

CONCLUSIONS

This project has sought to describe the variety of fractures that can develop within laminated paint films as the result of specific types of damage to the paint film, and to the underlying panel. The controlled damage-inducing contacts conducted for this project have verified earlier field observations that contact-damaged paint films often develop one- or more features that can provide a reliable record of both the nature- and the directionality of the damaging event. These tests also produced damage patterns that had not previously been recognized in earlier field examinations—but have since been documented in damage occurring in on-the-road accidents.

The moderately brittle nature of the paints applied to rigid panels causes them to fail in a limited range of patterns, which closely resemble the failure patterns of non-crystalline solid materials when subjected to similar stresses. The elasticity of the paints applied to resilient panels can conceal fracture damage, but dye-penetrant methods are available to reveal these patterns.

Further investigation and standardized testing may identify additional directional indicators in contact-damaged paint. Additional study is warranted to better understand the mechanisms that produce specific types of paint damage, and such study could further refine our understanding of the causes of individual damage patterns.

ACKNOWLEDGEMENTS

The authors wish to acknowledge the debt of gratitude owed to Hugh Harrison ("Harry") Hurt III, Professor Emeritus of the School of Safety and Systems Management at the University of Southern California, for his ceaseless encouragement to bring this project to publication, and his assistance in achieving that goal. We also wish to thank Mr. Stein E. Husher for his technical assistance in the production of this document.

REFERENCES

1. U.S. Department of Transportation, National Highway Traffic Safety Administration. "Traffic Safety Facts 1999: A Compilation of Motor Vehicle Crash Data from the Fatal Analysis Reporting System and the General Estimates System". Washington D.C., December 2000.

2. Clark, Roger E. "Investigation into the Characteristics and Interpretation of Paint Damage Patterns Created During Motor Vehicle Collisions." Unpublished term research project, Institute of Safety and Systems Management, University of Southern California, Los Angeles, 1977.

3. Yanik, Anthony J. "Color Era Dawned with '24 Oakland True Blue Six." Automotive News, America at the wheel – 100 years of the Automobile in America. September 21, 1993. (p.80). Crain Communications, Inc., Ann Arbor, Michigan.

4. Sawyers, Arlene. "Automotive Color Tips its Hat to the Fashion Runways of the World." Automotive News, America at the Wheel – 100 Years of the Automobile in America, September 21, 1993 (p.56). Crain Communications, Inc., Ann Arbor, Michigan.

5. Crewdson, Frederick M. "SPRAY PAINTING – Industrial and Commercial." Frederick J. Drake & Co., Chicago, Illinois, 1957.

6. Fuller, Wayne R. "Understanding Paint." THE AMERICAN PAINT JOURNAL COMPANY, St. Louis, Missouri, 1965.

7. "Tri-coat refinishing." DuPont Refinishing product brochure, DuPont Company, Wilmington, Delaware.

8. MAZDA MOTOR OF AMERICA, INC. "ENVIRONMENTAL PAINT DAMAGE REPAIR", Japan, July 1993.

9. "Fundamental Painting Procedures." TOYOTA MOTOR CORPORATION/LEXUS DIVISION. Overseas Service Division, Japan, 1989.

10. "TECHNOLOGY OF PAINTS, VARNISHES AND LACQUERS". Martens, Charles R., Editor. REINHOLD BOOK CORPORATION, New York.

11. Hurt, H.H. Jr. "Fundamentals of Helicopter Structures – Part II", University of Southern California, November, 1967.

12. Hurt, H.H. Jr. Personal communications, March 1994 through December 1994.

13. MAGNAFLUX CORPORATION. "Principles of Stresscoat – Brittle Coating Stress Analysis", Chicago, Illinois. 1971.

14. Geology Explorer, Iverson Software Co., Fridley MN USA, http://www.iversonsoftware.com/geology/c/conchoidal_fracture.htm

15. "Webster's Ninth New Collegiate Dictionary", Merriam-Webster Inc., Springfield, Massachusetts, 1988.

16. Goodsel, Don. "Dictionary of Automotive Engineering – 2nd Edition", – Society of Automotive Engineers, Inc., Warrendale, Pennsylvania, 1995.

17. Tomsic, Joan L. "SAE DICTIONARY OF AEROSPACE ENGINEERING – 2nd Edition.", Society of Automotive Engineers, Inc., Warrendale, Pennsylvania, 1998

TERMINOLOGY

Several "everyday" terms for panel damage appear throughout this paper. Since universally accepted classifications for these features were not found within the Accident Reconstruction literature, we developed our own working definitions of these terms, along with comparative references to other damage features [14,15,16]. The terms are defined in order of the size of the feature and/or the severity of the contact they represent, rather than alphabetically:

SCUFF: A superficial, often broad abrasion of the outermost layer of the finish, which typically alters the color or reflectivity of the contacted area, without penetrating the paint film. *The least-penetrating form of linear contact damage.*

SCRATCH: A narrow- and shallow furrow that enters one- or more of the upper layers of the finish itself, and may expose the under-coat--but does not reach the underlying panel material. *More penetrating- and usually more localized than a scuff.*

SCRAPE: A moderately deep- or broad abrasion that may penetrate several layers of the paint film to expose the panel, but does not cause damage to the panel itself. *A scrape can be composed of numerous, parallel scratches.*

GOUGE: A deep furrow that penetrates- and displaces multiple layers of the paint film, and causes damage to the underlying panel. *The most severe linear contact damage—deeper- but often more localized than a scrape.*

CRACK or FRACTURE: A thin, linear fissure in one- or more layers of the paint film, resulting from stress. *Cracks often develop in furrowed paint—straight fractures diverge from the furrow in the direction-of-motion, while curved fractures "bulge" in the direction of motion.*

CHIP: A small fragment of the paint film (or the pit produced by the loss of that fragment) due to concentrated damage from a localized contact—such as a strike by a small projectile, or dislodged during the formation of a **scratch** or **scrape**. A chip through the top-coat often carries with it material from one- or more of the underlying layers. *A chip is a "point" feature, while scratches, scrapes and gouges are "line" features.*

DENT: A permanent, concave indentation of a panel surface, caused by direct impact or -force. *A dent may- or may not be accompanied by paint damage.*

PAINT FAILURE TERMINOLOGY

We have coined the following terms to describe specific paint damage features that provide directional information on the damage-producing event. They are listed relationally:

FAN CHIP: A generally quarter-round flake of paint material detached from the paint film by the formation of straight- and curved fractures along furrowing damage features; also, the correspondingly-shaped hole left in the paint film by the loss of the Fan Chip. The orientation of a Fan Chip, or either its component fractures, identifies the direction of the damaging event.

SIDE FRACTURE: A straight fracture that originates at the edges of furrowing contact, and propagates forward- and outward from the line of contact in the direction of motion. Side Fractures typically develop at 45 degrees to the line of motion, and often form in mirror-image pairs on opposite sides of the furrow.

FACE FRACTURE: A curved fracture that spans the terminal ends of the side fractures, as the apparent final event in the formation of a Fan Chip. The convex Face Fracture bulges in the direction of motion of the damaging event.

ADHESION: Material remnants from the underside of a departed Fan Chip, left on the exposed surface by the detachment of the Face edge of the chip. The convex shape of an Adhesion is identical to that of the departed Face Fracture—convex in the direction of motion.

STEPPED DELAMINATION: Enlarged detachment of progressively higher layers of material during development of a Fan Chip, with the Steps being broadest in the direction of motion. Stepped Delamination occurs when the line of force rises out of the paint film—due to changes in the path of the impactor, or changes in the contours of the panel.

CONCHOIDAL CHIP: An eccentric, ovoid flake of two- or more layers of paint material, detached by an impactor rising out of a multi-layered paint film; also, the correspondingly-shaped hole left in the paint film by the loss of the Conchoidal Chip. The direction of the offset of the progressively higher layers identifies the direction of motion.

CONTACT

ROGER CLARK ASSOCIATES,
6407 Alondra Boulevard,
Paramount, CA 90723-3759 USA
Tel: 562 529 6900, Fax: 562 529 6914
E-mail: info@rc-recon.com

Reviewer Discussion

Reviewers Name: Philip V. Hight P.E.
Paper Title: Identification and Interpretation of Directional Indicators in Contact-Damaged Paint Films
Paper Number: 2002-01-0542

SAE #2002-01-0542
Identification and Interpretation of Directional Indicators in Contact-Damaged Paint Films
Neil Clark, Roger E. Clark, Authors

I should like to congratulate the authors on bringing yet another new tool into the field of accident reconstruction. This comprehensive paper attempts to explain the different impact dynamics and mechanisms that cause visible damage to the paint at impact sites and adjacent sites. Many of the tests were conducted using a pendulum. The laminated layers of automobile paint were on conventional panels, semi resilient panels and flexible panels. Some of the paint had flex agents to allow the paint to remain intact in minor impacts.

One important factor of this research may be as a tool to evaluate the magnitude of rear end impacts. Dynamic crush is not easily available from inspecting the resilient bumpers, which return to their normal position post collision.

A test on a small, non-deformed area of the bumper could duplicate the paint fractures that are seen on the collision area of the bumper. This may lead to another damage scale.

Further test work is required to duplicate paint damage to accident vehicles where the direction and magnitude of impact is well established. The pendulum may be changed to a portable impactor. There may be three different impact faces in order to best coorelate the localized impact damage. I look forward to follow up papers in this pioneering field.

Reviewer Discussion

Reviewers Name: Chris Armstrong
Paper Title: Identification and Interpretation of Directional Indicators in Contact-Damaged Paint Film
Paper Number: 2002-01-0542

Chris Armstrong Reviewer Comments:

The paper "Identification and Interpretation of Directional Indicators in Contact-Damaged Paint Film" is an important addition to the lexicon of accident reconstruction literature. Mr. Clark has engaged in a detailed and illuminating study of automotive paint structures. I believe his definition of terminology will allow reconstructionists to communicate in a coherent manner when addressing the issue of paint film damage.

Chris Armstrong
Staff Engineer, KEVA Engineering
601 Daily Dr. Suite 225
Camarillo, CA 93010
805-388-6016

Evaluating the Uncertainty in Various Measurement Tasks Common to Accident Reconstruction

Wade Bartlett
Mechanical Forensics Engineering Services, LLC

William Wright and Oren Masory
Florida Atlantic Univ.

Raymond Brach
Notre Dame Univ.

Al Baxter
Suncoast Reconstruction

Bruno Schmidt
Southwest Missouri State Univ.

Frank Navin
University of British Columbia

Terry Stanard
Klein Associates

Copyright © 2002 Society of Automotive Engineers, Inc.

ABSTRACT

When performing calculations pertaining to the analysis of motor vehicle accidents, investigators must often select appropriate values for a number of parameters. The uncertainty of the final answers is a function of the uncertainty of each parameter involved in the calculation.

This paper presents the results of recent tests conducted to obtain sample distributions of some common parameters, including measurements made with tapes, measurements made with roller-wheels, skidmark measurements, yawmark measurements, estimation of crush damage from photographs, and drag factors, that can be used to evaluate the uncertainty in an accident reconstruction analysis. The paper also reviews the distributions of some pertinent data reported by other researchers.

INTRODUCTION

When performing calculations pertaining to the analysis of motor vehicle accidents, investigators often must establish values for a number of parameters, such as drag factor, distances along roadways, crush depths, skid lengths, and yaw marks. Some of these are established by making measurements, some through the use of tables, and others by using judgement. The uncertainty of the final answers is a function of the variations of each parameter involved in the calculation. Though it has long been recognized that variations exist in all measured data, including those related to accident analysis [1, 2, 3, 4] very little published information exists to assist the investigator in assigning realistic input parameter variations.

There are at least three different ways to estimate the uncertainty of reconstruction calculations. One approach is simply to combine the parameter ranges in such a way as to generate extreme high and low results. Though possible, this situation is unlikely to exist in the real world. A more analytical approach requires taking partial derivatives of the constitutive equations involved in the analysis. With these derivatives, the overall uncertainty of a calculation can be estimated using the Root-Sum-of-Squares (RSS) method outlined by the NIST [5] and demonstrated by Tubergen. [6] However, given the complicated nonlinear equations typically used in accident reconstructions, this can become impractical. Another approach, Monte Carlo simulation, often provides a more convenient means of quantifying the combined uncertainty. Monte Carlo analyses require the assignment of probability distributions and their parameter values to the input variables in order to determine the uncertainty of a reconstruction result to a particular confidence level. Several researchers have demonstrated the application of Monte Carlo Simulation techniques in accident analysis [2, 6, 7,8], but none has discussed which probability distributions should be used, nor have they suggested ranges that should be used for variations that exist in commonly used parameters.

This paper presents the results of recent tests conducted to obtain sample distributions of some common parameters, including measurements of well-defined distances made with tapes, roller-wheels, and a laser transit, measurements of skidmarks, arcs, and yawmarks, measurement of crush damage, estimation of crush damage from photographs, and measurement of drag factors by several methods. This paper also reviews the distributions of pertinent data reported by other researchers in the accident reconstruction field. Prudent application of proper distributions and ranges can lead to more accurate accident reconstructions, regardless of which approach is used.

STATISTICS AND PROBABILITY DISTRIBUTIONS

Notation in this paper follows proper statistical practice in that $\bar{x}$ and s are used to indicate the mean and standard deviation of a sample (comprised of a set of measurements) and μ and σ are the mean and standard deviation of a population.

After systematic errors and blunders have been accounted for in a group of measurements with one variable, random measurement errors will remain. In practice, random errors often follow a normal distribution, or "bell-curve." The frequency function describing this type of distribution is written as:

$$f(x) = \frac{e^{-(x-\mu)^2/2\sigma^2}}{\sigma\sqrt{2\pi}}$$ EQ. 1

Where μ = population mean
  σ = population standard deviation

In a normally distributed population, the measured value falls within one standard deviation of the mean value ($\mu \pm \sigma$) 68.3% of the time, as shown in Figure 1. The measured value can be expected to fall within two standard deviations of the mean ($\mu \pm 2\sigma$) 95.5% of the time and within three standard deviations ($\mu \pm 3\sigma$) 99.7% of the time. [9] There is no absolute upper or lower bound to the value. This function has a total area under the curve from -∞ to +∞ of 1 square unit. Unless

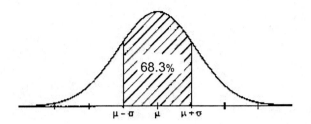

Figure 1: Normal distribution, showing the area encompassed by the mean plus or minus one standard deviation ($\mu \pm \sigma$)

otherwise stated, all statistical analyses in this paper will assume a normal distribution.

Two ways to determine if a particular set of data follows a normal distribution are the use of a normal probability axis and a Chi-squared "goodness-of-fit" test. In the former method, one determines if the data follows a straight line when plotted on a normal probability axis, as shown in Figure 7. The latter test is used to determine to a particular level of confidence if the data is a sample from a normal distribution. A comprehensive description of this method is beyond the scope of this paper, but can be found in most statistics books such as reference [10].

A second type of distribution commonly encountered in accident reconstruction analyses is the uniform or rectangular distribution. This type can be used in the absence of evidence to suggest that the value is more likely to be near the center of the range. This distribution is more conservative than a normal distribution, as it gives equal probability to all values in the specified range. At the same time, it precludes values outside the specified range, so the bounding terms, ($\mu - a$) and ($\mu + a$), must be selected with great care. The standard deviation for this type of distribution is equal to [$a / SQRT(3)$]. The range covered by ($\mu \pm \sigma$) includes 57.74% of all values. [10]

For a review of the terminology of uncertainty and statistical methods associated with evaluating uncertainty in measurements and calculations, the reader is directed to the references. [1, 5, 10]

LINEAR MEASUREMENTS TASKS

A series of linear measuring tasks were devised to test the repeatability of measurements that required little or no judgment regarding the location or dimensions of the item being measured. Uncertainty from these measurement tasks forms the basis for better understanding the errors involved in more complex tasks.

Many of these measurement tasks were conducted during the World Reconstruction Exposition 2000 (WREX2000) held in September 2000, at College Station, Texas, USA.

SHORT LINEAR MEASUREMENT

Participants were asked to measure the distance between two parallel lines printed on an 8-1/2 x 11-inch piece of paper using a standard 25 foot carpenter's tape measure labeled in units of feet and inches, with 1/16 inch graduations. Of the 28 participants, 24 reported the length to be 4-3/16 inches. One participant interpolated between graduations on the scale to report a length of 4.2-inches. Three participants, who reported they were comfortable measuring in units of feet and inches, apparently misread the scale. Two of these reported a length of 4-3/8 inches, and one reported 4-3/32 inches.

FLEXIBLE MEASURING TAPES

Two sets of three targets were arranged as shown in Figure 2. Participants were asked to measure the distances in the first set using a flexible tape measure graduated in feet and inches. On the second set a tape graduated in feet and tenths was used. Grass growing between the joints in the otherwise flat concrete surface was removed. The short measurement was nominally 36 feet (11 m) while the longer distance was nominally 90 feet (27 m). The "zero" end of the tape was held under the measurer's supervision by another random participant. There were twenty-nine participants. One participant reported a distance of 36.63 feet in the short-feet/tenths exercise which was more than 3 standard deviations over the mean, and was discarded for this analysis. The "actual" distances were measured with the Sokkia total station described later in this paper.

The distribution of these measurements appeared to be normal. The summary statistics for all four tape-measurement tasks are shown in Table 1.

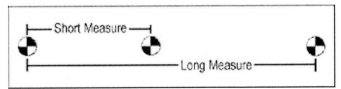

Figure 2: Target arrangement for short/long linear measurement task.

ROLLER WHEELS

On a set of flat, clean areas similar to those used for the tape measurement task above, participants were asked to measure the distance between targets using a single-wheel or a dual-wheel small-wheeled roller tape. Each task in this series had 20 to 30 participants. Figure 3 shows the distribution of results for the short-distance dual-roller wheel measurement task. The results from the other three tasks had similar normal-appearing results. Table 2 shows the summary statistics for four small-wheeled roller-tape measurement tasks.

One data point in each of the single wheel tasks and in the dual-wheel short measurement task was discarded for being more than three standard deviations from the

	Feet-inches, Short (ft.)	Feet-inches, Long (ft.)	Feet-tenths, Short (ft.)	Feet-tenths, Long (ft.)
"Actual"	38.50	90.60	36.06	91.60
Mean (ft.)	38.51	90.57	36.07	91.69
St. Dev. (ft.)	0.025	0.061	0.017	0.060
Coefficient of Variation	6.5×10^{-4}	6.7×10^{-4}	4.7×10^{-4}	6.5×10^{-4}
Minimum (ft.)	38.43	90.46	36.04	91.54
Maximum (ft.)	38.61	90.78	36.13	91.85
Count	29	29	28	29

Table 1: Results of measurements made with two types of fiberglass tape measures: feet/inches and feet/tenths.

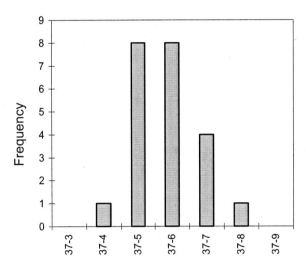

Figure 3: Short dual-wheel "roller-tape" measurement task results. ($n=20$, $\bar{x}=37.56$, $s=0.08$)

	Single wheel short	Single wheel long	Dual wheel short	Dual wheel long
"Actual"	35.08	91.47	37.50	89.56
Mean (ft.)	35.24	91.17	37.56	89.71
St. Dev. (ft.)	0.081	0.116	0.076	0.160
Coefficient of Variation	2.3×10^{-3}	1.3×10^{-3}	2.0×10^{-3}	1.8×10^{-3}
Minimum (ft.)	35.00	90.96	37.42	89.42
Maximum (ft.)	35.42	91.33	37.75	90.00
Count	30	28	22	20

Table 2: Results of measurement tasks using single-wheel and dual-wheel roller tapes.

mean. In the latter case the reported length was 35.50 feet, which appeared to have been a transcription error on the part of the participant. The other two data points were just slightly outside the high 3-sigma boundary.

The dual roller wheel was also used to measure an extended distance over a jointed concrete surface, with some small tufts of grass scattered along the path. The 23 participants in this exercise reported an average distance of 124.98 feet, with a standard deviation of 0.19 feet, giving a coefficient of variation of 1.5×10^{-3}.

The distributions of measurements made with roller wheels appear to be normal. The standard deviation for all measurements taken with several different roller wheels was larger than those taken with the fiberglass tape but was still very small. A linear regression curve fit using only the data obtained on the unobstructed paved surface indicates that the standard deviation (in feet) of similarly taken measurements can be approximated by:

$$s_x = 0.0011x + 0.039 \qquad \text{EQ. 2}$$

where x = the distance measured in feet.

MEASURING ARCS

At IPTM's 2001 Special Problems conference, [11] an experiment was conducted to assess participants' ability to determine an arc-radius using the chord and middle-ordinate measurement method. In this experiment, an arc with a constant radius of approximately 182 feet was scribed in white chalk on black asphalt, using a steel wire pinned at one end and pulled with a steady force. The intent of this process was to reduce the variability associated with selecting points to be measured on a less-than ideal tire mark. Each participant directed where a random attendee should hold the zero end of the tape, and then measured the chord and middle ordinate of the arc, as shown in Figure 4.

The radius of the arc was then calculated using:

$$R = \frac{C^2}{8M} + \frac{M}{2} \qquad \text{EQ. 3}$$

where R = radius, meters or feet
C = chord, meters or feet
M = middle ordinate, meters or feet

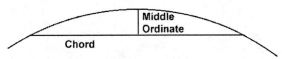

Figure 4: Chord and middle ordinate for radius determination.

Eighteen participants carried out the measuring task. Nine participants chose a chord length of 50 feet. The longest chord utilized was 57.5 feet (the largest allowed by the arc) and the shortest was 30 feet. The reported middle ordinates ranged from 0.625 to 2.25 feet. The calculated radius ranged from 173.5 feet to 193.1 feet. The frequency distribution of the radius calculated from the measurements is shown in Figure 5.

The data from this experiment appears to be normally distributed. Since only one arc was measured insufficient data exists to model standard deviation as a function of the curve radius.

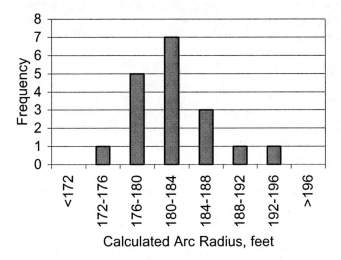

Figure 5: Measuring a chalked constant-radius arc. ($n=18$, $\bar{x}=182.3$ ft, $s=4.5$ feet)

MEASURING YAWMARKS

A yawmark was created on a dry concrete surface by a sport-utility vehicle. Participants were asked to measure the chord and radius of the mark with little guidance as to how to execute the task. Most participants used a 30-meter (100-foot) tape to measure the chord and a 7.6-meter (25 foot) tape to measure the middle ordinate. Participants were also asked to report the frequency with which they had performed this type of measurement in the past 12 months: zero times, one to five times, or more than five times. The radius was calculated using Equation 3, and a histogram of the results is plotted in Figure 6.

Two data points (one high and one low) were discarded because they were approximately 3 or more standard deviations from the mean. The standard deviation of the 30 remaining measurements was 4.6% of the mean.

To evaluate the normality of the data, the 30 data points were plotted on a normal probability axis, as shown in Figure 7. The data appears to follow a straight line, indicating they are normally distributed. Additionally, a Chi-Square Goodness-of-Fit test (not shown) revealed that at the 95% confidence level, the 30 points are from a normal distribution. No significant variation occurred in the measured quantities as a function of the time of day.

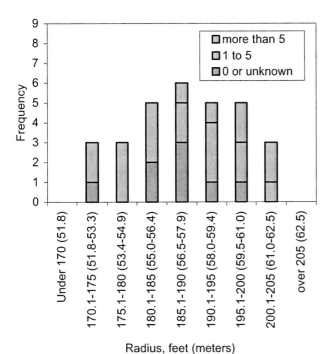

Figure 6: Measured radius of yawmark on concrete as a function of how often the participant had performed similar measurements in the past year, $n=30$, $\bar{x}=188.4$ ft (57.4 m), $s=8.7$ ft (2.7 m).

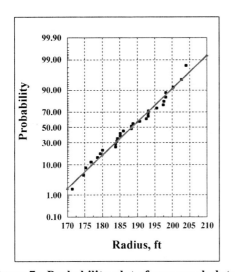

Figure 7: Probability plot of yaw-mark data.

MEASURING ANGLES

Measuring an angle with two linear measuring devices is a common field task. This process involves defining the two sides of a right triangle, with the angle to be measured between one side and the hypotenuse, as shown in Figure 8. An arbitrary point is selected along one side, and the distance from the angle of interest to the point is measured (shown as Side 1 in Figure 8). A line crossing through that point perpendicular to Side 1 is then visually approximated, defining Side 2 as shown in Figure 8. The length of Side 2 is then measured, allowing the angle of interest to be calculated with trigonometry.

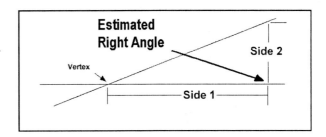

Figure 8: Layout for angle-measurement task.

The setup for this experiment involved striking two chalk lines on clean pavement. The seventeen participants were supplied two 25-foot carpenter's tapes. Linear distances selected on Side 1 ranged from 7 to 15 feet. The frequency distribution of the angles calculated from this task is illustrated in Figure 9. One data point was discarded for being approximately 3 standard deviations from the overall mean.

The data from this experiment appear to be normally distributed. Since only one angle was measured insufficient data exists to model standard deviation as a function of angle. The average of the measures correlates well with the angle as measured by a total station of 38.40 degrees.

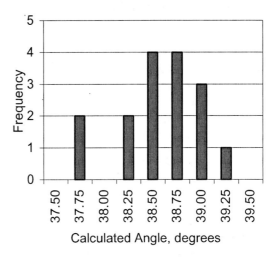

Figure 9: Angle calculated using two visually selected perpendicular measurements.

Mean, deg.	38.38
Standard Deviation, deg.	0.56
Coefficient of Variation	0.014
Minimum, deg.	37.06
Maximum, deg.	39.24
Count	16

Table 3: Summary statistics for angle-measuring task.

MEASURING DISTANCE WITH TOTAL STATIONS

Laser transits or "total stations" have long been a surveyor's tool but have gained popularity in scene mapping for accident analysis in the last decade. One such device, the Sokkia Set 6E total station, was used to collect data to allow an estimate of the uncertainty in the sighting of a total station shot. This measuring instrument locates points via an optically sighted electronic distance measurement device combined with a theodolite that accurately measures the vertical and horizontal angles to the sighted point. The resulting measurements describe a point in a spherical coordinate system. Rectangular coordinates are calculated with trigonometry from the spherical coordinates.

The unit was set up at one end of a clear area. Each of the 20 to 23 participants had previously used a similar instrument. They were asked to sight to two stationary prisms located nominally 74 and 536 feet from the instrument. Then they were asked to hold a pole-mounted prism over a target similar to those shown in Figure 2 that was on the ground near each of the stationary prisms while the total station's owner sighted the shot. Standard 31-mm prisms were used at all positions. Results were thus recorded for four groups: a stationary near prism with many operators, a stationary distant prism with many operators, a near hand-held prism with many holders and a common operator, and a distant handheld prism with many holders and a common operator.

The distribution of both distances and angles recorded appeared to be normally distributed; however, the effect of transferring the points from spherical to rectangular coordinates rendered the point distribution as an arc as illustrated in Figure 10, which reflects the recorded locations for the stationary prism around the nominal mean location (approximately 74 feet from the instrument). The tangent to the arc is perpendicular to the direction in which the shot was sighted. The distance measuring resolution of the instrument can be seen as the spacing between the three arcs, while the angular resolution can be seen as the smallest repeated space between points in each arc. The variation introduced by the operators in selecting their shot position can be seen as the total spread of points along the direction of the three rows.

The Sokkia instrument used was approximately 10 years old, and was capable of resolving distances in feet to two decimal places and angles to 10 seconds. Many newer models offer tighter resolution capabilities which would affect the affect of instrument variability on the spread of collected data. Even in this older instrument, though, this instrument-induced variability was clearly less than the variability introduced by the human operator sighting to a stationary prism, which in turn was less than the variability introduced by using a handheld prism.

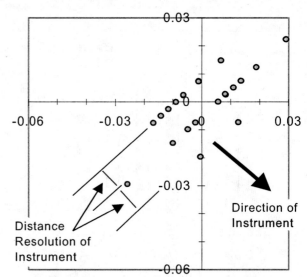

Figure 10: Scatter (inches) over a nominal distance of 74 feet with many operators and a single stationary prism.

When the direction of the sight (and thus the original spherical coordinate system orientation) is unknown the standard deviation of the distance to the actual point (forming a circle around the measured point) can be expressed as:

$s_x = 0.00005x + 0.0204$ (handheld prism)   EQ. 4

$s_x = 0.00006x + 0.0123$ (stationary prism)   EQ. 5

where x is expressed in feet.

These relationships are shown graphically in Figure 11. This standard deviation was calculated by resolving the rectangular coordinates for each point, determining the standard deviation for both dimensions in the plane of the roadway, and then taking the square root of the sum of their squares.

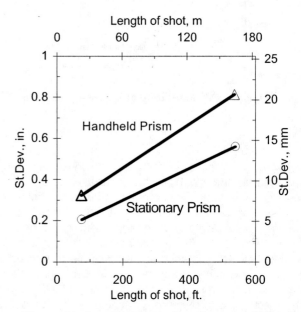

Figure 11: Standard deviation of near and distant points, with stationary and hand-held prisms.

Figure 12 displays the recorded data point locations for the distant hand-held prism, showing a circle with a radius of one standard deviation around the group's mean location.

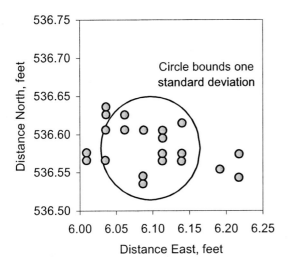

Figure 12: Recorded location for the distant hand-held prism, $n=23$, $\bar{x}=536.6$ ft (163.5 m), $s=0.068$ ft (0.021 m).

MEASURING FRICTIONAL DRAG

The frictional drag acting between a vehicle and the roadway is often an important aspect of an accident analysis. This parameter can be measured in many ways, including with whole vehicles, various types of drag sleds, and "skid trailers." The results of several methods are presented here.

USING VEHICLES

The American magazine Road and Track regularly provides a summary of the peak lateral accelerations measured during their vehicle tests using a 100-foot (30.5 meter) radius circle. Data from one such summary were examined. [12] In order to accurately reflect vehicles which might reasonably be found on public highways, the analysis was limited to models with more than 500 new US-registrations in the preceding year as reported by Automotive News. [13] The histogram for the 78 vehicles meeting that requirement is shown in Figure 13.

Eubanks, et al, reported the results of a large number of skid tests in which five or six tests were conducted with each of eight vehicles. [14] Figure 14 shows the average decelerations for each stop (identified as mu-braking by some researchers [15]) as calculated using data generated by the McInnis fifth wheel system. [16]

A large number of skid tests reported by Wallingford, et al, reflected standard deviations of between 0.03 and 0.08 for particular vehicle/tire/surface/speed test configurations. [17]

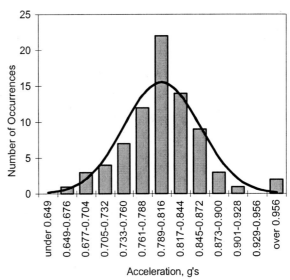

Figure 13: Peak lateral acceleration on skidpad for 78 common vehicles ($n=78$, $\bar{x}=0.80$, $s=0.056$)

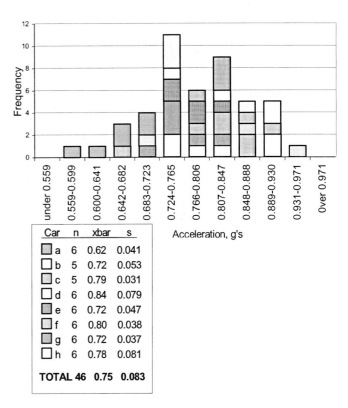

USING A DOT-SKID TRAILER

As part of a bus-accident investigation in 1985, the National Transportation Safety Board measured the "skid number" at 40 miles per hour (64 kilometers per hour) of wet bituminous concrete and wet Portland cement concrete on a section of Maryland highway using a "DOT-skid-trailer." The mean of the 6,448 tests conducted on wet bituminous concrete was 0.50, with a standard deviation of 0.09. The results for 767 tests on wet Portland cement were 0.54 в 0.05. [18] Other large data sets of dry roadway friction testing have yielded standard deviations of approximately 0.07g's. [2]

-63-

USING DRAG SLEDS

At two separate gatherings, participants brought their own drag sleds for comparison on a common surface.

Data entitled "MD-uniques" collected by 20 participants on a dry one-year old asphalt street. [19] A Maryland D.O.T. skid-trailer reported a drag factor on the same roadway of 0.829 at 40 mph (64 kph). The data set entitled "TX-uniques" included 10 participants working on a dry concrete surface, which had an ASTM skid-trailer-measured drag factor of 0.846 at 22 mph (35 kph), 0.802 at 31 mph (50 kph), and 0.821 at 41.5 mph (66 kph). [20] Additionally, at the TX-event, 35 participants were asked to measure the drag factor of the concrete surface using the same concrete-filled tire drag sled. In all three sets of data, participants performed five pulls. The drag factor was calculated as an average of the five pulls from each sled. Figure 15 shows the histogram of this analysis.

Experiment administrators noted a wide variety of drag sled pulling techniques, including different pull angles, pull lengths, and scale-reading points. The TX-generic tests employed an electronic "fish scale" that is widely used in practice. It was noted that participants had difficulty obtaining a stable reading from this device. Another significant trend revealed in the data was the reduction of the standard deviation per pull number during successive pulls. The standard deviation of all fourth pulls was much lower than the standard deviation of all the first pulls. This result is consistent with "learned" human performance.

Subsequent testing with the same generic drag sled has revealed that by reducing the variations in technique noted above through a brief training, and using a quality spring scale, the variation in measured values is dramatically reduced. Additional research is ongoing to determine the feasibility of generating consistent results with drag sleds.

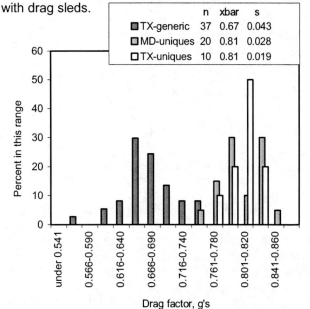

Figure 15: Distribution of drag factors, after averaging over five pulls for each participant.

Table 4 shows a summary of the recorded measurements and standard deviations presented in this section.

Measurement Method	No. of Samples	Average g's	Std.Dev. g's
Vehicles, skid pad, lateral accel. [12]	78	0.80	0.06
Vehicles, locked wheels [14]	46	0.75	0.08
Vehicles, locked wheel, 50mph, asphalt [17]	~100	0.76	0.08
Vehicles, locked wheel, 30mph, asphalt [17]	~100	0.77	0.08
Vehicles, locked wheel, 50mph, concrete [17]	~100	0.77	0.05
Vehicles, locked wheel, 30mph, concrete [17]	~100	0.75	0.06
Skid trailer, wet bitum. concrete [18]	6,448	0.50	0.09
Skid trailer, wet portland cement [18]	767	0.54	0.05
Generic drag sled, dry concrete [20]	37	0.67	0.04
Unique drag sleds, dry concrete [19]	20	.81	0.03
Unique drag sleds, dry asphalt [20]	10	0.81	0.02

Table 4: Descriptive statistics for drag factor measurements discussed in this section.

SKID MARK MEASUREMENT

A locked-wheel stop was executed on a concrete surface with a 1995 Ford extended-cab pickup truck with no ABS. The truck was traveling approximately north-to-south, and was equipped with a VC2000 which reported the stopping distance to be 67-feet (20.4-m). Participants were advised which direction the vehicle had been travelling and where it had stopped, and were asked to measure the skidmark made by the front left tire using a flexible 200-foot (60-m) tape measure.

Ten participants measured the front skidmark as requested, while 14 participants included the mark made by the rear wheel, yielding a length approximately one wheelbase longer than intended. The placement of the tape at the terminal end of the mark was observed to be fairly consistent (within an 8-inch span) while the selection of the mark's beginning point was less consistent. Table 5 contains the descriptive statistics for the measurement results. The measured length of the lighter rear tire mark was observed to decrease as the day progressed, while the measured length for the fairly dark front mark did not change significantly. Results are summarized graphically in Figure 16.

	Front+Rear ft. (m)	Front mark ft. (m)
Minimum	72.0 (21.9)	64.0 (19.5)
Maximum	77.4 (23.6)	68.6 (20.9)
Mean	75.2 (22.9)	67.0 (20.4)
Standard Deviation	1.7 (0.5)	1.7 (0.5)
Count	14	10

Table 5: Descriptive statistics of skidmark measurements.

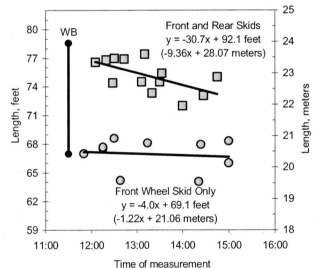

Figure 16: Measured skid length as a function of time of day. Trendlines use "x" equal to the fraction of the day, noon=0.5.

MEASURING VEHICLE CRUSH

The amount of deformation a vehicle sustains is often of interest to a reconstructionist conducting an energy-based accident analysis. While the vehicle itself is sometimes available for measurement, reconstructionists are occasionally asked to determine deformation depths from one or more photographs. Experiments examining the variation involved in each of these methods of crush measurement and estimation are described below.

DIRECT MEASUREMENT OF DAMAGED VEHICLE

Participants were provided a Chevrolet Astro van, shown in Figure 17, which had been involved in a partial overlap frontal impact. They were given a data sheet (Shown in APPENDIX A) on which to record their measurements which included a description of the

Figure 17: Damaged Astro-van with measurement jig for direct crush measurement experiment.

CRASH 3 measurement protocol [21], a description of the parameters to be measured, the vehicle's original overall length, overall width, and wheelbase.

The equipment made available was an adjustable, rectangular frame on the ground that could be used as a reference, if desired. Two measuring devices were provided: a 16-foot (5-m) steel rollup tape measure and a 6-foot (2-m) folding carpenter's ruler. A plumb bob was also available.

A review of the data showed that three participants recorded the data backwards. After reversing the order of their data points, the 17 crush profiles are shown in Figure 18. Using a weighted average which considers only half the value of the end points, the average crush depth recorded by the 17 participants ranged from 11.8 to 34 inches, with an average of 19.4 inches and a standard deviation of 5.2 inches. The reported length of the damage area ranged from 45 to 78 inches, with an average of 62.4 inches and a standard deviation of 9.9 inches.

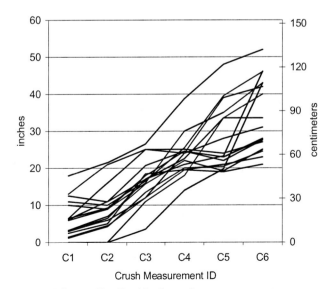

Figure 18: Residual crush measurement profiles recorded by 17 participants.

CRUSH ESTIMATES FROM PHOTOGRAPHS

Participants were asked to estimate the crush and or Equivalent Barrier Speed (EBS) of a vehicle from a single photograph (Figure 19) or a set of two photographs (Figures 20a, and 20b) without any additional information. In the single-photograph exercise, 57 participants provided 11 crush estimates and 49 EBS estimates. The average estimated crush was 13.6 inches with a standard deviation of 4.2 inches. The EBS results are given in Table 6.

Figures 20a and 20b: Two photographs used together for one crush estimation exercise.

Figure 19: Single photograph used for speed/crush estimate.

EBS, mph (kph)	# of responses
6 – 10 (9.7-16.1)	2
11 – 15 (17.7-24.2)	12
16 - 20 (25.8-32.2)	16
21 - 25 (33.8-40.3)	14
26 - 30 (41.9-48.3)	5

Table 6: Crush estimates from a single photograph.

In the two-photograph exercise, 52 participants provided 8 crush estimates and 51 EBS estimates. The average estimated crush was 13.1 inches with a standard deviation of 3.3 inches. The EBS results are given in Table 7.

EBS, mph (kph)	# of responses
11 – 15 (17.7 - 24.2)	5
16 - 20 (25.8 - 32.2)	12
21 - 25 (33.8 - 40.3)	18
26 - 30 (41.9 - 48.3)	12
31 – 35 (49.9 - 56.4)	3
36 – 40 (58.0 - 64.4)	0
41 - 45 (66.0 - 72.5)	1

Table 7: Crush depth estimates from two photographs.

Most of the participants in this exercise reported that they would not attempt to use information generated in this fashion in a reconstruction without additional details and analysis.

CONCLUSIONS

This paper presents the results of a variety of experiments designed to quantify the uncertainty in some measuring tasks commonly faced by accident reconstructionists. Modelling the probability of many accident reconstruction variables as normally distributed appears to be appropriate. The standard deviations from many common measurement tasks were presented. These data can help determine appropriate ranges for variables used in accident reconstruction uncertainty analyses.

When making measurements that have been shown here to have very small standard deviations, an investigator may be comfortable making a single measurement. When making measurements that have been shown to have large standard deviations, though, an investigator may wish to consider making several repeated measurements to generate a distribution from which a centered value can be selected.

In the one measurement task with a large standard deviation in which the root causes were studied (drag sleds), lack of standardized training was found to be a significant issue that affected the tool's repeatability. It seems apparent from the results presented in this paper that repeatability of some other tasks common to accident reconstruction would improve with standardized protocols.

ACKNOWLEDGMENTS

The authors wish to acknowledge the invaluable assistance provided by Chenry Baughman in allowing us to use his Sokkia total station, and operating the unit at WREX2000.

CONTACT

Wade Bartlett can be contacted by post at P.O. Box 1958, Dover NH, 03867, USA, or via email at wade@mfes.com.

REFERENCES

1. Y.Beers, *Introduction to the Theory of Error*, Addison-Wesley, Reading MA, 1957

2. R.Brach, "Uncertainty in Accident Reconstruction Calculations," SAE Paper 940722, Warrendale PA, 1994

3. L.Fricke, Traffic Accident Investigation, Northwestern University Traffic Institute Section 28

4. A.Fonda, "Nonconservation of Momentum During Impact", SAE Paper 950355, Warrendale PA, 1995

5. B.Taylor, C.Kuyatt, "Guidelines for Evaluating and Expressing the Uncertainty of NIST Measurement Results," National Institute of Standards and Technology, Technical Note 1297, U.S. Government Printing Office, Washington D.C., 1994

6. R.Tubergen, "The Technique of Uncertainty Analysis as Applied to the Momentum Equation for Accident Reconstruction," SAE Paper 950135, Warrendale PA, 1995

7. G.Kost, S.Werner, "Use of Monte Carlo Simulation Techniques in Accident Reconstruction," SAE Paper 940719, Warrendale PA, 1994

8. D.Wood, S.O'Riordan, "Monte Carlo Simulation Methods Applied to Accident Reconstruction and Avoidance Analysis," SAE Paper 940720, Warrendale PA, 1994

9. *CRC Standard Mathematical Tables*, 11th Edition, pages 239-241, Chemical Rubber Publishing Company, Cleveland OH, 1957

10. D.Montgomery, G.Runger, "Applied Statistics and Probability for Engineers", 2nd Ed, John Wiley, 1999

11. Institute for Police Technology and Management, Special Problems in Accident Reconstruction, April 2001

12. Road and Track Magazine, pages 162-163, September 1995

13. Automotive News, April 1995

14. J.Eubanks, W.Haight, R.Malmsbury, D.Casteel, "A Comparison of Devices used to Measure Vehicle Braking Deceleration," SAE Paper 930665, Warrendale PA, 1993

15. D.Goudie, J.Bowler, C.Brown, B.Heinrichs, G.Siegmund, "Tire Friction During Locked Wheel Braking," SAE 2000-01-1314, Warrendale PA, 2000

16. McInnis Engineering Associates, Ltd., 11-11151 Horseshoe Way, Richmond, BC Canada, VTA 4S5

17. J.Wallingford, B.Greenlees, S.Christoffersen, "Tire-Roadway Friction Coefficients on Concrete and Asphalt Surfaces Applicable for Accident Reconstruction," SAE paper 900103, Warrendale PA, 1990

18. "Intercity Bus Loss of Control and Collision with Bridge Rail on Interstate 70 Near Frederick Maryland August 25, 1985," National Transportation Safety Board Report PB93-916202, as reported in Accident Investigation Quarterly, Winter 1994

19. MdATAI Joint Conference in Waldorf Maryland, September 1998

20. ASTM Trailer Friction Test and drag sled results, as reported in the preliminary results from WREX-2000

21. CRASH3 User's Guide and Technical Manual, NHTSA, DOT Report HS 805 732, February 1981

ID No:

WREX2000 *Measurement Task Activity*
Residual Frontal Crush Measurement Data Sheet

CRASH3 measurement protocol:
6 side-by side, equidistant, *w*, measurements of induced and direct crush, C_i, in a horizontal plane at a distance *h* above the ground

551. C_1 _____ in
552. C_2 _____ in
553. C_3 _____ in
554. C_4 _____ in
555. C_5 _____ in
556. C_6 _____ in
557. L _____ in
558. w _____ in
559. h _____ in
560. REF _____ in
561. RWB _____ in
562. LWB _____ in

Please complete the following ☒ ☑ ■:
563. In the past 12 months I have made measurements such as these:

    0 ☐   1-5 ☐   >5 ☐ times,

564. Normally, I make measurements using the
    SI (metric) system ☐   US (in, lb, ft, . . .) System ☑

565. date:_____
    565. time:_____ ☐ am ☐ pm of measurements.

crash3.dat

Reviewer Discussion

Reviewers Name: Greg Anderson
Paper Title: Reconstruction Low Speed Impact Collision Severity: Comparison between Full Scale Testing and Analytical Prediction.
Paper Number: 2002-01-0546

Greg Anderson Comments
Proposed comments on Wade Bartlett's paper as follows.

This paper is an interesting look at the variability of some of the everyday measurements that are dealt with during crash investigation and analysis. The increase in standard deviation when the measurement in question is of an actual yaw mark versus a chalk line is an interesting comparison of an actual field exercise to the pure measurement. It is not surprising that as the level of judgment required increases, so does variability, reaching a perhaps alarming level with the crush measurement task. It should also be noted that, in addition to improving the repeatability of drag sled measurements, it would be useful to verify their effectiveness at duplicating actual at-speed pneumatic-tire coefficients. Overall, a lot of worthwhile information.

Greg Anderson
Scalia Safety Engineering

2002-01-0548

Accelerations and Shock Load Characteristics of Tail Lamps From Full-Scale Automotive Rear Impact Collisions

Lindsay "Dutch" Johnson and Jeffrey Croteau
Exponent® Failure Analysis Associates

Joseph Golliher
Origin Engineering, LLC

Copyright © 2002 Society of Automotive Engineers, Inc.

ABSTRACT

An estimated fifty percent of automobile accident fatalities occur at night. When investigating these accidents, the question often arises as to whether the lamps were on or off at impact. One approach to answering this question is to inspect the lamp for damage evidence that can be compared to research or previous investigations. Articles addressing automotive lamp examination can be found in the literature dating back prior to 1960; however, only a relatively few of these articles have incorporated experimental work. The articles that have experimentally studied the shock load characteristics of automotive tail lamps have: 1) based conclusions on acceleration data far removed from the lamps, or 2) have made assumptions as to what the "actual" accelerations are at the tail lamp itself. In order to gain insight into what accelerations, and associated shock load characteristics, are actually "seen" by tail lamps during an automotive collision, four full-scale, instrumented, rear impact crash tests were conducted. The results indicate that the accelerations at the directly contacted tail lamp housing can reach values an order of magnitude greater than those "seen" at the vehicle CG. Pre- and post-impact lamp condition comparisons were made for selected lamps. Comparisons were also made between lamps with respect to their location relative to the direct contact zone, as well as with respect to their filament state at impact (i.e., on or off).

INTRODUCTION

The most common type of tail lamp is the traditional 1157-style lamp with an offset pin base. The 3157-style lamp, as used in the Ford Taurus, is essentially a 1157 lamp with a "wedge" base. Even though these lamps have been in use for many decades, and post-collision filament characteristics are commonly used for accident reconstruction purposes, no experimental studies have been conducted that have measured the "actual" accelerations "seen" at the lamp impact zone. Conclusions regarding impact zone lamp accelerations have been based on 1) acceleration data far removed from the lamps, or 2) assumptions as to what the "actual" accelerations are at the tail lamp itself.

Articles addressing tail lamp and headlamp examination can be found in the literature dating back prior to 1960. A number of papers addressing lamp examination for use in accident reconstruction describe procedures for analyzing the lamp following impact [1, 2, 9, and 12-17]. Others describe lamp analysis procedures using microscopic techniques such as SEM [7 and 10], as well as light microscopy [6]. Only four articles addressing tail lamp examination, that incorporated experimental work, were found [3, 4, 8, and 11].

Severy, 1966, performed full-scale crash tests and analyzed the traditional incandescent headlights and taillights in an attempt to classify collision-induced bulb filament deformation. Accelerations, which were measured eight feet aft of the front bumper, ranged from 4 to 27 g. Even though up to eight headlights and eight tail lamps were mounted to the same vehicle, and subjected to the same collision, the majority of these bulbs could not have experienced the same decelerations due to their respective locations to the impact zone. Severy concluded that white hot filament deformation is proportional to the magnitude of impact directed to the headlight unit, however, in order to obtain more reliable filament deformation data, testing of bulbs subjected to the same deceleration will be required. Dolan, 1971, performed single bulb tests to simulate various types of 1157 tail-lamp failures. The impact testing consisted of dropping a pivoted metal bar from a measured height onto the illuminated lamp being tested (bulb broken tests), and tapping the bulb against a wooden block until plastic deformation was observed (bulb intact tests). Although diagrams are provided in an attempt to classify the shape of the bulb filaments following both cold and hot shock loading, the uncontrolled nature of the experiments (especially the cold shock experiments) creates uncertainty in the results based on the method in which the data was collected. Keskin et al., 1988, performed barrier impact and centrifuge loading tests on 1157 taillights to determine threshold values of acceleration necessary to plastically deform the filaments. The first barrier test subjected the bulbs to a fairly realistic, albeit severe, pulse of 100 milliseconds in duration with a peak deceleration between 50 and 60 g, however, only two bulbs, in different orientations, were subjected to the same loading. Dydo et al., 1988, performed pendulum impact tests using 1157 taillights to record filament response to low speed impact. Each test consisted of mounting three bulbs, in three different orientations, onto a carrier mounted to the pendulum. The impact pulses ranged from 8.0 milliseconds down to 0.8 milliseconds in duration with respective peak decelerations ranging from 100 to 3800 g. As indicated by Dydo et al., more typical collision pulses range in duration from 30 to 70 milliseconds with peak decelerations on the order of 20 to 50 g, especially at the vehicle CG.

Based on the above review of the literature, it can be seen that a need exists to experimentally investigate the accelerations actually "seen" by the tail lamp in the impact zone. In order to address this need, the work presented in this paper consisted of performing four full-scale, instrumented, rear impact crash tests using both energized and non-energized tail lamps. This work is the first to identify the "actual" accelerations and associated shock load characteristics of tail lamps in the impact zone of actual automobile collisions.

MATERIALS AND METHODS

Exponent's crash rail facility in Phoenix, Arizona was used to perform four full-scale, instrumented, rear impact crash tests. These tests were conducted not only to analyze filament deformation, but also to determine closing speeds in rear impact collisions, the results of which were presented at the 2001 SAE Congress by Croteau et. al. The bullet vehicle was a Class 8 truck in two tests and a flat-faced barrier in the other two tests. A different 1989 Ford Taurus was the stationary target for each of the four tests.

The bullet vehicle used in the truck tests was a Class 8 1960 Kenworth 6x4 agricultural farm truck. The Kenworth had an overall width of 96 inches, clearance between the bumper (bottom edge) to ground of 22 inches, and a test weight of 20,060 lbf. Each Taurus had an overall width of approximately 71 inches, a rear bumper (top edge) to ground height of 21 inches and a test weight of approximately 3,450 lbf.

Instrumentation for each Taurus consisted of a damped 1000-g Entran accelerometer, model EGA -1000 (with a shock load limit of /-10,000 g), located at approximately the center of the forward-facing surface of the left and right rear tail lamp housings. The accelerometer was oriented so that its sensitive axis was aligned with the longitudinal direction of the vehicle. Additionally, one triaxial accelerometer set was located at the base of the left and right A-pillars. Instrumentation for the Kenworth consisted of one triaxial accelerometer set located on the vehicle centerline just aft of the cab.

For the first truck test, the vehicles were positioned such that the front bumper of the Kenworth impacted the rear structure above the rear bumper of the Taurus. The longitudinal centerlines of both vehicles were coincident with each other at impact, as shown in Figure 1.

Figure 1: Still photograph of full engagement test between truck and car.

The Kenworth was traveling forward at a speed of 25.6 mph when it struck the rear of the stationary Taurus. At impact the Taurus' side marker, parking, turn indicator, brake, and reverse lamp filaments were in either the energized or non-energized state as indicated in Table A1 in the Appendix. The 3157-style lamp was used in both the reverse and parking/turn/brake socket and is a two-filament lamp. Vehicle design indicates that only the larger (wire gage and coil diameter) filament can be energized in the reverse socket and both filaments can be energized in the other socket depending on the driver settings. The photograph in Figure 2 depicts a tail-lamp housing assembly with the lamp positions and the accelerometer location labeled.

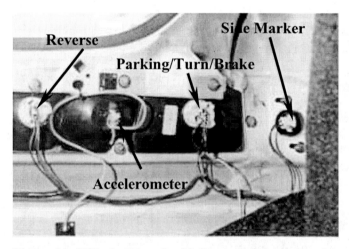

Figure 2: Still photograph of left rear tail-lamp housing from inside the trunk area.

In the offset truck test, the vehicles were positioned such that the right front of the Kenworth impacted the left rear of the Taurus. The longitudinal centerline of the Taurus was 45 inches to the right of the Kenworth's centerline at impact. The Kenworth was traveling forward at a speed of 25.8 mph when it struck the rear of the stationary Taurus. At impact the Taurus' side marker, parking, turn indicator, brake, and reverse lamp filaments were in either the energized or non-energized state as indicated in Table A2 in the Appendix.

Two additional single-moving tests were conducted where a flat-faced moving barrier impacted two more 1989 Ford Taurus's. The Barrier had a front platen width of 81 inches, and a test weight of 4,000 lbf. The test condition and instrumentation of each Taurus in the barrier tests was of the same configuration as in the truck tests. Instrumentation on the barrier consisted of one triaxial accelerometer set located on the vehicle centerline at approximately amidship.

For the first barrier test, the vehicles were positioned such that the Barrier face impacted the rear bumper of the Taurus. As in the full engagement test with the truck, the longitudinal centerlines of both vehicles were coincident with each other at impact. The barrier was traveling forward at a speed of 31.6 mph when it struck the rear of the stationary Taurus. At impact the Taurus' side marker, parking, turn indicator, brake, and reverse lamp filaments were in either the energized or non-energized state as indicated in Table A3 in the Appendix. The second barrier test was configured to be an offset test such that the right front of the barrier face impacted the left rear of the Taurus, as depicted in Figure 3

Figure 3: Still photograph of offset barrier test between barrier and car.

The longitudinal centerline of the Taurus was 40 inches to the right of the Barrier's centerline at impact and was traveling forward at a speed of 31.6 mph when it impacted the stationary Taurus. At impact the Taurus' side marker, parking, turn indicator, brake, and reverse lamp filaments were in either the energized or non-energized state as indicated in Table A4 in the Appendix.

RESULTS AND DISCUSSION

The speeds in each test were based on the energy available for deformation (65,000 ft-lbf) given a 30 mph rear impact between the FMVSS 301 flat-faced barrier and a 3,450 lbf Taurus [16]. The delta-V of the Taurus center of gravity was approximately 23 mph in the truck tests and approximately 18 mph in the barrier tests. The ground clearance of the truck bumper was more than the height from the ground to the top of the rear bumper in the Taurus, resulting in direct contact between the truck bumper and the tail-lamp housing assemblies. Each tail lamp used in the reported tests was photographically documented prior to and after each test. This was done to permit the characterization of deformation, if any, induced by the tests.

Selected acceleration data plots, from the offset truck test, are shown in Figures A1-A5 in the Appendix. All data was acquired at a sample rate of 10,000 Hz. Per SAE Recommended Practice J211, all CG acceleration data was filtered with a class 60 (100 Hz) filter; and all lamp housing data was filtered with a class 600 (1000 Hz) filter. Shown in Figure A1 is the acceleration at the Taurus' CG, which shows a peak of approximately 12 g with a pulse duration of approximately 270 milliseconds (ms). Shown in Figures A2 and A3 is the acceleration data from the right rear (non-direct impact side) tail-lamp housing of the Taurus, with a peak acceleration of approximately 195 g and a pulse duration of approximately 5 ms. Figures A4 and A5 show the acceleration data (SAE filter class 600) from the left rear (direct impact side) tail-lamp housing of the Taurus. Two anomalies were encountered during acquisition of the data that are shown in Figures A4 and A5: 1) There was a positive DC shift in the data of 425 g at approximately 13 ms, which was due to an unintended voltage shift in the data acquisition system, and 2) a few of the peaks, in the upward-shifted data beyond 13 ms, were "clipped" (A/D converter saturation) at 1500 g. Even though the data shown in Figures A4 and A5 was filtered over the entire time range (-50 to 400 ms), it was considered valid only out to approximately 13 ms due to the clipping that occurred beyond this point. The peak occurring at approximately 10 ms occurred prior to the data shift and was considered to be the maximum. This peak had a magnitude of approximately 910 g with a pulse width of approximately 1 ms.

Unfortunately, the left and right tail-lamp housing accelerations saturated the A/D converter at the 300 g full-scale setting in both the full engagement tests. The unfiltered acceleration data reached levels of at least 300 g for both the left and right side housings for both full engagement tests. The CG and the lamp housing accelerations for the Taurus for each test are summarized in **Table 1**.

Table 1: Peak acceleration levels measured on left and right tail-lamp housings as compared to center of gravity.

Peak Longitudinal Accelerations in Taurus			
	Left Side [g]	**CG [g]**	**Right Side [g]**
Barrier Tests			
Full Engagement	300	19	300
50 Offset	760	15	360
Truck Tests			
Full Engagement	300	14	300
50 Offset	910	12	195

Figure 4: Lamp from left side tail-lamp housing from the full engagement truck test. Neither filament was incandescent at impact.

Figure 5: Lamp from reverse socket in left tail-lamp housing from the full engagement truck test. Only the larger filament was incandescent at impact.

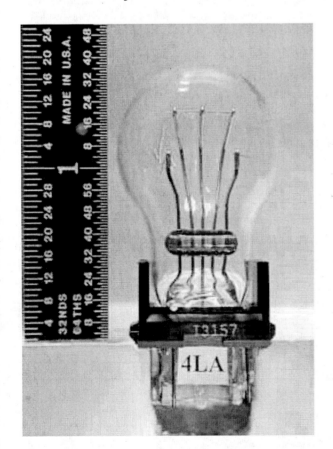

Figure 6: Reverse lamp from left tail-lamp housing from the offset barrier test. Neither filament was incandescent at impact.

Figure 7: Reverse lamp from right tail-lamp housing from the offset barrier test. Only the larger filament was incandescent at impact.

A total of sixteen 3157 lamps were subject to crash level accelerations, half were incandescent at impact and twelve were in the area of direct contact. The acceleration field was three-dimensional, however due to cost considerations given the likelihood of transducer damage, only the longitudinal acceleration was measured.

Figures 4 and 5 depict the post impact condition of the adjacent 3157 lamps from the left side tail-lamp housing in the full engagement barrier test. Both of these lamps were received with the vehicle. The lamp in the parking/turn/brake socket (Figure 4) was not energized and the larger filament in the reverse socket (Figure 5) was incandescent at the time of impact. Note that neither filament in Figure 4 was energized and both were fractured.

Only the larger filament in Figure 5 was incandescent, yet the smaller filament exhibits evidence of distortion, indicating that radiant heat from the larger filament was sufficient to keep the filament from fracturing.

Replacement lamps were installed in both reverse sockets in the offset barrier test, where the left side was not energized but the right side was incandescent at the moment of impact. Note that the posts are different in the replacement 3157 lamps than the as-received posts, such as those shown in Figures 4 and 5. Figure 6 depicts the post impact condition of the non-energized reverse lamp on the left side from the offset barrier test. Note that the large filament (two post support) was fractured; yet the smaller filament that is supported by three posts was not.

Figure 7 shows the post-impact condition of the right side reverse lamp from the offset barrier test. The peak acceleration is within 10 of the left side however it was not in the area of direct contact with the barrier. Despite the similar acceleration peaks and that the larger filament was incandescent at impact, there was no apparent deformation to either filament.

Out of the eight 3157 lamps tested in the barrier-to-car tests, only one lamp was not recovered and only 2 of the seven recovered globes were fractured. In the truck-to car tests, all of the lamps were recovered and there were three globes that were fractured.

Figure 8 depicts the condition of the recovered lamp from the full engagement truck test. It is clear that the globe is broken and the discoloration on the small filament (parking) is consistent with the fact that the filament was hot, but not incandescent at impact. There is also distortion of the large filament (brake and turn). The right turn indicator was activated yet we know from the high-speed film that it was approx. 0.2 seconds pre-fully incandescent at impact. (The turn indicator cycle time for each test was approximately 0.7 seconds.) Filament deformation alone may lead to an improper conclusion by an investigator as to the state of a lamp. In the case of a broken globe, the oxidation or discoloration of the filament itself is an indicator. The transient to/from incandescence must also be considered. Analysis of adjacent filaments in the same lamp may be necessary to conclusively determine, if possible, the on/off state of a lamp.

Figure 9 shows the damaged state of the left park/turn/brake lamp from the offset truck test. Note that all five posts are deformed as a result of direct contact damage. The parking filament was on and the left turn signal was approx. 0.2 seconds post-fully incandescent at impact. Even with a broken globe, an accident investigator would be able to learn something about the state of the filament for this lamp.

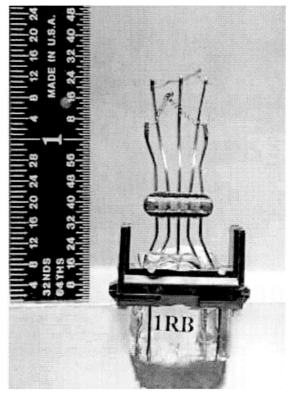

Figure 8: Park/Turn/Brake lamp from the right tail-lamp housing from the full engagement truck test. The parking filament was incandescent and the right turn signal was approx. 0.2 seconds pre-fully incandescent at impact.

CONCLUSIONS

1) Greater than an order of magnitude of difference was found between the contact zone accelerations (SAE Class 600 filtered) and the CG accelerations (SAE Class 60 filtered) for the Taurus in each test. For the offset tests, the differences in acceleration between the left tail-lamp housing (directly impacted) and the Taurus' CG ranged from 745 g to 898 g and from 183 g to 345 g for the right tail-lamp housing (non-directly impacted). These differences will be smaller if the same SAE filter class is used for both the CG and tail-lamp housing acceleration data.

2) Attempts to predict contact zone accelerations using measured vehicle CG accelerations only, will result in a high degree of uncertainty due to the large differences between, and the variability associated with the measured accelerations

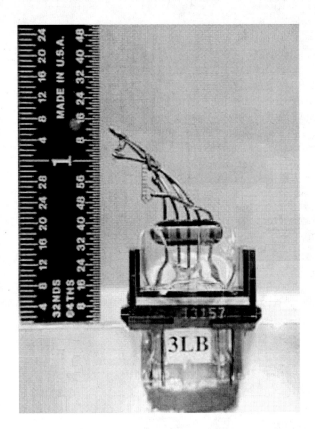

Figure 9: Park/Turn/Brake lamp from the left tail-lamp housing from the offset truck test. The parking filament was incandescent and the left turn signal was approx. 0.2 seconds post-fully incandescent at impact.

at the contact zone and the vehicle's CG. Additionally, the complex nature of the vehicle's stiffness between the CG and the contact zone, and the huge number of possible impact configurations will make accurate prediction extremely difficult using only vehicle CG acceleration data.

3) The reported post-collision lamp characteristics are typical of those observed by others in the field, however, from the present work, these characteristics can now be associated with actual automobile collision contact zone acceleration pulses.

Based on this work, the authors recommend that the following three instrumentation guidelines be followed when attempting to acquire direct impact zone accelerations: 1) set the transducer full scale range in the A/D converter to at least 2000 g, 2) use a transducer with a high shock load limit (e.g. /-10,000 g, as was used in this work), and 3) use a critically damped transducer with a frequency response of at least 2 kHz (at less than 1 dB).

This work is the first to measure actual automobile collision contact zone accelerations and to relate the post-collision characteristics of tail lamps, in the same zone, to these accelerations. This work is one step closer to an improved understanding of how actual automobile collision induced accelerations contribute to post-collision lamp characteristics. Additional work will be required to determine the relationship between acceleration pulse peak, duration, and filament deformation.

Future work will include additional measurements of contact zone accelerations during full-scale automotive crash tests, and controlled studies using these measured accelerations to induce identifiable lamp shock load characteristics.

REFERENCES

1. Baker, J.S., "Lamp Examination for On or Off in Traffic Accidents", in Traffic Accident Reconstruction, Northwestern University Traffic Institute, Vol. 2, 1st Ed., 1990.
2. Bleyl, R.L., "Who Had the Green Signal ", *Journal of Police Science and Administration*, Vol. 8, No. 4, 1980.
3. Dolan, D.N., "Vehicle Lights and Their Use as Evidence", *Journal of the Forensic Science Society*, Vol. 11, No. 2, London, 1971.
4. Dydo, J.R., Bixel, R.A., Wiechel, J.F., Stansifer, R.L., and Guenther, D.A., "Response of Brake Light Filaments to Impact", *SAE Accident Reconstruction*, Paper No. 880234, February 1988.
5. Fries, T.R., Lapp, R.O., "Accident Reconstruction – Response of Halogen Light Filaments During Vehicle Collisions", *SAE Motor Vehicle Accident Reconstruction*, Paper No. 890856, February 1989.
6. Haas, M.A., Camp, M.J., and Dragen, R.F., "A Comparative Study of the Applicability of the Scanning Electron Microscope and the Light Microscope in the Examination of Vehicle Light Filaments", *Journal of Forensic Sciences*, Vol. 20, No. 1, January 1975.
7. Hayes, T.L., "Comparison of Tungsten Filaments by Means of the Scanning Electron Microscope", *Journal of the Forensic Science Society*, Vol. 11, No. 3, London, 1971.
8. Keskin, A.T., Reed, W.S., and Friedrich, R.L., "Brake Light Filament Deformation Analysis for Vehicular Collisions", *SAE Accident Reconstruction*, Paper No. 880233, February 1988.
9. Mathyer, J., "Evidence Obtained From the Examination of Incandescent Electric Lamps, Particularly Lamps of Vehicles Involved in Road Accidents, Parts 1 and 2", *International Criminal Police Review*, No. 284, Paris, France, 1975.
10. Powell, G.L.F., "Interpretation of Vehicle Globe Failures: The Unlit Condition", *Journal of Forensic Sciences*, Vol. 22, No. 3, February 1977.
11. Severy, D.M., "Headlight-Taillight Analysis From Collision Research", *SAE Proceedings of the 10th Stapp Car Crash Conference*, Paper No. 660786, November 1966.
12. Stone, I.C., "Forensic Laboratory Support to Accident Reconstruction", *SAE Accident Reconstruction*, Paper No. 870427, February 1987.
13. Thiele, R., "Examination of Car Lights After a Road Accident", *International Criminal Police Review*, No. 116, Paris, France, 1958.
14. Thompson, J.W., "Switched On ", *Journal of the Forensic Science Society*, Vol. 11, No. 3, London, 1971.
15. Woods, J.D., "Headlights are Tools in Traffic Accident Investigation", *Law and Order*, June 1977.
16. Kawakami, A., Sekimori, H., Shinohara, A., "Accident Information for Traffic Accident Reconstruction – The Role of the Automobile Lamp Filament," SAE Paper 930661, Accident Reconstruction: Technology and Animation III, SP-946, March 1993.
17. Collins, J.C., "Accident Reconstruction," Charles C. Thomas, Pub., Springfield, IL, 1979.
18. Croteau, J.J., Werner, S.M., Habberstad, J.L., and Golliher, J., "Determining Closing Speed in Rear Impact Collisions with Offset and Override," SAE Paper, 2001-01-1170, SAE 2001 World Congress, Detroit, MI, March 5-8, 2001.

APPENDIX

Table A1. Lamp location, identifying label, age and filament state at impact, for the full engagement truck test. Key for the table is given following Table A4 below.

Lamp And Location	Lamp Label[2]	Lamp Filament	Filament State / Lamp Age[3]
Left Rear Side Marker	1LC	Side Marker	On / AR
Left Rear Outboard	1LB	Parking	Off / AR
		Turn Indicator	Off / AR
		Brake	Off / AR
Left Rear Inboard[1]	1LA	Reverse	On / AR
		Not Used	Off / AR
Right Rear Inboard[1]	1RA	Reverse	Off / New
		Not Used	Off / New
Right Rear Outboard	1RB	Parking	On / New
		Turn Indicator	On-0.2 / New
		Brake	Off / New
Right Rear Side Marker	1RC	Side Marker	On / New

Table A2. Lamp location, identifying label, age and filament state at impact, for the offset truck test. Key for the table is given following Table A4 below.

Lamp And Location	Lamp Label[2]	Lamp Filament	Filament State / Lamp Age[3]
Left Rear Side Marker[1]	3LC	Side Marker	On / New
Left Rear Outboard	3LB	Parking	On / New
		Turn Indicator	On 0.2 / New
		Brake	Off / New
Left Rear Inboard[1]	3LA	Reverse	Off / New
		Not Used	Off / New
Right Rear Inboard[1]	3RA	Reverse	Off / New
		Not Used	Off / New
Right Rear Outboard[1]	3RB	Parking	Off / New
		Turn Indicator	Off / New
		Brake	Off / New
Right Rear Side Marker	3RC	Side Marker	On / New

Table A3. Lamp location, identifying label, age and filament state at impact, for the full engagement barrier test. Key for the table is given following Table A4 below.

Lamp And Location	Lamp Label[2]	Lamp Filament	Filament State / Lamp Age[3]
Left Rear Side Marker	2LC	Side Marker	On / AR
Left Rear Outboard	2LB	Parking	Off / AR
		Turn Indicator	Off / AR
		Brake	Off / AR
Left Rear Inboard	2LA	Reverse	On / AR
		Not Used	Off / AR
Right Rear Inboard	2RA	Reverse	Off / New
		Not Used	Off / New
Right Rear Outboard	2RB	Parking	On / New
		Turn Indicator	On-0.1 / New
		Brake	Off / New
Right Rear Side Marker	2RC	Side Marker	On / New

Table A4. Lamp location, identifying label, age and filament state at impact, for the offset barrier test. Key for the table is given below.

Lamp And Location	Lamp Label[2]	Lamp Filament	Filament State / Lamp Age[3]
Left Rear Side Marker[1]	4LC	Side Marker	On / New
Left Rear Outboard[1]	4LB	Parking	On / New
		Turn Indicator	On 0.3 / New
		Brake	Off / New
Left Rear Inboard[1]	4LA	Reverse	Off / New
		Not Used	Off / New
Right Rear Inboard	4RA	Reverse	On / New
		Not Used	Off / New
Right Rear Outboard	4RB	Parking	Off / New
		Turn Indicator	Off / New
		Brake	Off / New
Right Rear Side Marker	4RC	Side Marker	On / New

Key for Tables A1 – A4:

[1] Indicates lamp was dislodged from socket during impact
[2] Lamp label code:
  L = Left side   R = Right side
  A = inboard lamp (reverse lamp filament)
  B = outboard lamp (parking and brake/ turn indicator filaments)
  C = side marker lamp
[3] Filament state and lamp age key:
  AR = As Received lamp used
  New = New replacement lamp used
  On = Lamp energized
  Off = Lamp not energized
  On+#.# = filament #.# seconds post-fully incandescent
  On-#.# = filament #.# seconds pre-fully incandescent

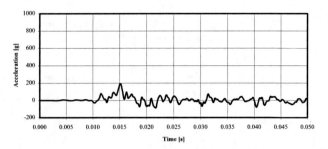

Figure A3. Right tail-lamp housing (non-direct impact side) acceleration v time plot (partial time range) for the 50% offset truck test.

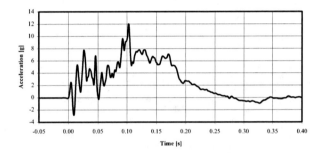

Figure A1. Taurus CG acceleration v time plot for the 50% offset truck test.

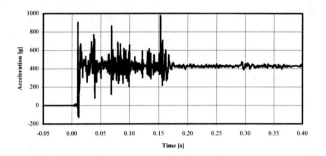

Figure A4. Left tail-lamp housing (direct impact side) acceleration v time plot (full time range) for the 50% offset truck test.

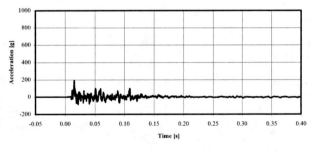

Figure A2. Right tail-lamp housing (non-direct impact side) acceleration v time plot (full time range) for the 50% offset truck test.

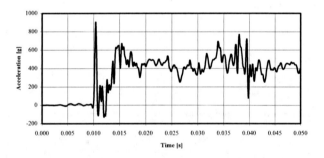

Figure A5. Left tail-lamp housing (direct impact side) acceleration v time plot (partial time range) for the 50% offset truck test.

Reviewer Discussion

Reviewers Name: Tom Pearl
Paper Title: Light Filament Deformation Paper
Paper Number: 2002-01-0548

Tom Perl Discussion
Re: Light Filament Deformation Paper

This paper makes a valuable first step to determine the acceleration pulse necessary to deform light filaments in vehicular accidents. Obviously the conclusions from this testing were limited due to the clipping that occurred with the acceleration data, but this information is valuable for establishing the data acquisition parameters for future testing. Future testing should include instrumentation necessary to define not only the magnitude, but also the shape of the acceleration pulse experienced by the light fixtures. Hopefully, the authors or others will have the opportunity to further this work in the future.

2002-01-0549

Possible Errors Occurring During Accident Reconstruction Based on Car "Black Box" Records

Marek Guzek and Zbigniew Lozia
Warsaw University of Technology

Copyright © 2002 Society of Automotive Engineers, Inc.

ABSTRACT

The authors simulate the vehicle motion in situations where typical defensive maneuvers in road traffic are being carried out, namely straight braking, singular lane-change and "J-turn". They reproduce the vehicle motion (i.e. velocity, vehicle's C.G. trajectory and yaw angle) by using the simulated "black box" outputs. Thus, they operate in a way that is similar to that employed by forensic experts who analyze the "black box" outputs. The employed simulation method provides means of comparing the simulated exact values of velocity and of the vehicle's C.G. trajectory to the readings obtained by integrating the "black box" outputs simulation. Estimation of the error produced by the expert employing the "black box" outputs can be made in this way. The body pitch and roll are sources of longitudinal and lateral acceleration recording errors because a gravitational acceleration component for the respective direction is added in simplified version of the device. The results show a possibility of occurrence of significant values of the error, in particular, in situations where maneuvers producing a large share of lateral movement are made.

INTRODUCTION

Development of technology results in lowering the production costs of high-tech equipment capable of recording and storing quantities that identify pre-crash and crash situations for various means of transportation. Now, "black boxes" designed for installation in automobiles, trucks and buses are available. These devices are intended to record quantities that can be useful for forensic experts in identifying the accident/crash sequence and determining its parameters (e.g. initial car velocity, its position on the road). However, incomplete information can result in generation of errors in the accident/crash identification and, in consequence, a significant change in the final analysis outcome.

CAR "BLACK BOXES"

"Black boxes" used in motorization are named ADR – Accident Data Recorder, or EDR – Event Data Recorder. In Europe another abbreviation is also used, namely UDS – Unfall Daten Speicher (in English: Accident Data Recorder). Those devices register variables that help car technical and court experts reconstruct the moment of the accident and estimate the value of its parameters, like, for example, the car's initial velocity and its initial position on the road.

We can distinguish two types of "black boxes" (BB) according to their measuring possibilities:

- the first one (being hereinafter referred to as BB1) being a solution similar to the one used in aircraft; registered values: three components of acceleration - longitudinal a_w, lateral a_p and "vertical" a_z, three values describing the angular position of the car body: the yaw angle ψ_1, the pitch angle φ_1 and the roll angle ϑ_1, or analogous angular velocities ($\dot{\psi}_1, \dot{\varphi}_1, \dot{\vartheta}_1$) - fig. 1a,

- devices of the second type (being hereinafter referred to as BB2) used in motorization, being a simplified version of the aircraft type solution; their construction differs in the fact that the "vertical" acceleration a_z, pitch angle φ_1 and roll angle ϑ_1 (or analogous angular velocities - $\dot{\varphi}_1, \dot{\vartheta}_1$) are not being registered - fig. 1b.

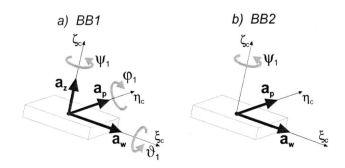

Figure 1. Two types of "black box": BB1 i BB2

An example for the "black box" registration is illustrated on figure 2. The device constantly monitors and analyses the parameters with the frequency of 500 Hz in a 30-second time loop. The final registration begins in the moment of collision (exceeding the threshold values of accelerations). The history preceding the collision is also registered: 100 ms with the frequency of 500 Hz and the preceding 30 s with the frequency of 25 Hz. This is followed by the current registration: the next 100 ms with the frequency of 500 Hz and then with the frequency of 25 Hz. The registration process ends 15 seconds after the moment of collision, unless it is turned off by the user. Based upon the registered values, the working of the control mechanisms of the car (indicators, "stop" lights, outside lighting, etc.) is checked. One can also work out the time history of the car velocity in the longitudinal and side directions (and thus the resultant velocity) and the trajectory of the car's motion. To evaluate this the so-called quadrature procedures are used. These are methods of numerical integration, enabling the integration of the registered accelerations (and, optionally, the angular velocities) of the car in adequate conditions. The calculations are conducted "backwards" starting in a specified point in time and space (for example starting from the moment and location of the collision).

Figures 3 to 6 represent examples of time history of the simulation of the BB1 device registration process, during the maneuver of a singular lane change (1LCM) carried out by a passenger car medium size (gross weight) with initial velocity of 90 km/h = 25 m/s (the amplitude of the steering wheel angle α_{km}=1.15 rad, period T_α=2 s). Figure 3 shows the acceleration of the installation point of the BB1 in the longitudinal direction a_w, lateral a_p and vertical a_z. Figure 4 presents the time histories of the three angles describing the angular position of the car body: the yaw angle ψ_1, the pitch angle φ_1 and the roll angle ϑ_1 An optional solution includes also the registration of the according angular velocities (the yaw velocity $\dot\psi_1$, the pitch velocity $\dot\varphi_1$ and the roll velocity $\dot\vartheta_1$).

Knowing all the six values (a_w, a_p, a_z, ψ_1, φ_1, ϑ_1) enables the evaluation of the acceleration components in the Oxyz system fixed with the road (details in Appendix A and B). Figures 5 and 6 present the results of the integration of the time histories shown by figures 3 and 4 (after adequate transformations) starting from the end time t_k= 5 s backwards (the fact being marked by an arrow). In this way the time history of the car velocity V has been obtained (optionally the two components: the longitudinal velocity V_w and the lateral velocity V_p) as well as the trajectory $y_{O_1}(x_{O_1})$ of the car mass centre O_1 (C.G.). The results show that the initial velocity of the maneuver V_0=V(t=0) (that is at the time when the car was still moving by a rectilinear movement at constant velocity) was indeed 90 km/h = 25 m/s, and the car was

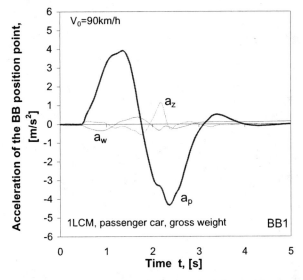

Figure 3. Time histories of longitudinal a_w, lateral a_p and "vertical" a_z acceleration. Singular lane-change manoeuvre

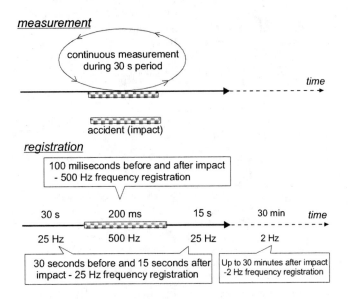

Figure 2. Registration of parameters using device UDS 2165 [3, 8, 14]

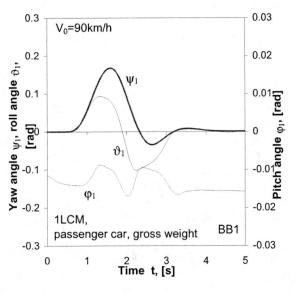

Figure 4. Time histories of yaw angle ψ_1, pitch angle φ_1 and roll angle ϑ_1. Singular lane-change manoeuvre

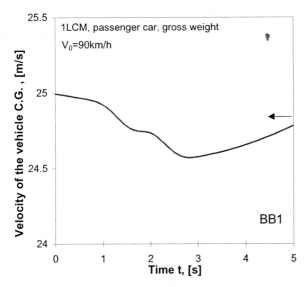

Figure 5. Reconstructed time history of vehicle motion velocity based on BB1 records. Singular lane-change maneuver

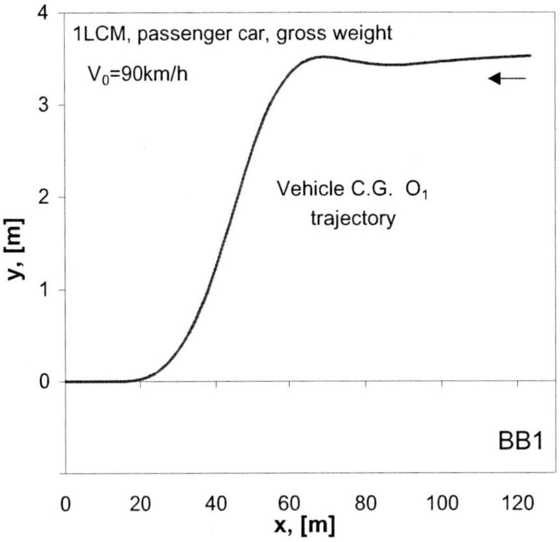

Figure 6. Reconstructed vehicle body C.G. O_1 trajectory on the road plane Oxy based on BB1 records. Singular lane-change manoeuvre

located in the position $x_{O_1}(t=0) = y_{O_1}(t=0) = 0$ on the Oxyz system fixed with the road. The exact location of the point O and the Ox axis (and therefore also the Oy and the Oz axes) is defined by an expert. This may be, for example, the point in which the car was found after the accident, and the direction of the Ox axis might be the direction of the road or the direction shown by the final location of the car. In the presented results, for clear presentation, we have assumed that t=0 is the time of starting the maneuver, and that the $x_0=x(t=0)=y_0=y(t=0)=0$ location is the initial location of the maneuver.

The information presented above is very helpful during appraisals made by court (forensic) experts identifying the circumstances of the car accident.

POTENTIAL SOURCES OF ERRORS IN THE "BLACK BOXES'" MEASUREMENT

It was assumed that we have perfect sensors measuring the accelerations, angular velocities and angular positions. Thus, their interior errors have been omitted.

The focus was set upon the influence of the car body angular movements. It was assumed that the use of gyroscopically stabilized foundations of installing the sensors is not possible for serial solutions of BB.

Another source of error is the incorrect leveling of the set of sensors and the orientation of the angular sensor in relation to the car body plane of symmetry. Authors include those errors in their calculations as fixed errors. For simple presentation of the authors' reasoning and clarity of the article, the authors decided to omit them in the presented calculations.

Another source of incorrectness of the vehicle's movement reconstruction not covered by this article is the registration frequency.

The point of installing BB was marked P. Acceleration sensors, in addition to the values of the components of the acceleration resulting from the dynamics of the mass a_w, a_p, a_z ($a_w = a_{P\xi_c}$, $a_p = a_{P\eta_c}$, $a_z = a_{P\zeta_c}$; see Appendix B, dependence (B.21)), measure also the components of the vector of acceleration of gravity g_{ξ_c}, g_{η_c}, g_{ζ_c} (see Appendix B, dependence (B.22)). With the exception of the vertical component of acceleration, the value of these additional components may be regarded as disturbing factors, being the source of error for the registered acceleration values. Appendix B gives the formal description of the transformation from the system fixed with the Oxyz to the system fixed with the car $O_1\xi_1\eta_1\zeta_1$ and with the BB $P\xi_c\eta_c\zeta_c$ sensors.

As a result of integrating the registered BB values we identify the location and the velocity in the Oxyz system fixed with the road. Thus, one should determine the values of the components of the acceleration in this system. They have the form shown below and refer to the horizontal values of acceleration at a certain point in the longitudinal $a_{\xi h}$ direction, lateral $a_{\eta h}$ direction and the vertical component $a_{\zeta h}$:

$$\mathbf{a} = [\ddot{x}, \ddot{y}, \ddot{z}]^T = \mathbf{A}_\psi \cdot \mathbf{a}_h = \mathbf{A}_\psi \cdot [a_{\xi h}, a_{\eta h}, a_{\zeta h}]^T \quad (1)$$

The components $a_{\xi h}$, $a_{\eta h}$, $a_{\zeta h}$ are sometimes referred to as components of acceleration in the "leveled system".

If the measurements are performed using inertial sensors, their readings include an error originating from measuring the component of the acceleration of gravity (see Appendix B, point B1a):

$$a_w^c = a_w - g \cdot \sin\varphi_1$$
$$a_p^c = a_p + g \cdot \cos\varphi_1 \cdot \sin\vartheta_1 \quad (2)$$
$$a_z^c = a_z + g \cdot \cos\varphi_1 \cdot \cos\vartheta_1$$

The upper index "c" represents the value shown by the sensor. The measurement is performed in directions matching the $O_1\xi_1\eta_1\zeta_1$ system axis.

For the device of the BB1 type an exact description of the components of vector $\mathbf{a}_h$ is possible, basing on the readings of the sensors (see also Appendix B). Disregarding the errors of the sensors (2) we get:

$$\mathbf{a}_h = [a_{\xi h}, a_{\eta h}, a_{\zeta h}]^T = \mathbf{A}_\varphi \cdot \mathbf{A}_\vartheta \cdot [a_w, a_p, a_z]^T \quad (3)$$

$$a_{\xi h} = a_w \cdot \cos\varphi_1 + a_p \cdot \sin\varphi_1 \cdot \sin\vartheta_1 + a_z \cdot \sin\varphi_1 \cdot \cos\vartheta_1$$
$$a_{\eta h} = \phantom{a_w \cdot \cos\varphi_1 + {}} a_p \cdot \cos\vartheta_1 \phantom{{} \cdot \sin\vartheta_1} - a_z \cdot \sin\vartheta_1$$
$$a_{\zeta h} = -a_w \cdot \sin\varphi_1 + a_p \cdot \cos\varphi_1 \cdot \sin\vartheta_1 + a_z \cdot \cos\varphi_1 \cdot \cos\vartheta_1$$
$$(4)$$

Taking into consideration the errors of the sensors (2), we replace the values a_w, a_p, a_z in relation (4) by the values a_w^c, a_p^c, a_z^c and after transformations we obtain, for the device of the BB1 type:

$$a_{\xi h}^c = a_{\xi h}$$
$$a_{\eta h}^c = a_{\eta h} \quad (5)$$
$$a_{\zeta h}^c = a_{\zeta h} + g$$

If the measurement system is calibrated in such a way that we measure only the relative values of the vertical acceleration, then the constant values of the acceleration of gravity g is dropped in the $a_{\zeta h}^c$ element.

The above shows that the working conditions of the of the BB1 type device do not influence the registration accuracy of the car acceleration (taken that the angles $\psi_1, \varphi_1, \vartheta_1$ are properly determined), and therefore they are not a source of error in reconstructing the motion of the car.

If we use a device of the BB2 type, because the angles φ_1, ϑ_1 are unknown, we are forced to assume in the relation (4) $\cos\varphi_1 = \cos\vartheta_1 = 1$, $\sin\varphi_1 = \sin\vartheta_1 = 0$, which leads to the relation:

$$a_{\xi h}^c = a_w^c$$
$$a_{\eta h}^c = a_p^c \quad (6)$$
$$a_{\zeta h}^c \approx 0 \quad \text{(because } a_z \text{ isn't measured)}$$

The values of a_w^c, a_p^c are subjected to an error described by equation (2). These errors "are unknown" for device of the BB2 type.

Figure 7 shows the simulations of time histories of the sensors' readings based on the example shown by figures 3 to 6. For comparison, the acceleration values of $a_{\xi h}, a_{\eta h}$ in the point of installing the sensors have been included. The differences between accelerations in the leveled system $a_{\xi h}, a_{\eta h}$ and the readings a_w^c, a_p^c of the sensors were marked as reading errors $\Delta a_w^c, \Delta a_p^c$ of the sensors.

In the studied example this leads to the following calculation results, based on the records of the BB2 device. Figures 8 and 9 show the time history of the car motion velocity and the trajectory of motion, calculated using the quadratures method. The estimation of the value of the initial velocity of 87.8 km/h = 24.4 m/s has been obtained, at the assumed velocity being 90 km/h (error -2.4%) and the initial position $x_{O_1}(t=0) = 1.69$ m, $y_{O_1}(t=0) = -0.88$ m. Taking into consideration that the estimated initial position of the car motion is $x_{O_1} = y_{O_1} = 0$ and that the value of the lateral movement (in direction y) during the simulation of the maneuver was about 3.5 m, the error in this direction reaches 25%. The differences are then meaningful. So, it explains the reason for more research.

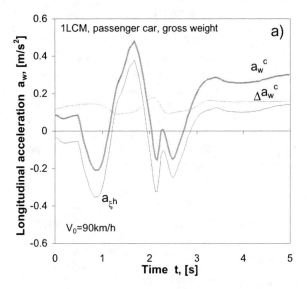
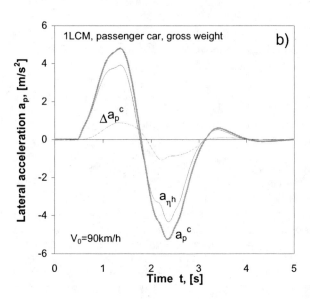

Figure 7. Time histories of longitudinal (fig. a) and lateral (fig. b) acceleration: accurate values $a_{\xi h}, a_{\eta h}$, sensors indications a_w^c, a_p^c and their errors $\Delta a_w^c, \Delta a_p^c$. Singular lane-change maneuver. Device BB2

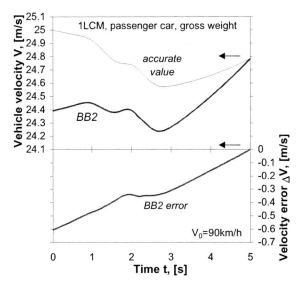

Figure 8. Reconstructed time history of vehicle motion velocity based on BB2 records. Singular lane-change maneuver

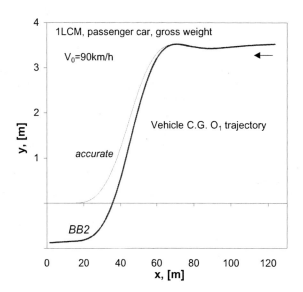

Figure 9. Reconstructed vehicle body C.G. O_1 trajectory on the road plane Oxy based on BB2 records. Singular lane-change maneuver

THE ACCEPTED METHOD OF RESEARHING THE POTENTIAL ERRORS WHEN USING THE "BLACK BOXES" OF THE BB2 TYPE

To estimate the results of the errors in functioning of the BB2 type "black boxes", connected with the relative motion of the car body, one needs to have an exact recording of the analyzed values (the accelerations in chosen directions, angular velocities or the angles of the turn of the car body around the known axes). The use of the most sublime measuring techniques on a real object (highest class sensors, gyroscopically stabilized foundations of the acceleration sensors) will not protect us against measuring errors (internal errors of the sensors, errors in the placement of installation of the sensors). Furthermore, it is a very expensive method of research.

It is much easier (especially for learning purposes and initial researches leading to quantity conclusions) to use an experimentally verified model of simulation of the motion and the dynamics of a vehicle, expanded by a model simulating the readings of the BB2 type devices.

Appendix A presents shortly two models simulating the motion and the dynamics of a two-axle vehicles, used in this work. Those models were experimentally verified for a passenger car of medium class (of total mass up to 1550 kg) and a truck of medium loading capacity (total mass up to 11500 kg), see [9, 10].

Appendix B describes the method of determining the readings of the sensors, including the errors coming from the motion of the car body, the installation of the BB in the body, and from calibration.

The calculation procedure was the following. First, the car motion simulation was performed. Based on its results the registration process of the BB2 was simulated. This registration was the foundation for reconstruction of the previously simulated motion. The procedure of numerical integration – the Newton-Cotes quadrature [13] - was used. The scheme of the car motion reconstruction is presented on figure 10.

The comparison of the simulation results of a given value (for example the motion velocity) and the results based on the BB2 registration, helps to estimate how accurate the results of the car motion reconstruction may be, using the device.

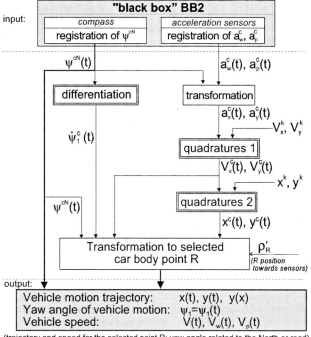

Figure 10. The scheme of processing the BB2 data (index "c" represents the sensors' reading or a value calculated on their basis, index "k" – the final values, index "x", "y" – the components in the Oxy system; R – any point on the car body, ψ^{cN} – the yaw angle, relative to the North)

The absolute value of the error of this parameter was accepted to be equal to the difference between its value estimated by the use of BB2, and the one calculated in the simulation of the motion. The relative error is the absolute value of the error compared to the value accepted as the accurate value that is the one calculated in the car motion simulation experiment.

The authors assumed that considered "black box" has the same parameters as actual product existing on the European market [3, 8, 14]; see also figure 2. As the result of above, the important problem of recording frequency was not treated. However, the presented models may be useful to analyze this problem.

EXAMPLARY CALCULATIONS FOR TYPICAL DRIVER'S DEFENCE MANOEUVRES

As examples of maneuvers that may lead to an accident, and thus to the use of the BB1 and BB2 devices' recording, the following were chosen:
- the maneuver of braking during rectilinear motion (Straight Line Braking, SLB);
- the singular lane change maneuver (1LCM);
- the entering a curve maneuver (the Ramp Input on the Steering Wheel, RISW; "J-turn" in US).

The calculations were performed for two types of vehicles:
- a passenger car of medium size in two load conditions:
  - partially loaded (total mass m_c=1285 kg)
  - gross weight (total mass m_c=1545 kg)
- a two-axle truck in two load conditions:
  - with no loading (marked P0, total mass m_c=6900 kg)
  - partially loaded (marked P3, total mass m_c=9165 kg)

Because the calculations for the "black box" of the BB1 type show (see chapter 3) a possibly high accuracy of the car motion reconstruction, the results presented below concern only the second type of BB device (BB2). It was assumed in the calculations that BB2 is placed underneath the driver's seat.

BRAKING DURING RECTILINEAR MOTION - The process of a rectilinear braking (SLB) was forced by a temporary course of the pressure force P_N exerted on the brake pedal: constant value preceded by a time of reaction t_0 (P_N=0) and a period of linear increase t_n (P_N grows from zero to the nominal value). In all calculations the reaction time was taken as 1 s, and the increase time as 0.2 s. Generally, conditions of the test meet the ISO 6597 [4] standards requirements.

Figure 11 presents an example of braking in a rectilinear motion of a passenger car from an initial velocity of V_0= 90 km/h, at a nominal value of the pressure force exerted on the brake pedal P_N=300 N on a dry asphalt surface. Figure 11a presents the time history of the readings of the sensor showing the longitudinal acceleration a_w^c as compared to the simulated real value of that acceleration $a_{\xi h}$ and the difference between them – the error of the sensor Δa_w. Figures 11 b and c present the consequences on the example of BB2: the reconstructed time history of the car motion (b) and the change of position (c). The initial value V_0 was overstated by approximately 5.1 km/h (5.7% compared to the real value, 90km/h). Also the change of position is – according to the analysis based on the BB2 records – overstated: by approximately 3.8m (5.9% relative to 64.2 m). It is a result of an overstatement of the absolute values of longitudinal acceleration (see fig.11a). This "overstatement" is mainly due to an increase of the pitch angle during the process of braking, and the related component of the acceleration of gravity. Figure 12 presents collectively the results for a few dozens of examples for braking of the same vehicle, differing in the value of the pressure force exerted on the brake pedal P_N, and in the value of the initial velocity V_0 (50 and 90 km/h). They include the absolute and relative values of the error in evaluating the initial velocity using the BB2 records (a), the braking distance S_h in the simulation and reconstruction of the car motion, and the absolute and relative values of the error in calculating S_h by using the BB2 records. Both errors ΔV_0 and ΔS_h are equal to about 5 - 7%, with a small increasing tendency for larger pressure forces exerted on the brake pedal (for blocked wheels). In absolute values, the initial velocity error ΔV_0 may reach the level of a few km/h: about 3 km/h for V_0=50 km/h and about 5 - 6 km/h for V_0=90 km/h. The error in calculating the braking distance reaches from about 1.5 - 2 m for V_0=50 km/h to about 5 - 6 m for V_0=90 km/h.

THE SINGULAR LANE CHANGE MANOEUVRE - The singular lane change maneuver (1LCM) was forced by the use of temporary course of the steering wheel angle α_k in the form of one period of the sinusoid curve. The conditions of those experiments are similar to the test suggested by ISO [5, 6]. The amplitude α_{km} and the period T_α were chosen specifically, so that the car's motion in a lateral direction at the end of the maneuver equals to about 3.5 m (the standard width value of a traffic lane). The time of reaction (time from the initial moment to the moment of exerting the forcing – a non-zero steering wheel angle α_k) in all experiments was the same: 0.5 s.

The research was performed for a passenger car and a truck. The main goal was to define the possible errors in evaluating the trajectory. It was assumed that the direction of the road was marked by the Ox axis. The standard value of the period T_α for the passenger car was 2 s, and for the truck – 3 s (unless stated otherwise). The maneuver was treated as finished after 5 s from the initial time (t_k=5 s).

Figure 13 presents the results of the evaluation of the centre of mass O_1 motion trajectory y(x) for four different combinations of parameters for a passenger car, differing only in the velocity of performing the maneuver, the car's load, and the amplitude of the steering wheel angle.

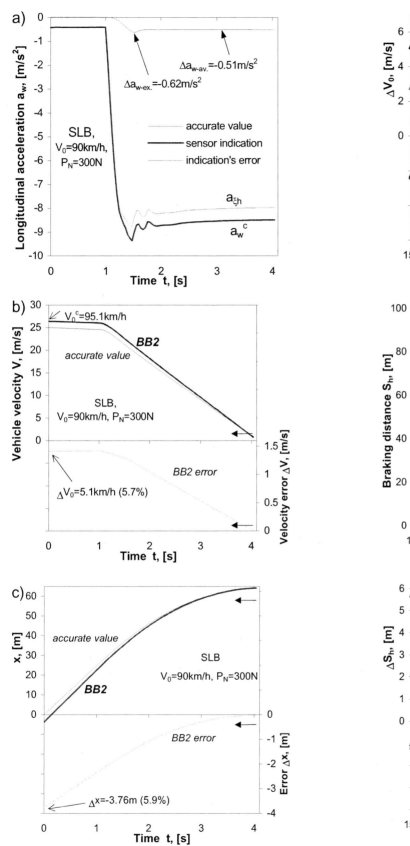

Figure 11. Time histories for straight-line braking: longitudinal acceleration $a_{\xi h}$, a_w^c, Δa_w^c (fig. a) and reconstructed vehicle velocity V (fig. b) and traveled distance of vehicle C.G. O_1 (fig. c). Partially loaded passenger car

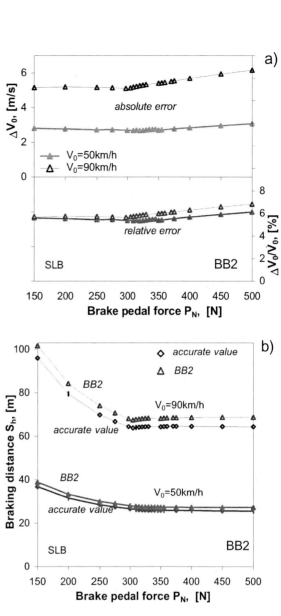

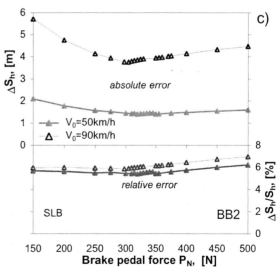

Figure 12. Reconstructed vehicle motion parameters: initial velocity error ΔV_0 (fig. a), braking distance S_h (fig. b), braking distance error ΔS_h (fig. c) as a function of brake pedal force P_N; for two assumed initial velocities 50 and 90 km/h. Partially loaded passenger car

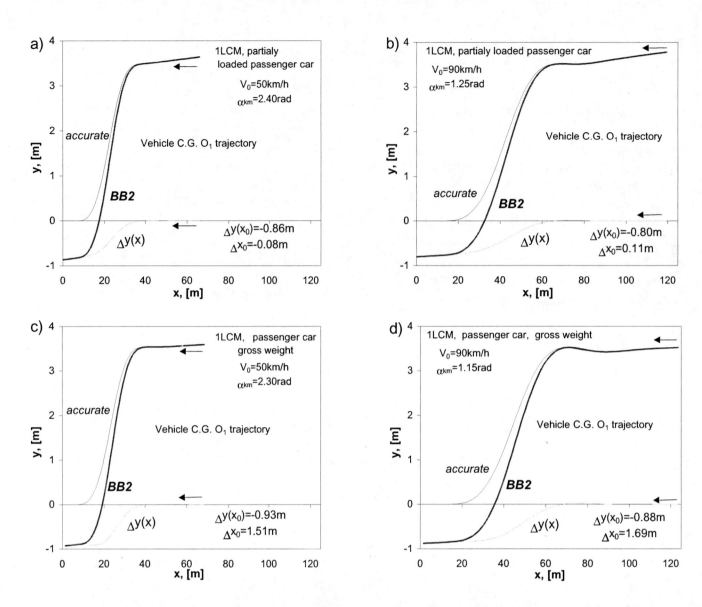

Figure 13. Singular lane-change maneuver (1LCM). Vehicle body C.G. O_1 trajectories on the road plane Oxy for different vehicle load and velocity. Passenger car

The results include the exact progress (based on the simulation of the motion), the one obtained from the BB2 registration and its error. It is visible that irrespective from the chosen combination of parameters, the error in evaluating the trajectory $\Delta y_0(x_0)$ for the initial position equals about 0.8 - 0.9 m, although it is a bit smaller for a partially loaded car. It is also worth noticing that the estimation error of the longitudinal position Δx_0 (at the initial time) was much bigger for the totally loaded car.

Figure 14 presents analogical results for four combinations of parameters for the case of a truck. In this case the results are different. For the a, b and c vehicles the centre of mass O_1 location error at the initial moment is small, although in the middle phase the error $\Delta y(x)$ reaches about 0.5 m. (figure 14 a and b). The small value of this error is mainly the result of the "symmetry" of the maneuver, that is a similarity (in a higher extent than in the case of a passenger car) of the lateral acceleration "half period" history. The trajectory error, increasing in the first phase of the calculations, is compensated by the change in the sign of the error of the lateral acceleration in the second phase.

In case d (a shorter time of forcing, $T_\alpha = 2$ s) the trajectory error is very large (exceeds 3 m). This is caused by a large value of the error of indication of the sensor Δa_p^c (reaches absolute values above 2 m/s^2 at nominal value $a_{\eta h}$ equal roughly to the same). The source of such behavior of the sensor is the high value of the roll angle ϑ_1 (reaching extreme values of about –0.2 rad). At this moment the internal wheel looses its contact with the road surface.

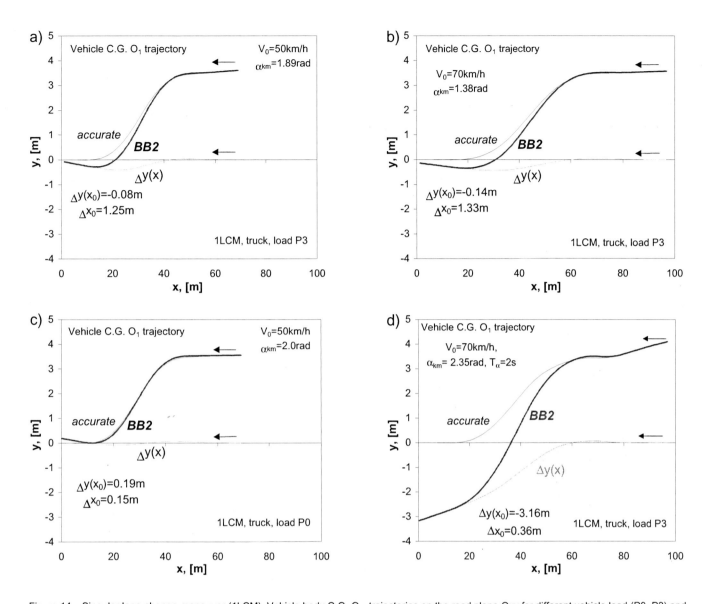

Figure 14. Singular lane-change maneuver (1LCM). Vehicle body C.G. O_1 trajectories on the road plane Oxy for different vehicle load (P0, P3) and velocity. Truck

RAMP INPUT ON THE STEERING WHEEL ("J-TURN") - The maneuver, known also as "entering the curve" as well as "the J-turn" has been carried out in conditions similar to the test suggested by ISO [5]. The test is based on a step jump on the steering wheel angle α_k preceded by a period of linear growth. The maximum value of the steering wheel rotation angle was marked α_{km} and the time of growth t_α. The rate of growth α_{km}/t_α fall within the values suggested by ISO. The calculations of the simulation ended, that is the final parameters of the motion of the car were set when the lateral displacement referring to the axis of the road (described as before, by the direction of the Ox axis) exceeded 7 m (this corresponds to the situation, when the tested vehicle falls out of the two-lane road having a width of 7 m and a sideway of about 1.75 m, in a situation when it was driving in the middle of his lane and then was forced to perform the maneuver, driving past the opposite lane and leaving the road).

Tests were carried out for the case of a passenger car and a truck. The main goal of the tests was to establish the possible errors in evaluating the trajectory.

Figure 15 shows the results of evaluating the trajectory of the centre of the mass O_1 motion y(x) for four combination of parameters for a passenger car, differing from each other in the speed of carrying out the maneuver, the load of the car, the maximum value of the steering wheel angle. The results include the exact progress (obtained from the simulation of the motion), the one obtained from the BB2 registration and its error. It is visible that the error in evaluating the trajectory $\Delta y_0(x_0)$ for the initial position equals about 2.5 - 2.8 m, and similar to the case of the singular lane change maneuver it is slightly smaller in the case of the partially

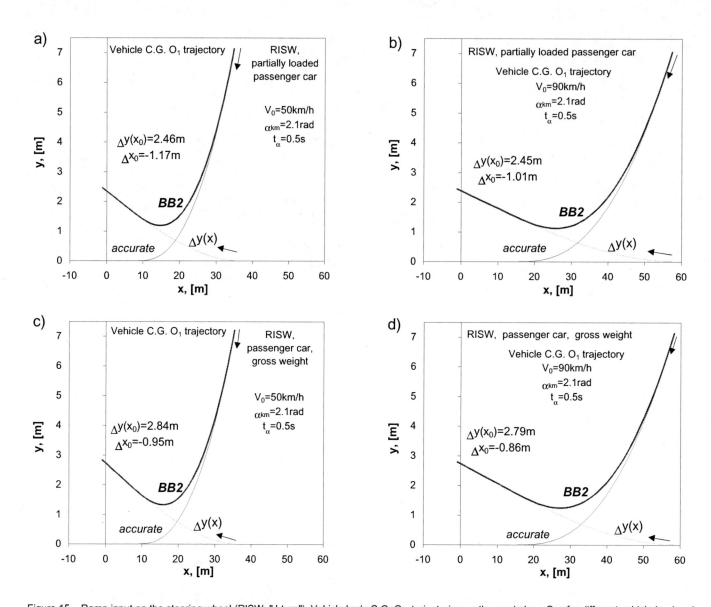

Figure 15. Ramp input on the steering wheel (RISW; "J-turn"). Vehicle body C.G. O_1 trajectories on the road plane Oxy for different vehicle load and velocity. Passenger car

loaded car. Taking into consideration the standard width of the traffic lane (3.5 m) these are large values, as they may result in wrongly estimating the lane used by the car at the beginning of the analyzed situation.

Figure 16 presents analogical results for four combinations of parameters for the case of a truck. The results are similar to the case of the passenger car, although in cases a and b the values of the errors in evaluating the trajectory for the initial position are smaller ($\Delta y_0(x_0)$ equals to about 1.75 m). In the case presented by figure 16c the error is significantly larger (exceeds 6 m). For this maneuver we have an analogical situation to the maneuver 1LCM, case P3, V_0=70 km/h (illustrated by fig. 14d) – large values of the car body roll angle, lost contact of the internal wheel with the road surface, and subsequently large values of the error in evaluating the acceleration values.

Estimating the reconstruction of the pre-accident situation having the character of "entering the curve" we need to note, that as opposed to the singular lane change maneuver, and similar to the braking in rectilinear motion, there isn't any "compensation" of errors in evaluating the acceleration values, resulting from the change in the sign of the acceleration.

CONCLUSION

This paper presents the observations of the car movement simulation in situations before the accident, using experimentally verified models of the vehicle dynamics, including the simulations of "black boxes" recordings. The reconstruction of the vehicle motion, based upon the registered vehicle motion parameters was taken into consideration, using two different solutions: BB1 – with its characteristics similar to the aircraft type, BB2 – with its characteristics typical for the solutions offered on the car market. The focus was put

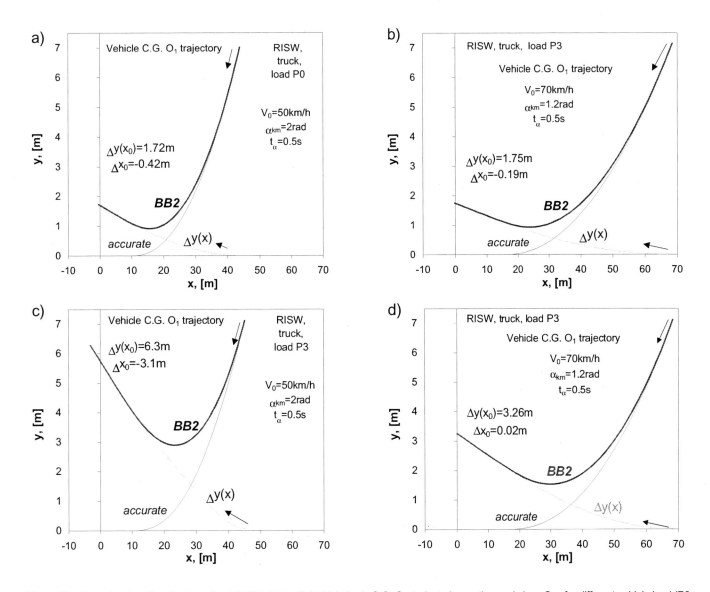

Figure 16. Ramp input on the steering wheel (RISW; "J-turn"). Vehicle body C.G. O_1 trajectories on the road plane Oxy for different vehicle load (P0, P3) and velocity. Truck

mainly on the influence of the angular movements of the car body on the correctness in reconstructing the vehicle motion.

The characteristics of the BB2 type device results in such a way, that the motion of the vehicle is regarded to be a flat movement. BB2 does not register the angles of the pitch and roll motion of the car body, neither the "vertical" acceleration. As a consequence, the accelerations considered when reconstructing the progress of the velocity and trajectory of the vehicle motion are affected by an error resulting from measuring the components of the acceleration of gravity. This may reach considerable values (even up to few dozens of percent related to the real values). In result, significant errors in reconstructing the motion of the car may be obtained when using the BB2 records. This relates especially to the trajectory of the movement. It is different in the case of devices of the type BB1. In this case, the knowledge of the pitch angle and roll angle enables to compensate the error and thus obtain a reconstruction matching the real progresses.

In summary, the calculations presented in this paper show a significant importance of recordings made by "black boxes" used in cars, showing the angular positioning of the body (pitch and roll angles). They have a significant influence on the quality of the vehicle motion reconstruction. It is worth noting, that similar suggestions have been reached by the R&D EDR WG (Research and Development Event Data Recorder Working Group) group working within the NHTSA, which had as a goal to describe the present and future possibilities and the scope of EDR devices implementation in car vehicles [2].

ACKNOWLEDGMENTS

The research work presented in the paper was partly supported by the state committee for scientific research (KBN), grants Nos 9T12C05219, 8T12C02821.

REFERENCES

1. Arczewski K.P., *The Kinematics of Discrete Systems.* Oficyna Wydawnicza Politechniki Warszawskiej (Publishing House of the Warsaw University of Technology). Warsaw, 1994. (in the Polish language).
2. http://dms.dot.gov/ (ID: 5218, Organization: NHTSA).
3. http://www.vdokienzle.com/pkw
4. ISO 6597: *Straight Line Braking Test.* 1988.
5. ISO 7401: *Road Vehicles - Lateral Transient Response Test Method.* (DIS: 1986, TR: 1988).
6. ISO TR 8725: *Road Vehicles - Transient Open-Loop Response Test Method with One Period of Sinusoidal Input.* 1988.
7. Kaminski E., Pokorski J., *The Theory of the Car. The Dynamics of the Suspensions and Driving Systems of Car Vehicles.* WKiL. Warsaw 1983. (in the Polish language).
8. Lehman G., Reynolds T., *The Contribution of Onboard Recording Systems to Road Safety and Accident Reconstruction.* Proceedings of the NTSB International Symposium on Transportation Recorders. May, 3 - 5, 1999.
9. Lozia Z., *The Analysis of the Movement of a Two-axle Car on the Background of Modelling its Dynamics.* Monograph. Zeszyty Naukowe Politechniki Warszawskiej (Warsaw University of Technology Scientific Papers). Warsaw 1998. (in the Polish language).
10. Lozia Z., *Vehicle dynamics and motion simulation versus experiment.* 1998 SAE International Congress and Exposition. Detroit, Michigan USA. February 23 - 26, 1998. SAE TP 980220. (Published also in the Special Publication SP-1361 "Vehicle Dynamics and Simulation, 1998" and in the SAE Transactions 1998).
11. Maryniak J., *The Dynamic Theory of Moving Objects.* Politechnika Warszawska (Warsaw University of Technology). Prace naukowe (Scientific Papers). Mechanika (Mechanics). No 32. WPW. Warsaw 1976. (in the Polish language).
12. Osiecki J., *The Fundamentals of the Analysis of Mechanical Vibrations.* Politechnika Swietokrzyska. Kielce 1979. (in the Polish language).
13. Stoer J., *Introduction to the Numerical Methods.* Vol. 1. PWN, Warszawa, 1979 (in original *Einführung in die Numerische. Mathematik 1*, Springer Verlag, Berlin , New York, Heidelberg 1972, 1976). (in the Polish language).
14. *UDS 2165. The Accident Data Recorder.* Information sheet provided by DRABPOL company, Czestochowa; http://www.drabpol.pl/tachografy

APPENDIX A. SIMULATION MODELS OF CAR MOTION AND ITS DYNAMICS

Two types of models simulating the vehicle motion and its dynamics [9, 10] have been used. The first one (fig. A1) describes the characteristics of a two-axle passenger car or van having an independent front suspension and a dependent rear one. The second one describes a two-axle truck or van having a dependent front and rear suspension. Figure A1 presents the major characteristics of both models.

The following co-ordinate systems were chosen:

- $Oxyz$ – the inertial system fixed with the road; the Ox and the Oy axis are horizontal, the vertical Oz axis is orientated upwards;
- $O_1x_1y_1z_1$ – the non-inertial system with axis parallel accordingly to the Ox, Oy and the Oz axis and the origin of the system lying in the mass centre of the car body O_1;
- systems fixed with model masses: the car body ($O_1\xi_1\eta_1\zeta_1$), the rear axle mass ($O_4\xi_4\eta_4\zeta_4$), the front axle mass ($O_9\xi_9\eta_9\zeta_9$) and four wheels ($O_5\xi_5\eta_5\zeta_5$, $O_6\xi_6\eta_6\zeta_6$, $O_7\xi_7\eta_7\zeta_7$, $O_8\xi_8\eta_8\zeta_8$);
- auxiliary systems, enabling the evaluation of the transformation matrices.

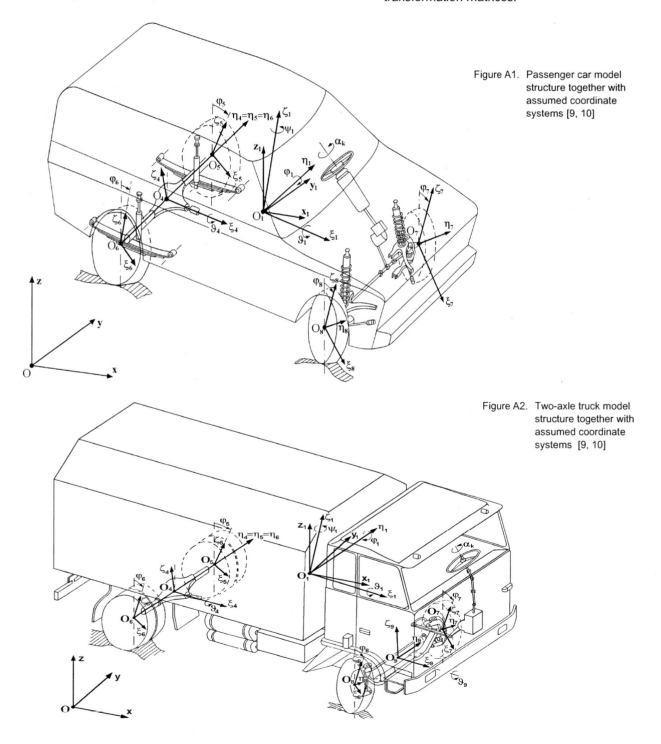

Figure A1. Passenger car model structure together with assumed coordinate systems [9, 10]

Figure A2. Two-axle truck model structure together with assumed coordinate systems [9, 10]

The translatory movement of bodies and material points in the model are described by using the position of the centers of the mass of the bodies (O_1, O_4, O_9, O_5, O_6, O_7, O_8) and points O_2 and O_3, treated as material points (front "unsprung masses" for model shown by figure A1; they lie on the crossing of the rotation and king-pin axes). The axes $O_i\xi_i$, $O_i\eta_i$, $O_i\zeta_i$ (i = 1, 4, 5, 6, 7, 8, 9) are treated as the main central axes of inertia of the corresponding stiff bodies.

The spherical motion of the vehicle body against the pole O_1 has been described using the "airplane angles" [7, 11, 12] known also as "quasi-Euler" [7, 11], "Cardan" or "Bryant" [1]:

- the yaw angle ψ_1 (rotation around the axis $O_1\zeta_1$),
- the pitch angle φ_1 (rotation around the axis $O_1\eta_1$),
- the roll angle ϑ_1 (rotation around the axis $O_1\xi_1$).

The succession of rotations corresponds to the succession of their description.

The transformation of the system $O_1\xi_1\eta_1\zeta_1$ to the system $O_1x_1y_1z_1$ is described by the relation:

$$[x_1, y_1, z_1]^T = \mathbf{A} \cdot [\xi_1, \vartheta_1, \zeta_1]^T \qquad (A.1)$$

where $[a, b, c]^T = \begin{bmatrix} a \\ b \\ c \end{bmatrix}$;

"T" means the transposition of the matrix or the vector.

Table A1. VEHICLE MODEL STRUCTURE

The two-axle car model	The two-axle truck model
➢ Two-axle vehicle with independent front wheels and dependent rear wheels.	➢ Two-axle vehicle with dependent front and rear wheels.
➢ 8 mass-elements treated as rigid bodies (body, rear axle beam, two front unsuspended masses, four rotating wheels).	➢ 7 rigid mass elements bodies (body, front and rear axle beams, four rotating wheels).
➢ Vehicle motion is described in fixed coordinate system Oxyz connected with the road and in many local systems connected with the model bodies.	
➢ 14 DOF (degrees of freedom): 3 co-ordinates of vehicle body centre of gravity O_1 position in fixed system Oxyz (x_{O_1}, y_{O_1}, z_{O_1}), 3 angular co-ordinates of body position (yaw ψ_1, pitch φ_1 and roll ϑ_1), 4 co-ordinates of unsprung masses relative motion (ζ_{10_2}, ζ_{10_3}, ζ_{10_4}, ϑ_4), 4 angular co-ordinates of wheels' rotation (φ_5, φ_6, φ_7, φ_8).	➢ 14 DOF (degrees of freedom): 3 co-ordinates of vehicle body centre of gravity O_1 position in fixed system Oxyz (x_{O_1}, y_{O_1}, z_{O_1}), 3 angular co-ordinates of body position (yaw ψ_1, pitch φ_1 and roll ϑ_1), 4 co-ordinates of unsprung masses relative motion (ζ_{10_4}, ϑ_4, ζ_{10_9}, ϑ_9), 4 angular co-ordinates of wheels' rotation (φ_5, φ_6, φ_7, φ_8).
➢ Non-linear elastic-damping suspension characteristics (limiters, dry friction, asymmetry of shock-absorbers).	
➢ Suspension compliances are taken into account but in very general way, without inclusion of bushing elements models.	
➢ The steering system model, apart from geometric properties, describes also its elastic properties.	
➢ It is assumed lack of geometrical roll-steer (resulting from geometrical constrains existing in the steering system as well in the suspension).	
➢ Three-dimensional location of the king-pin axis is described in the body-fixed system $O_1\xi_1\eta_1\zeta_1$ (king-pin inclination and castor angles; they are functions of suspension deflection).	
➢ The pneumatic tire model describes interaction with even and uneven road surface: - elastic-damping properties in the radial direction, - elastic properties in the lateral and longitudinal direction.	
➢ The tire shear forces and aligning moment model is a semi-empirical model (**HSRI-UMTRI** or **Magic Formula**). It includes the influence of: - wheel centre velocity, normal reaction of road, camber angle (related to the road), - king-pin inclination, caster, and toe-in angles on forces and moments generated in the footprint. In low frequency analysis, transient properties of tire are neglected. When considering higher frequency phenomenon first order differential equation for lateral force is applied based on **von Schlippe-Dietrich** or **LPG-tire** theory.	
➢ Force and moment inputs result from: aerodynamic drag, wheel rolling resistance, braking, driving.	
➢ The steering wheel angle is treated as an external input.	
➢ Two steering formulae (taken from literature) describe driver steering activity for typical motion cases.	
➢ The road surface irregularities are also inputs. They have a determined 3D form. They may be a result of real road measurements or the realization of stationary Gaussian random process describing real road according to ISO.	

Matrix **A** has the form:

$$\mathbf{A} = \begin{bmatrix} \cos\psi_1 \cdot \cos\varphi_1 & \cos\psi_1 \cdot \sin\varphi_1 \cdot \sin\vartheta_1 - \sin\psi_1 \cdot \cos\vartheta_1 & \cos\psi_1 \cdot \sin\varphi_1 \cdot \cos\vartheta_1 + \sin\psi_1 \cdot \sin\vartheta_1 \\ \sin\psi_1 \cdot \cos\varphi_1 & \sin\psi_1 \cdot \sin\varphi_1 \cdot \sin\vartheta_1 + \cos\psi_1 \cdot \cos\vartheta_1 & \sin\psi_1 \cdot \sin\varphi_1 \cdot \cos\vartheta_1 - \cos\psi_1 \cdot \sin\vartheta_1 \\ -\sin\varphi_1 & \cos\varphi_1 \cdot \sin\vartheta_1 & \cos\varphi_1 \cdot \cos\vartheta_1 \end{bmatrix}$$

(A.2)

The transformation in the opposite direction (from $O_1 x_1 y_1 z_1$ to $O_1 \xi_1 \eta_1 \zeta_1$) is described by the inverse matrix $\mathbf{A}^{-1}$, where:

$$\mathbf{A}^{-1} = \mathbf{A}^T \qquad (A.3)$$

which is derived from their mutual orthogonality.

Both models have been positively verified by experiments. The first one for the case of a passenger car of medium size (total mass up to 1550 kg), the second one for a truck of medium load (total mass up to 11500 kg). Experimental verifications were performed on very good roads for maneuvers recommended by ISO. For a medium class car: straight line braking, one period of sinusoidal input on steering wheel (normally it is singular lane change maneuver). For a two-axle truck: circular steady state motion, step input on steering wheel ("J-turn"), double lane change maneuver. The results of the verification are shown in the paper [10].

APPENDIX B. THE KINEMATICS OF THE VEHICLE MOTION. THE "BLACK BOX" MODEL

The movement of the vehicle body is treated as a combination of the translatory movement of the centre of the mass of the body O_1 and the spherical movement of the body against point O_1. Thus we consider a movement having 6 degrees of freedom (3 displacements and 3 rotations). Figure B1 presents the assumed coordinate systems (the description conforming to the one included in Appendix A) and the vectors characterizing the kinematics of the movement.
The $P\xi_c\eta_c\zeta_c$ system is a moving system, fixed with BB.

The following markings for the vectors were chosen:
r – translatory location;
V – translatory movement velocity;
a – translatory movement acceleration;
ω – angular velocity;
ε – angular acceleration.

The description of vectors (notation in matrix form "T" means transposition):

$\bar{r}_{O_1} \equiv \mathbf{x}_{O_1} = [x_{O_1}, y_{O_1}, y_{O_1}]^T$ - the position of the centre of the mass of the car body O_1 in the inertial Oxyz system;

$\bar{r}_P \equiv \mathbf{x}_P = [x_P, y_P, y_P]^T$ - the position of point P in the inertial Oxyz system;

$\bar{\rho} \equiv \rho = [\xi_P, \eta_P, \zeta_P]^T$ - the position of point P in the $O_1\xi_1\eta_1\zeta_1$ system, connected with the vehicle (see Appendix A);

$\bar{\omega} \equiv \omega = [\dot{\psi}_1, \dot{\varphi}_1, \dot{\vartheta}_1]^T$ - angular velocity;

$\bar{\varepsilon} \equiv \varepsilon = [\ddot{\psi}_1, \ddot{\varphi}_1, \ddot{\vartheta}_1]^T$ - angular acceleration;

$\bar{V}_{O_1} \equiv \dot{\mathbf{x}}_{O_1} = [\dot{x}_{O_1}, \dot{y}_{O_1}, \dot{z}_{O_1}]^T$ - the velocity of point O_1;

$\bar{a}_{O_1} \equiv \ddot{\mathbf{x}}_{O_1} = [\ddot{x}_{O_1}, \ddot{y}_{O_1}, \ddot{z}_{O_1}]^T$ - the acceleration of point O_1;

$\bar{a}_P \equiv \ddot{\mathbf{x}}_P = [\ddot{x}_P, \ddot{y}_P, \ddot{z}_P]^T$ - the acceleration of point P.

The kinematics of point P is described as follows (**A** – the rotation matrix, described in Appendix A):

position: $\quad \mathbf{x}_P = \mathbf{x}_{O_1} + \mathbf{A} \cdot \rho \quad$ (B.1)

velocity: $\quad \dot{\mathbf{x}}_P = \dot{\mathbf{x}}_{O_1} + \dot{\mathbf{A}} \cdot \rho + \mathbf{A} \cdot \dot{\rho} \quad$ (B.2)

acceleration: $\quad \ddot{\mathbf{x}}_P = \ddot{\mathbf{x}}_{O_1} + \ddot{\mathbf{A}} \rho + 2 \cdot \dot{\mathbf{A}} \dot{\rho} + \mathbf{A} \ddot{\rho}$

Because point P is fixed the vehicle body, the relative velocity is equal zero ($\dot{\rho} = 0,0,0]^T$). So, the Coriolis acceleration is zero and the relative acceleration is equal zero ($\ddot{\rho} = [0,0,0]^T$). Thus:

velocity: $\quad \dot{\mathbf{x}}_P = \dot{\mathbf{x}}_{O_1} + \dot{\mathbf{A}} \cdot \rho \quad$ (B.4)

acceleration: $\quad \ddot{\mathbf{x}}_P = \ddot{\mathbf{x}}_{O_1} + \ddot{\mathbf{A}} \cdot \rho \quad$ (B.5)

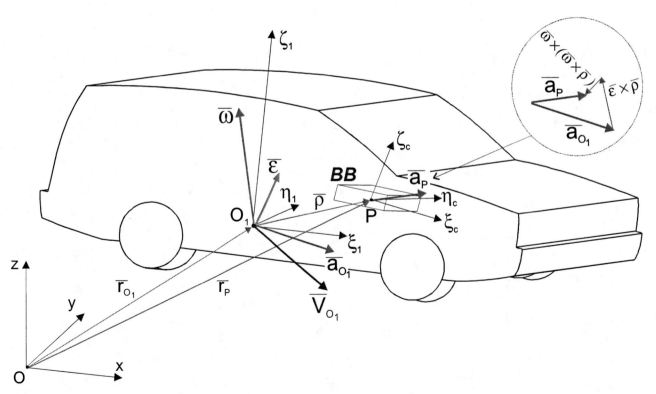

Figure B1. The model of the kinematics of the movement of the vehicle equipped with a "Black Box" BB fixed at point P

The acceleration in a leveled system in any point of the vehicle body may be described as:

$$\mathbf{a}_h = [a_{\xi h}, a_{\eta h}, a_{\zeta h}]^T = \mathbf{A}_\psi^{-1} \cdot \ddot{\mathbf{x}} \qquad (B.6)$$

The accelerations $a_{\xi h}$ and $a_{\eta h}$ lie on a horizontal surface, and $a_{\zeta h}$ has a vertical direction. Thus $\mathbf{a}_h$ is described in the Oxyz system, including the yaw angle ψ_1.

In order to obtain the velocities and accelerations of point P, the knowledge of relating derivatives of matrix $\mathbf{A}$ is necessary. These can be easily obtained, by using the following algorithm.

We present the rotation matrix $\mathbf{A}$ in a form of a product of matrices:

$$\mathbf{A} = \mathbf{A}_\psi \cdot \mathbf{A}_\varphi \cdot \mathbf{A}_\vartheta \qquad (B.7)$$

Rotation matrices $\mathbf{A}_\psi, \mathbf{A}_\varphi, \mathbf{A}_\vartheta$ have the form of:

$$\mathbf{A}_\psi = \begin{bmatrix} \cos\psi_1 & -\sin\psi_1 & 0 \\ \sin\psi_1 & \cos\psi_1 & 0 \\ 0 & 0 & 1 \end{bmatrix},$$

$$\mathbf{A}_\varphi = \begin{bmatrix} \cos\varphi_1 & 0 & \sin\varphi_1 \\ 0 & 1 & 0 \\ -\sin\varphi_1 & 0 & \cos\varphi_1 \end{bmatrix},$$

$$\mathbf{A}_\vartheta = \begin{bmatrix} 1 & 0 & 0 \\ 0 & \cos\vartheta_1 & -\sin\vartheta_1 \\ 0 & \sin\vartheta_1 & \cos\vartheta_1 \end{bmatrix} \qquad (B.8)$$

We can write:

$$\dot{\mathbf{A}} = \dot{\mathbf{A}}_\psi \cdot \mathbf{A}_\varphi \cdot \mathbf{A}_\vartheta + \mathbf{A}_\psi \cdot \dot{\mathbf{A}}_\varphi \cdot \mathbf{A}_\vartheta + \mathbf{A}_\psi \cdot \mathbf{A}_\varphi \cdot \dot{\mathbf{A}}_\vartheta \qquad (B.9)$$

$$\ddot{\mathbf{A}} = \ddot{\mathbf{A}}_\psi \cdot \mathbf{A}_\varphi \cdot \mathbf{A}_\vartheta + \mathbf{A}_\psi \cdot \ddot{\mathbf{A}}_\varphi \cdot \mathbf{A}_\vartheta + \\ + \mathbf{A}_\psi \cdot \mathbf{A}_\varphi \cdot \ddot{\mathbf{A}}_\vartheta + \\ + 2 \cdot \dot{\mathbf{A}}_\psi \cdot \dot{\mathbf{A}}_\varphi \cdot \mathbf{A}_\vartheta + 2 \cdot \dot{\mathbf{A}}_\psi \cdot \mathbf{A}_\varphi \cdot \dot{\mathbf{A}}_\vartheta + \\ + 2 \cdot \mathbf{A}_\psi \cdot \dot{\mathbf{A}}_\varphi \cdot \dot{\mathbf{A}}_\vartheta \qquad (B.10)$$

where:

$$\dot{\mathbf{A}}_\psi = \dot{\psi}_1 \cdot \mathbf{D}_\psi; \quad \ddot{\mathbf{A}}_\psi = \ddot{\psi}_1 \cdot \mathbf{D}_\psi - \dot{\psi}_1^2 \cdot (\mathbf{A}_\psi - \mathbf{1}_\psi)$$

$$\dot{\mathbf{A}}_\varphi = \dot{\varphi}_1 \cdot \mathbf{D}_\varphi; \quad \ddot{\mathbf{A}}_\varphi = \ddot{\varphi}_1 \cdot \mathbf{D}_\varphi - \dot{\varphi}_1^2 \cdot (\mathbf{A}_\varphi - \mathbf{1}_\varphi) \qquad (B.11)$$

$$\dot{\mathbf{A}}_\vartheta = \dot{\vartheta}_1 \cdot \mathbf{D}_\vartheta; \quad \ddot{\mathbf{A}}_\vartheta = \ddot{\vartheta}_1 \cdot \mathbf{D}_\vartheta - \dot{\vartheta}_1^2 \cdot (\mathbf{A}_\vartheta - \mathbf{1}_\vartheta)$$

and

$$\mathbf{D}_\psi = \begin{bmatrix} -\sin\psi_1 & -\cos\psi_1 & 0 \\ \cos\psi_1 & -\sin\psi_1 & 0 \\ 0 & 0 & 0 \end{bmatrix},$$

$$\mathbf{D}_\varphi = \begin{bmatrix} -\sin\varphi_1 & 0 & \cos\varphi_1 \\ 0 & 0 & 0 \\ -\cos\varphi_1 & 0 & -\sin\varphi_1 \end{bmatrix},$$

$$\mathbf{D}_\vartheta = \begin{bmatrix} 0 & 0 & 0 \\ 0 & -\sin\vartheta_1 & -\cos\vartheta_1 \\ 0 & \cos\vartheta_1 & -\sin\vartheta_1 \end{bmatrix},$$

$$\mathbf{1}_\psi = \begin{bmatrix} 0 & 0 & 0 \\ 0 & 0 & 0 \\ 0 & 0 & 1 \end{bmatrix}, \quad \mathbf{1}_\varphi = \begin{bmatrix} 0 & 0 & 0 \\ 0 & 1 & 0 \\ 0 & 0 & 0 \end{bmatrix}, \quad \mathbf{1}_\vartheta = \begin{bmatrix} 1 & 0 & 0 \\ 0 & 0 & 0 \\ 0 & 0 & 0 \end{bmatrix}$$

Transformations of angular velocities were made with accordance to the accepted description (see fig. B2).

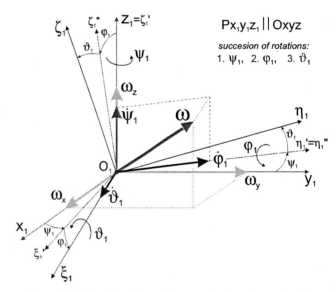

Figure B2. Transformations of the coordinate systems. Angular velocity

Accepting the formula: $\omega = [\dot{\psi}_1, \dot{\varphi}_1, \dot{\vartheta}_1]^T$
and the markings of:

$\omega_x = [\omega_x, \omega_y, \omega_z]^T$ - angular velocity in the Oxyz system;

$\omega_\rho = [\omega_\xi, \omega_\eta, \omega_\zeta]^T$ - angular velocity in the $O_1\xi_1\eta_1\zeta_1$ system;

and also $\omega_x = \mathbf{A} \cdot \omega_\rho$ we can write:

$$\omega_x = \mathbf{B}_x \cdot \omega, \qquad \omega_\rho = \mathbf{B}_\rho \cdot \omega \qquad (B.12)$$

where matrices $\mathbf{B}_x$ and $\mathbf{B}_\rho$ have the form:

$$\mathbf{B}_x = \begin{bmatrix} 0 & -\sin\psi_1 & \cos\psi_1 \cdot \cos\varphi_1 \\ 0 & \cos\psi_1 & \sin\psi_1 \cdot \cos\varphi_1 \\ 1 & 0 & -\sin\varphi_1 \end{bmatrix},$$

$$\mathbf{B}_p = \mathbf{A}^{-1} \cdot \mathbf{B}_x = \begin{bmatrix} -\sin\varphi_1 & 0 & 1 \\ \cos\varphi_1 \cdot \sin\vartheta_1 & \cos\vartheta_1 & 0 \\ \cos\varphi_1 \cdot \cos\vartheta_1 & -\sin\vartheta_1 & 0 \end{bmatrix} \quad (B.13)$$

B.1. MATHEMATICAL MODEL OF THE READINGS OF THE SENSORS BB

The position of the sensors BB is defined by the point of fixing P and the axes of the system $P\xi_c\eta_c\zeta_c$, fixed with the device:

- $P\xi_c$ - the longitudinal axis, showing the direction of the measurement of the longitudinal acceleration a_w^c and the roll angle ϑ_1^c (or the roll velocity $\dot{\vartheta}_1^c$);
- $P\eta_c$ - the lateral axis, showing the direction of the measurement of the transversal acceleration a_p^c and the pitch angle φ_1^c (or the pitch velocity $\dot{\varphi}_1^c$);
- $P\zeta_c$ - the "vertical" axis, showing the direction of the measurement of the "vertical acceleration" a_z^c and the yaw angle ψ_1^c (or the yaw velocity $\dot{\psi}_1^c$).

The $P\xi_c\eta_c\zeta_c$ system is obtained from the $O_1\xi_1\eta_1\zeta_1$ system by translation by a vector $\overline{p}$ and rotation described by matrix $\mathbf{C}$. Analogical rotations to the ones describing the angular position of the car body in relation to the road (the yaw ψ_1, the pitch φ_1, the roll ϑ_1) have been taken, but in the opposite sequence: the BB roll ϑ_c (rotation around the longitudinal axis ξ_c), the BB pitch φ_c (rotation around the lateral axis η_c), the BB yaw ψ_c (rotation around the "vertical" axis ζ_c) – see figure B3. Such a sequence of rotations has been taken because of the ease of leveling the sensors (orientated in relation to the vehicle). Their introduction enables any angular positioning of BB in relation to the body. This in turn enables to account for the related errors of the readings of the BB sensors.

The matrix $\mathbf{C}$ has the form:

$$\mathbf{C} = \begin{bmatrix} \cos\psi_c \cdot \cos\varphi_c & -\sin\psi_c \cdot \cos\varphi_c & \sin\varphi_c \\ \cos\psi_c \cdot \sin\varphi_c \cdot \sin\vartheta_c + \sin\psi_c \cdot \cos\vartheta_c & -\sin\psi_c \cdot \sin\varphi_c \cdot \sin\vartheta_c + \cos\psi_c \cdot \cos\vartheta_c & -\cos\varphi_c \cdot \sin\vartheta_c \\ -\cos\psi_c \cdot \sin\varphi_c \cdot \cos\vartheta_c + \sin\psi_c \cdot \sin\vartheta_c & \sin\psi_c \cdot \sin\varphi_c \cdot \cos\vartheta_c + \cos\psi_c \cdot \sin\vartheta_c & \cos\varphi_c \cdot \cos\vartheta_c \end{bmatrix}$$

(B.14)

The transformations from the system $P\xi_c\eta_c\zeta_c$ to the system $O_1\xi_1\eta_1\zeta_1$ have the form:

$$[\xi_1, \eta_1, \zeta_1]^T = \mathbf{C} \cdot [\xi_{1c}, \eta_c, \zeta_c]^T \quad (B.15)$$

The opposite transformation (from $O_1\xi_1\eta_1\zeta_1$ to $P\xi_c\eta_c\zeta_c$) is described by the inverse matrix $\mathbf{C}^{-1}$, orthogonal against $\mathbf{C}$:

$$\mathbf{C}^{-1} = \mathbf{C}^T \quad (B.16)$$

B1a. The readings of the acceleration sensors.

The inertial acceleration sensors show the value proportional to the sum of the components in the direction of the activity of the sensor: the force of inertia and the force of gravity. The sensor's indication is the sum of the components in the direction of the activity of the sensor of the real acceleration and the acceleration of gravity – see figure B4.

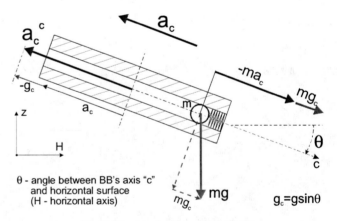

θ - angle between BB's axis "c" and horizontal surface (H - horizontal axis)

$g_c = g\sin\theta$

Figure B4. Acceleration measurement

Generally, the dependence of the sensor's readings in the "c" direction is the following:

$$a_c^c = \frac{-\sum F_c}{m} \quad (B.17)$$

where:

m — the seismic mass of the sensor,

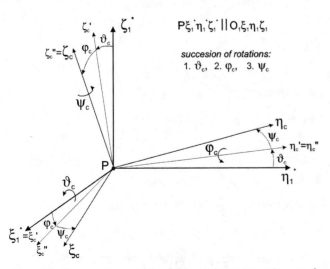

Figure B3. The angular positioning BB in relation to the main axes of the vehicle

$\sum F_c$ - the sum of the forces acting on the seismic mass working in direction "c":

$$\sum F_c = -ma_c + mg_c \quad (B.18)$$

Thus the value presented by the sensor on figure B4:

$$a_c^c = a_c - g_c \quad (B.19)$$

Accepting that, in general the acceleration sensor is three-axial, that is:

$$\mathbf{a}^c = [a_w^c, a_p^c, a_z^c]^T$$

we obtain the general vector relation for the readings of the sensor

$$\mathbf{a}^c = \mathbf{a}_{Pc} - \mathbf{g}_c \quad (B.20)$$

where

$$\mathbf{a}_{Pc} = [a_{P\xi_c}, a_{P\eta_c}, a_{P\zeta_c}]^T = \mathbf{C}^{-1} \cdot \mathbf{a}_P = \mathbf{C}^{-1} \cdot \mathbf{A}^{-1} \cdot \ddot{\mathbf{x}}_P \quad (B.21)$$

$$\mathbf{g}_c = [g_{\xi_c}, g_{\eta_c}, g_{\zeta_c}]^T = \mathbf{C}^{-1} \cdot \mathbf{g}_\xi = \mathbf{C}^{-1} \cdot \mathbf{A}^{-1} \cdot \mathbf{g} \quad (B.22)$$

and $\mathbf{g} = [0, 0, -g]^T$ – the vector of acceleration of gravity.

$\mathbf{a}_{Pc}$ and $\mathbf{g}_c$ represent the acceleration of point P and the acceleration of gravity, accordingly, described in the $P\xi_c\eta_c\zeta_c$ system.
The graphic illustration of the readings of the sensors in case of the BB2 type is shown on figure B5.

In the particular, when the axes of the sensor coincide with the axes of the vehicle ($O_1\xi_1\eta_1\zeta_1 \| P\xi_c\eta_c\zeta_c$), that is $\vartheta_c = 0$, $\varphi_c = 0$, $\psi_c = 0$, what is equivalent to $\mathbf{C} = \mathbf{C}^{-1} = \mathbf{1}$ (1 – the unit matrix) and the sensor is positioned in the centre of the mass O_1 ($P = O_1$) we obtain:

$$\mathbf{a}_{Pc} \equiv \mathbf{a}_{O_1} = [a_{O,\xi}, a_{O,\eta}, a_{O,\zeta}]^T = \mathbf{A}^{-1} \cdot \ddot{\mathbf{x}}_{O_1} \quad (B.23)$$

$$\mathbf{g}_c \equiv \mathbf{g}_\xi = \begin{bmatrix} g_\xi \\ g_\eta \\ g_\zeta \end{bmatrix} = \mathbf{A}^{-1} \cdot \mathbf{g} = \begin{bmatrix} g \cdot \sin\varphi_1 \\ -g \cdot \cos\varphi_1 \cdot \sin\vartheta_1 \\ -g \cdot \cos\varphi_1 \cdot \cos\vartheta_1 \end{bmatrix} \quad (B.24)$$

The calibration of the sensors is performed by achieving the zero sensor readings for the vehicle's condition when the acceleration of the vehicle body in P point are really equal to zero. From the physical point of view, it is the consequence of leveling the sensors in an immobilized car at the level of load, placed on a horizontal road surface. The easiest way of "calibration" in the mathematical BB model is analogical to the real one – the choice of proper angles of positioning the sensors ϑ_c and φ_c, so that the sensors' readings show zero.

To perform the quadrature procedures, the sensor readings are transformed to the system Oxyz related to the road. Transformations described by relations (B.15) and (A.1) are being performed. As the angle values, the reading values of the sensors are used, or instead, the quadrature of the readings of angular velocities sensors (i.e. angles).

B1b. Values of the angular position of the vehicle body

The measurement of values characterizing the angular movements of the vehicle may take place in different ways. When measuring the angular velocities, the readings will be the appropriate components related to the sensor axis:

$$\boldsymbol{\omega}^c = \boldsymbol{\omega}_c = [\omega_{\xi_c}, \omega_{\eta_c}, \omega_{\zeta_c}]^T = \mathbf{C}^{-1} \cdot \boldsymbol{\omega}_\rho = \\ = \mathbf{C}^{-1} \cdot \mathbf{A}^{-1} \cdot \mathbf{B}_\rho \cdot \boldsymbol{\omega} \quad (B.25)$$

where: $\boldsymbol{\omega}^c = [\dot{\psi}_1^c, \dot{\varphi}_1^c, \dot{\vartheta}_1^c]^T$ - angular velocity readings,

$\boldsymbol{\omega} = [\dot{\psi}_1, \dot{\varphi}_1, \dot{\vartheta}_1]^T$ - angular velocity of the car.

For typical solution of BB2 type device the only reading is the yaw angle using a compass operating on a surface related to BB2, and therefore in the general case, unleveled. The angle ψ^{cN} between the direction of the magnetic North N and the longitudinal axis of the sensor $P\xi_c$ is being measured. The sensor measures in the $P\xi_c\eta_c$ surface, fixed with the BB. Figure B6 illustrates the model of the compass described above.

There are 3 characteristic surfaces distinguished:
- the Π_H surface - the horizontal surface,
- the Π_{Nz} surface - the vertical surface including the North direction N,
- the Π_c surface - the "horizontal" surface of the BB2 - $P\xi_c\eta_c$.

ψ^{cN} is the angle between the N^c (trace of the vertical surface Π_{Nz} on the compass' operating surface Π_c) and the longitudinal direction ξ_c.

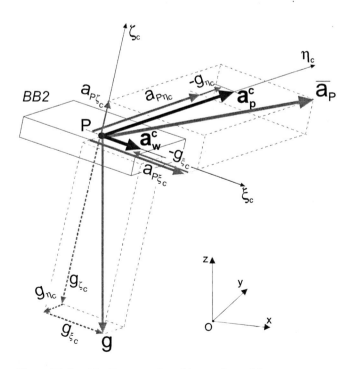

Figure B5. Graphical interpretation of the readings of the sensors measuring the longitudinal and the lateral accelerations (BB2 type device)

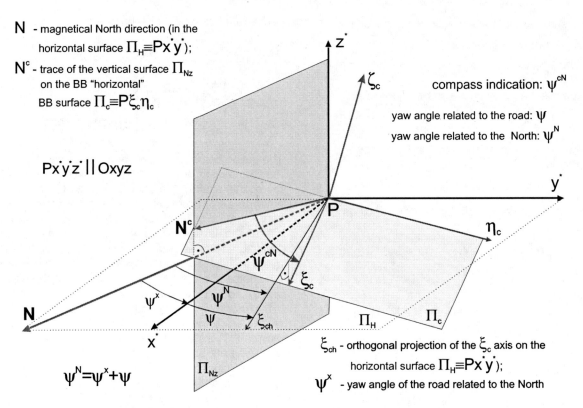

Figure. B6. The illustration of the operation of an unleveled compass

Accepting the markings:

$\overline{e}_{\xi_c}$, $\overline{e}_{\eta_c}$, $\overline{e}_{\zeta_c}$ - the versors of the axes in the $P\xi_c\eta_c\zeta_c$ system,

$\overline{N}$ - a vector enclosing in the North direction,

$\overline{N}^c$ - a vector enclosing in the N^c direction.

The angle ψ^{cN} may be described as the angle between the vectors $\overline{N}^c$ and $\overline{e}_{\xi_c}$, and thus:

$$\cos\psi^{cN} = \frac{\overline{N}^c \cdot \overline{e}_{\xi_c}}{|\overline{N}^c| \cdot |\overline{e}_{\xi_c}|} = \frac{\overline{N}^c \cdot \overline{e}_{\xi_c}}{|\overline{N}^c|} \quad (B.26)$$

The yaw angle relative to the road, based on the readings of the compass can thus be described as:

$$\psi_1^c = \psi^{cN} - \psi^x \quad (B.27)$$

(correction by ψ^x, for that this value may be treated as known)

In the case when $Ox || ON$ (the Ox axis related to the road and the North direction are parallel), that is $\psi^x=0$, the identity takes place:

$$\psi_1^c = \psi^{cN} \quad (B.28)$$

In a special case, that is when $\mathbf{C}^{-1}=\mathbf{1}$ (the axes of the BB sensors are parallel to the main axes of the car: $P\xi_c\eta_c\zeta_c || O_1\xi_1\eta_1\zeta_1$):

$$\cos\psi^{cN} = \frac{\overline{N}^c \cdot \overline{e}_{\xi_c}}{|\overline{N}^c|} = \frac{N_{\xi_c}}{|\overline{N}^c|} =$$

$$= \frac{\sin\psi_1 \cdot \sin\varphi_1 \cdot \sin\vartheta_1 + \cos\psi_1 \cdot \cos\vartheta_1}{\sqrt{(\sin\psi_1 \cdot \cos\varphi_1)^2 + (\sin\psi_1 \cdot \sin\varphi_1 \cdot \sin\vartheta_1 + \cos\psi_1 \cdot \cos\vartheta_1)^2}} \quad (B.29)$$

For the BB1 type device it was assumed that all the angles are known exactly, that is:

$$\psi_1^c = \psi_1, \qquad \varphi_1^c = \varphi_1, \qquad \vartheta_1^c = \vartheta_1 \quad (B.30)$$

2002-01-0550

Revision and Validation of Vehicle/Pedestrian Collision Analysis Method

Amrit Toor, Michael Araszewski, Ravinder Johal, Robert Overgaard and Andrew Happer
INTECH Engineering Ltd.

Copyright © 2002 Society of Automotive Engineers, Inc.

ABSTRACT

A comprehensive analysis method for assessing the vehicle impact speed in a vehicle/pedestrian collision was presented in an earlier publication (SAE #2000-01-0846) by the above authors [1]. This presented method provides a practical analytical approach for evaluating vehicle impact speed from the post-impact vehicle damage, pedestrian injuries and pedestrian throw distance. The applicability of this method to reconstructing real world vehicle/pedestrian collisions is examined. The results of this study indicate that the previously presented model can be used to assess a reasonable range for the vehicle impact speed in a vehicle/pedestrian collision. The regression model equations were revised to include the new data.

INTRODUCTION

Vehicle impact speed is an important parameter in the reconstruction of a vehicle/pedestrian collision. This parameter can be used to assess whether a vehicle was traveling at an unsafe speed, collision evasion potential, and other details surrounding the collision event.

Since 1970, several research studies have analyzed vehicle/pedestrian collisions and some have presented mathematical models for assessing vehicle impact speed from a pedestrian's post-impact trajectory. However, these models are only applicable for specific types of pedestrian trajectories and sufficient data to solve the models are rarely available.

The earlier publication by the authors [1] conducted an extensive review of existing pedestrian collision publications and published a comprehensive analysis method that enables an accident reconstructionist to evaluate a vehicle's impact speed from the post-impact vehicle damage, pedestrian injuries and pedestrian throw distance.

Since the original publication [1], unique sets of real world data have become available such that either the vehicle impact speed or the post-impact vehicle motion is known. In both cases, the vehicle impact speed can be ascertained with an acceptable degree of accuracy. The corresponding parameters necessary to assess the vehicle speed from the method presented in INTECH's paper are also accessible in the new data. Thus, a unique opportunity is available to examine the accuracy of the model presented in the original publication.

This paper tests the empirical model of the original publication [1] with data from real world pedestrian collisions that were not part of the original regression analysis. The previously presented analysis method was used to calculate vehicle impact speeds for several documented real world vehicle/pedestrian collisions. The actual impact speeds were then compared to the calculated values. The comparison revealed that the previously presented model accurately predicts vehicle impact speeds in real world pedestrian collisions.

The newly acquired real world collision data was also appended to the previously gathered data in order to re-evaluate or revise the original empirical equations.

SUMMARY OF SAE 2000-01-0846

An extensive review of the published literature pertaining to vehicle/pedestrian collisions led to three methods for evaluating vehicle speeds. The pre-requisite for each method of assessing the impacting vehicles speed were knowledge of:

- Pedestrian injuries
- Vehicle damage
- Pedestrian throw distance

All three methods were based on empirical data. The assessment of the vehicle speed from pedestrian injuries and vehicle damage was subjective and was intended to serve as a check for the regression analysis based on pedestrian throw distance. Two tables were presented to facilitate the subjective assessment of the vehicle speed. The original tables are reproduced in this paper (Tables 1,2 and 3).

Table 1 - Vehicle Damage Summary for Forward Projection Trajectories

Approx. Vehicle Impact Speed	General Damage Summary
< 20 km/h	Surface cleaning marks.
35 km/h	Leading edge of hood dented; deformation on front of vehicle.
60 km/h	Middle of hood dented.

Table 2 - Vehicle Damage Summary for Wrap Trajectories

Approx. Vehicle Impact Speed	General Damage Summary
< 20 km/h	Surface cleaning marks.
25 km/h	Head contact near bottom edge of windshield when pedestrian C.G. ~60 cm above low-fronted vehicle's bumper assembly; otherwise, head contact near middle of hood for average-sized vehicle and pedestrian. Body contact on roof when pedestrian C.G. ~85 cm above low-fronted vehicle's bumper assembly.
25 to 40 km/h	Head contact near trailing portion of hood or cowl; slight body panel deformation.
40 km/h	Head contact near bottom edge of windshield for impacts significantly below (~50 cm) pedestrian's C.G. (i.e. typical braking low-fronted vehicle).
40 to 50 km/h	Clearly defined dents on body panels.
50 km/h	Head contact near bottom edge of windshield when pedestrian C.G. ~40 cm above low-fronted vehicle's bumper assembly. Body contact on roof when pedestrian C.G. ~60 cm above low-fronted vehicle's bumper assembly.
50 to 55 km/h	Head contact near middle of windshield for typical braking low-fronted vehicle.
60 km/h	Head contact near bottom edge of windshield when vehicle's upper leading edge near pedestrian's C.G.
> 60 km/h	More probable body to roof contact.
70 km/h	Head contact near upper frame of windshield; significant deformation of body panels.
80 km/h	Pelvic contact with roof; roof deformation (unbraked vehicle).

Table 3 - Pedestrian Injury Summary

Approximate Vehicle Impact Speed	General Injury Description (Subjectively assessed)
<20 to 25 km/h	Minor injuries.
30 to 50 km/h	Moderate or major injuries; however, injuries likely non-fatal.
> 55 km/h	Severe injuries which are likely fatal.

The more useful aspect of the original paper was the regression analysis. It was found that there was a second order relationship between the pedestrian throw distance and vehicle speed. Three regression equations were published with 85th percentile pedestrian interval.

Forward Projection Trajectory Model:

$$V_v = 11.4(d_t)^{1/2} - 0.4 \quad [\pm 10.5 \text{ km/h}] \quad (1)$$

Wrap Trajectory Model:

$$V_v = 12.7(d_t)^{1/2} - 2.6 \quad [\pm 9.0 \text{ km/h}] \quad (2)$$

Combined Throw Distance Model:

$$V_v = 12.3(d_t)^{1/2} - 1.9 \quad [\pm 9.5 \text{ km/h}] \quad (3)$$

where: V_v = Vehicle impact speed [km/h]
d_t = Pedestrian throw distance [m]

In a forward trajectory, the pedestrian's C.G. is projected ahead of the vehicle. This type of trajectory is typically expected when the pedestrians C.G. is below the upper leading edge of the vehicle (typically below the hood height).

In a wrap trajectory, the pedestrians' C.G. moves rearward with respect to the vehicle. Wrap trajectory can occur when the pedestrians C.G. is at or above the upper leading edge of the vehicle.

In practice, there are many cases for which the trajectory cannot be accurately identified. For those cases, the combined regression equation was developed.

REAL WORLD DATA

The new data is taken from four different sources:

1. INTECH ENGINEERING – Three vehicle/pedestrian collision reconstruction cases from British Columbia were documented. Post-impact vehicle damage, vehicle displacement, vehicle deceleration, pedestrian injuries and pedestrian throw distance were recorded. These three cases are presented in some detail, as the authors have first hand knowledge of these cases.

2. HILL [2] – This paper provides post-impact vehicle skid distance and deceleration rates for twenty-six real world vehicle/pedestrian collisions that were documented by the West Midlands Police in Great Britain. The effective vehicle deceleration rates were assessed from skid tests utilizing a chalk gun. Pedestrian throw distances were also recorded for these incidents.

3. DETTINGER [3] – DEKRA Accident Research in Germany documented thirteen real world collisions that involved braking vehicles. The vehicle impact speeds and pedestrian throw distances were described.

4. RANDLES, ET AL. [4] – A video camera placed at an intersection in Helsinki, Finland, recorded several vehicle/pedestrian collisions. The authors of this paper interpreted the video footage and present vehicle impact speeds and pedestrian throw distances for nine applicable real world collisions.

DATA ANALYSIS

In total there are data from 51 real world vehicle/pedestrian collisions. In order to categorize the data to be consistent with the earlier INTECH paper, the post-impact trajectory of each struck pedestrian was classified as either a forward projection trajectory or a wrap trajectory.

In 22 of the documented real world collisions, it is known whether or not the impacting vehicle was not braking. However, insufficient information was available in the other 29 collisions to assess whether the vehicle was braking at the point of impact.

The previously presented comprehensive analysis method uses available post-impact information to assess a vehicle's impact speed. The three primary factors available to conduct an investigation into a pedestrian impact are:

1. Vehicle damage
2. Pedestrian injuries
3. Pedestrian throw distance

Although all of the analyzed real world collisions provide pedestrian throw distance data, only the three INTECH cases provide vehicle damage and pedestrian injury descriptions. Thus, the vehicle damage and pedestrian injuries in these three incidents were analyzed.

VEHICLE DAMAGE & PEDESTRIAN INJURIES

Tables A1 and A2 in Appendix A summarize the vehicle damage and pedestrian injuries sustained in these three incidents.

Case 1

This collision involved an impact between a 1989 Nissan Sentra 4-door sedan and a 6-year old boy. The vehicle was not braking at impact, but the driver applied the brakes shortly after the point of impact. The vehicle deposited about 23 metres skid marks on damp asphalt. The road/tire coefficient of friction was determined to be about 0.57. The impact speed was calculated to be about 58 km/h.

A fabric imprint was observed on the bumper and the grille was cracked. In addition, there were localized dents and scuff marks on the hood. The boy sustained a fractured leg and head lacerations/contusions.

The forward displacement of the boy's body conforms to the forward trajectory criteria. Thus, Table 1 was referenced. Table 1 suggests that deformation on the front of an impacting vehicle is expected at a speed of about 35 km/h. At speeds around 60 km/h, deformation near the middle of the hood is expected. However, in this case, the height of the boy would have prevented him from wrapping further onto the hood of the vehicle. Thus, if the impact speed was around 60 km/h, then damage may not be observed further rearward on the hood. Analysis of the damage sustained by the Nissan Sentra (and considering the lower stature of the pedestrian) indicates that the vehicle likely had an impact speed of about 35 km/h.

The boy's injuries were also compared to the general descriptions in Table 3. The injuries sustained in this incident were not minor as the boy fractured his leg and had head injuries. Furthermore, his injuries were not life threatening. Thus, his injuries may be described as "moderate". Comparing the pedestrian's injuries to Table 3, suggests, that the Nissan Sentra likely had an impact speed of about 30 to 50 km/h.

Both subjective analysis methods of using vehicle damage and pedestrian injuries resulting from this real world collision suggest that the Nissan Sentra was traveling about 30 to 50 km/h at the point of impact. This assessment is lower than the actual speed of about 58 km/h. It was cautioned in the original paper that the tables in this paper provide general damage and injury descriptions, which can be used to <u>approximate</u> vehicle impact speeds. Thus, the pedestrian throw distance should always be analyzed to provide a more accurate vehicle speed assessment.

The lower stature and physiology of the boy may also contribute to the inaccuracy of this analysis. The subjective criteria were primarily derived from collisions involving live adult humans, adult cadavers and adult-sized dummies. The results of this analysis indicate that care should be used when applying the comprehensive analysis method to vehicle damage and pedestrian injuries for collisions involving child pedestrians.

The post-impact displacement of the child pedestrian was about 35 metres. Using the regression equation (1), it is calculated that the average vehicle speed was about 67 km/h. While the actual impact speed is within the 85th percentile prediction limits (56.5 to 77.5 km/h) range, (i.e. the average speed is an over-estimate). This is not surprising as the empirical data used for the regression equation was also developed from data predominantly from adult sized pedestrian collisions. Furthermore, it is also known that the vehicle in this case was not braking at impact, but braked shortly after impact.

Case 2

A 4-door GMC Suburban impacted an adult male, which caused dents on the leading edge and near the middle of the hood. Due to the high frontal geometry of this vehicle, the pedestrian was projected forward. The GMC was braking at impact and deposited 19.35 metre long post-impact skid marks on a dry road surface. The measured coefficient of friction at the accident scene was 0.75. The calculated impact speed from the skidding evidence was about 61 km/h.

The pedestrian sustained fatal injuries including:

- Multiple contusions and abrasions.
- Multiple fractures to his legs, wrists and ribs.
- Chest trauma with excessive bleeding.

With reference to Table 1, at an impact speed of about 60 km/h, deformation is expected in the middle of the hood. The GMC sustained damage consistent with this description. Thus, analysis of the vehicle damage indicates that the GMC was traveling about 60 km/h.

The adult pedestrian was about 38 years old. His injuries were severe and fatal. Comparison of his injuries with the general descriptions provided in Table 3 indicate that the GMC Suburban was likely traveling in excess of 55 km/h at the point of impact.

The vehicle damage and pedestrian injuries resulting from this real world collision suggest that the GMC Suburban had an impact speed of about 60 km/h. This assessment is an excellent estimate of the actual speed of about 61 km/h.

The throw distance for this pedestrian was about 25.1 metres. The regression equation yields that the average vehicle speed was about 56.7 km/h. The average calculated speed is in good agreement with the actual speed of 61 km/h.

The analysis results for this incident are much more accurate than those for Case 1. This accuracy can be partially attributed to the pedestrian's stature being more representative of the height of an average-sized adult.

Case 3

The third INTECH case involved a 50-year old woman who was impacted by a braking 4-door Ford Crown Victoria sedan. The vehicle sustained superficial marks on the front bumper with no visible deformation. The lady wrapped onto the front of the vehicle and sustained a compound leg fracture. The vehicle deposited about 1 meter long post-impact skid marks on a damp road surface. The coefficient of friction at the incident scene was determined to be about 0.52.

Analysis of the vehicle's post-impact braking indicates an impact speed of about 12 km/h.

Table 2 was referenced as the woman pedestrian had a wrap type trajectory. The damage summary presented in the table describes that only surface cleaning marks are expected for a vehicle impact speed less than 20 km/h. The impact between the Ford Crown Victoria and the woman was insufficient to cause the woman's head to contact the vehicle and only cosmetic marks on the bumper were observed. Thus, analysis of the vehicle damage indicates that the Ford Crown Victoria was likely traveling less than 20 km/h.

The woman pedestrian sustained a leg fracture, which may be described as a relatively minor injury. Comparison of her injury with the general descriptions provided in Table 3 indicate that the Ford Crown Victoria was likely traveling less than 20 km/h at the point of impact.

The results of the vehicle damage and pedestrian injuries in this real world collision consistently indicate that the Ford Crown Victoria likely had an impact speed less than 20 km/h.

In this case, the throw distance was small, about 1.84 metres. The regression equation yields an average vehicle impact speed of about 14.6 km/h. Comparing this actual impact speed of about 12 km/h, suggests that the regression equation can be applied to lower impact speeds with a reasonable degree of confidence.

PEDESTRIAN THROW DISTANCE

Table A3 in Appendix A summarizes the throw distance and vehicle impact speed data derived from the four sources. The presented vehicle impact speeds were either supplied by the data sources or calculated from post-impact vehicle displacement and deceleration.

Of the documented real world collisions, 6 incidents involved forward projection trajectories and the remainder involved wrap trajectories.

Forward Projection Trajectories

Figure B1 in Appendix B illustrates the raw data that was obtained from the documented real world collisions involving forward projection trajectories. The vehicle impact speeds in all six collisions were within the 85th percentile prediction interval of the previously presented regression relationship.

Case numbers 1 and 45 lie close to the lower 85th percentile prediction limit. Case number 1 involves the only known unbraked vehicle in this data set. Thus, the vehicle in this incident may have likely carried the pedestrian further. Case number 45 is one of the Helsinki impacts presented by Randles, et al., and it is not known whether the vehicle was braking in this case. Thus, if the vehicle was not decelerating at the point of impact, then the position of the data close to the lower 85th percentile prediction limit may be explained.

Wrap Trajectories

Figure B2 shows the real world data from 45 documented collisions involving wrap trajectories. With the exception of five cases, all of the data lies within the 85th percentile prediction interval of the regression relationship. The five exceptions can be described as follows:

1. Case number 46 involves the only vehicle that is known to not have been braked at the point of impact. This incident was one of the documented Helsinki collisions presented by Randles, et al. The authors describe that the vehicle was observed to accelerate after colliding with the pedestrian. However, the regression relationship for wrap trajectories is only applicable for braking vehicles, the data from case number 46 does not meet this criteria.

2. Case number 43 also occurred in Helsinki and it was noted that the vehicle braking was not known and the pedestrian was carried further than usual. Thus, the vehicle may not have been braking in this case.

3. Dettinger documented case number 41 where a vehicle impact speed of about 97 km/h was recorded. The maximum document speed in the data used for the regression analysis was about 85 km/h. Discussion in the original paper indicates that *"as there is limited data at the lower and upper ends of the throw distance data ranges, the reconstructionist should use caution when applying the derived regression equations near the limits of the data ranges."* This caution would be applicable to the circumstances of case number 41.

4. Case numbers 6 and 22, from Hill, had vehicle impact speed of about 70 km/h. Thus caution mentioned in item #3, above, would also apply to these cases. In addition, the vehicles were braked to a stop after impact. However, it is not known whether they were braking at the point of impact or some time after impact.

The regression equations (1) to (3) cannot be applied to the exceptions described above. Thus, elimination of this non-applicable data from Figures B1 and B2 shows that the regression relationships can provide an accurate indicator of a vehicle's impact speed in a real world collision when the pedestrian throw distance is known.

Data Scatter

The real world throw distance data for both forward projection and wrap trajectories was scattered about the models' means. Figure B3 shows the residual differences between the measured pedestrian throw distances and the pedestrian throw distances calculated from the regression relationships.

This scatter is likely attributable to specific collision parameters that are not accounted for by the regression models and to some degree cannot be identified in real world collisions (e.g. vehicle frontal profile, pedestrian size, pedestrian pre-impact path, pedestrian walking speed, pedestrian orientation to the vehicle, etc.).

MODEL REVISION

The regression relationships for assessing vehicle impact speed from pedestrian throw distance in the original paper were derived from review of real world collisions and staged impact tests. The real world data obtained from the four new sources provide additional data that can be combined with the existing throw distance database. Using the complete dataset the regression equations are revised to be:

Forward Projection Trajectory Model:

$$V_v = 11.3 \times (d_t)^{1/2} - 0.3 \quad (1a)$$

85th Percentile Prediction Interval: ± 10.5 km/h
95th Percentile Prediction Interval: ± 14.0 km/h

Wrap Trajectory Model:

$$V_v = 13.3 \times (d_t)^{1/2} - 4.6 \quad (2a)$$

85th Percentile Prediction Intervals: ± 9.0 km/h
95th Percentile Prediction Intervals: ± 12.0 km/h

Combined Throw Distance Model:

$$V_v = 12.8 \times (d_t)^{1/2} - 3.6 \quad (3a)$$

85th Percentile Prediction Intervals: ± 9.5 km/h
95th Percentile Prediction Intervals: ± 12.8 km/h

Figures B4, B5 and B6 graphically illustrate the revised regression equations for each set of data are recalculated.

The 85th percentile prediction intervals presented within the original regression equation were chosen to encompass a majority of the pedestrian input data. The INTECH Case 1 illustrates that this prediction interval produces reasonable results, even for child pedestrians.

However, there may be cases for which 85th percentile prediction limits may not be suitable. Typically, these cases will most likely result from data uncertainty. In addition there will be cases (for example criminal cases) when the vehicle speed must be specified with more certainty. Thus, 95th percentile prediction intervals are also included.

Discussion of Model Review

As expected, with the introduction of the new data, the regression equations differed slightly from the original paper. The most notable change is in the wrap trajectory equation. The primary reason for this is that the new data included vehicle speeds higher than previously considered.

Conversely, the data points for the Forward Projection Trajectory were within the previously considered vehicle impact speeds. Consequently, the regression equations have a very minor change.

The combined equation encompasses both the forward projection and wrap trajectory. The major change in the equation is also due to high speed wrap trajectory data, not previously encompassed in the regression analysis.

The data used to derive the regression equation was limited to vehicles braking at or near impact. Thus, users of this model should exercise caution when applying this method to non-braking vehicles.

CONCLUSIONS

1. The comparison revealed that the majority of the known real world vehicle impact speeds were within the 85th percentile prediction interval presented in the original paper. However, as expected, the impact speeds for non-decelerating vehicles were outside the 85th percentile prediction interval. These are the exceptions that must be considered by the users of this method.

2. The original regression equations have been revised to include the new data.

3. With reference to the INTECH case 1, it is important to realize that the throw distance overestimated the vehicle impact speed, while the subjective analysis (based on vehicle damage and pedestrian injuries) under-estimated the vehicle impact speed. This case demonstrates the need to consider each case carefully. The analysis methods in this paper and the original paper are primarily based on data from adult pedestrian collisions.

RECOMMENDATION

The regression model is an excellent starting point for evaluating vehicle/pedestrian collisions. However, this model should be periodically re-evaluated to include new data. In particular data for child sized pedestrians should be included. There is a strong possibility that a separate set of regression equations may be necessary to accurately analyze child pedestrian collisions.

DEDICATION

This paper is dedicated to our friend and colleague, Eric Roenitz, who passed away in April 1999. His hard work and determination resulted in the initiation of our research program. He is missed greatly and his leading example provides a benchmark to follow.

ACKNOWLEDGEMENT

This paper would not have been possible without the efforts of the INTECH Engineering staff. In particular, the authors are appreciative of the assistance of Tina Coon.

ABOUT THE MAIN AUTHOR

Amrit Toor, Ph.D., P.Eng., is employed as a mechanical engineer specializing in accident reconstruction and vehicle dynamics testing services at INTECH Engineering Ltd., in Vancouver, British Columbia, Canada.

Questions or comments on the paper are welcomed and can be forwarded to:

INTECH Engineering Ltd.
#24 – 7711 128th Street
Surrey, BC V3W 4E6
Canada
Tel: 604 572-9900
Fax: 604 572-9901
Web: http://www.intech-eng.com
Email: intech@intech-eng.com

REFERENCES

1. Happer, A., Araszewski, M., Toor, A., Overgaard, Johal, R. "Comprehensive Analysis Method for Vehicle/Pedestrian Collisions." SAE #2000-01-0846, Society of Automotive Engineers, Inc. Warrendale, PA. 2000.

2. Hill, G.S. "Calculations of Vehicle Speed from Pedestrian Throw," IMPACT, Spring 1994. pp. 18-20.

3. Dettinger, J. "Methods of Improving the Reconstruction of Pedestrian Accidents – Development Differential, Impact Factor, Longitudinal Forward Trajectory, Position of Glass Splinters," [English translation] Verkehrsunfall und Fahrzeugtechnik. Volume 12, December 1996 and Volume 1, January 1997.

4. Randles, B.C., et al. "Investigation and Analysis of Real-Life Pedestrian Collisions." SAE #2001-01-0171, Society of Automotive Engineers, Inc., Warrendale, PA. 2001.

APPENDIX A

Table A1 – Vehicle Damage Summary

Case ID Num	Source	Vehicle			
		Description	Damage	Braking	Impact Speed [km/h]
1	INTECH	1989 Nissan Sentra – 4 door	Fabric imprint on bumper, grille cracked, dents & scuffs on hood	No	58
2	INTECH	1990 GMC Suburban – 4 door	Dents on leading edge & middle of hood	Yes	61
3	INTECH	1992 Ford Crown Victoria – 4door	Superficial marks on bumper (no deformation)	Yes	12

Table A2 – Pedestrian Injury Summary

Case ID Num	Source	Pedestrian		
		Description	Injuries	Trajectory
1	INTECH	6 yr old male	Fractured leg, head scrapes & bruises	Forward Projection
2	INTECH	38 yr old male	Fatality: multiple contusions, abrasions, fractures & chest trauma	Forward Projection
3	INTECH	50 yr old female, 1.68 m	Compound leg fracture	Wrap

Table A3 – Pedestrian Throw Distance Summary

Case ID Num	Source	Pedestrian Trajectory	Pedestrian Throw [m]	Vehicle Braking	Vehicle Impact Speed [km/h]
1	INTECH	Forward projection	35.00	No	58
2	INTECH	Forward projection	25.10	Yes	61
3	INTECH	Wrap	1.84	Yes	11
4	HILL	Wrap	67.90	N/A	100
5	HILL	Wrap	9.80	N/A	40
6	HILL	Wrap	23.50	N/A	72
7	HILL	Wrap	25.50	N/A	67
8	HILL	Wrap	13.50	N/A	46
9	HILL	Wrap	10.90	N/A	44
10	HILL	Wrap	21.00	N/A	57
11	HILL	Wrap	20.80	N/A	63
12	HILL	Wrap	16.10	N/A	51
13	HILL	Wrap	6.50	N/A	32
14	HILL	Wrap	19.30	N/A	53
15	HILL	Wrap	14.50	N/A	52
16	HILL	Wrap	15.20	N/A	49
17	HILL	Wrap	17.00	N/A	52
18	HILL	Wrap	11.30	N/A	45
19	HILL	Wrap	12.40	N/A	39
20	HILL	Wrap	14.00	N/A	47
21	HILL	Wrap	32.00	N/A	67
22	HILL	Wrap	23.00	N/A	69
23	HILL	Wrap	29.20	N/A	70
24	HILL	Wrap	16.00	N/A	56
25	HILL	Wrap	24.00	N/A	60
26	HILL	Wrap	18.40	N/A	56
27	HILL	Wrap	11.30	N/A	40
28	HILL	Wrap	17.90	N/A	47
29	HILL	Wrap	16.00	N/A	54
30	DETTINGER	Wrap	9.60	Yes	39
31	DETTINGER	Forward Projection	13.00	Yes	40
32	DETTINGER	Wrap	13.10	Yes	47
33	DETTINGER	Wrap	14.00	Yes	46
34	DETTINGER	Forward Projection	15.30	Yes	44
35	DETTINGER	Wrap	17.90	Yes	53
36	DETTINGER	Wrap	22.60	Yes	62
37	DETTINGER	Wrap	28.90	Yes	72
38	DETTINGER	Wrap	29.40	Yes	73
39	DETTINGER	Wrap	29.90	Yes	74
40	DETTINGER	Wrap	30.90	Yes	73
41	DETTINGER	Wrap	47.20	Yes	97
42	DETTINGER	Wrap	51.30	Yes	92
43	RANDLES	Wrap	10.38	N/A	27
44	RANDLES	Wrap	5.42	Yes	25
45	RANDLES	Forward Projection	7.41	N/A	20
46	RANDLES	Wrap	4.03	No	9
47	RANDLES	Wrap	4.73	Yes	20
48	RANDLES	Wrap	8.52	Yes	28
49	RANDLES	Wrap	4.45	Yes	23
50	RANDLES	Wrap	8.48	Yes	32
51	RANDLES	Forward Projection	7.90	N/A	33

APPENDIX B

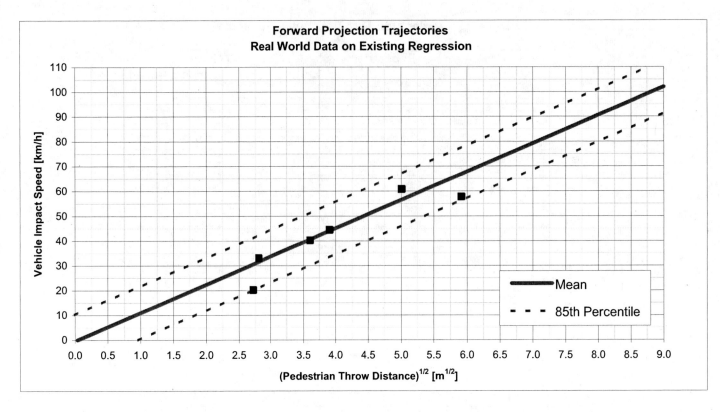

Figure B1 – Real World Forward Projection Trajectory Data

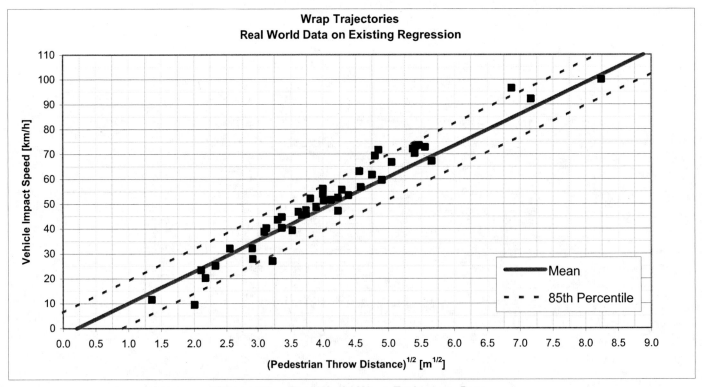

Figure B2 – Real World Wrap Trajectory Data

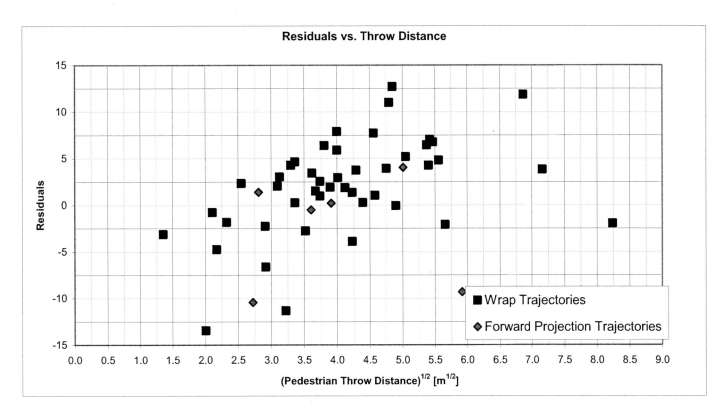

Figure B3 – Residual Differences Between Measured & Calculated Values

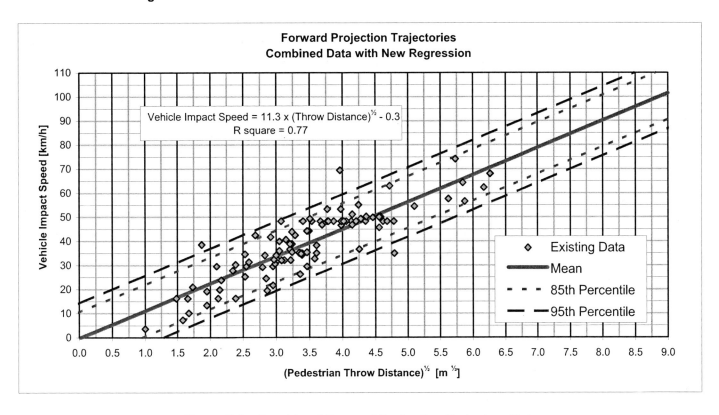

Figure B4 – Combined Forward Projection Trajectory Data

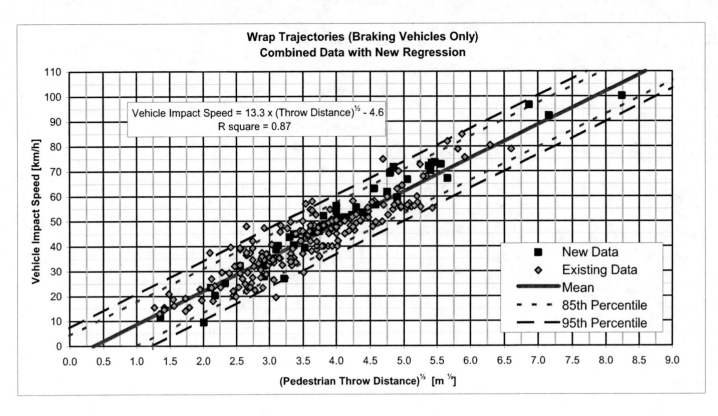

Figure B5 – Combined Wrap Trajectory Data

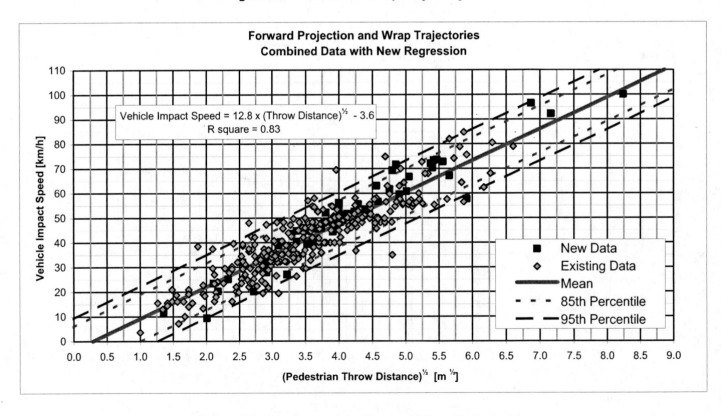

Figure B6 – Forward Projection and Wrap Trajectories

2002-01-0551

Seventeen Motorcycle Crash Tests into Vehicles and a Barrier

Kelley S. Adamson
Unified Building Sciences & Engineering, Inc.

Peter Alexander
Raymond P. Smith & Associates

Ed L. Robinson and Gary M. Johnson
Robinson & Associates, LLC

Claude I. Burkhead, III
Advanced Engineering Resources

John McManus
Consulting Engineering Services

Gregory C. Anderson
Scalia Safety Engineering

Ralph Aronberg
Aronberg & Associates Consulting Engineers, Inc.

J. Rolly Kinney
Kinney Engineering, Inc.

David W. Sallmann
Rudny & Sallmann Engineering Ltd.

Copyright © 2002 Society of Automotive Engineers, Inc.

ABSTRACT

Staged motorcycle-to-car and motorcycle-to-barrier collisions were conducted with seventeen early 1990's models Kawasaki 1000 motorcycles. The impact speeds into the barrier and cars were varied between 10 and 49 MPH. The purpose was to observe the change in motorcycle wheelbase, and characterize motorcycle-to-car and motorcycle-to-barrier crush profiles. These crash tests will expand the existing motorcycle crash test database.

The vehicles were instrumented with tri-axial accelerometers to facilitate the analysis of forces, speed change, and stiffness. Some of the crash tests were recorded by high-speed video cameras. This paper characterizes the data collection system, summarizes the data collected, and lists the parameters that characterize the collision. Crush data and vehicle rest positions were recorded by typical reconstruction methods.

INTRODUCTION

Seventeen staged motorcycle crash tests were conducted at the World Reconstruction Exposition 2000 (WREX2000) held at College Station, Texas September 25-30, 2000. The objective of the tests was to evaluate the characteristics of a heavy motorcycle involved in collisions with two stationary targets: a rigid heavy concrete block, and an automobile. Seven crash tests were conducted using the concrete barrier while ten crash tests were conducted using two 1989 Ford Thunderbirds.

TEST DESIGN

The four-cylinder air-cooled Kawasaki 1000 police motorcycles were donated by the New Orleans Police Department. The 1989 to 1993 motorcycles had been taken out of service but were generally complete with minimal damage and free movement of both the wheels and the head bearing. It was intended that the motorcycles would be upright and aligned perpendicular to the target surface at impact. Each motorcycle was individually weighed. The motorcycle weights ranged from 546 to 633 pounds. Figure 1 shows a typical crash test motorcycle.

Figure 1: Typical motorcycle employed in crash tests.

The VIN, initial wheelbase, dimensions and weight were recorded for each unit. Some motorcycles had fiberglass fairings, some had crash guard bars, and some had saddlebags and/or a cargo box. The front tires were size 18x65H and were inflated to 28 psi prior to the test. The wheels were seven spoke, cast aluminum. The rear tires were also the same size, on the same type of rim, and were inflated to 25 psi.

Motorcycle and vehicle weights were measured to +/- 0.1 % with four electronic scales. Vehicle dimensions were measured by steel tape to an accuracy of +/- 1/4 inch.

The test area was an abandoned concrete runway at the Texas Transportation Institute facility, with ample distance for vehicle acceleration, collision, and tow vehicle deceleration. The runway provided adequate room for test personnel plus observers to be safely located with respect to the crash tests.

Figure 2: Concrete barrier and motorcycle following test #5

Figure 2 shows the 11,080-pound concrete target block. Its dimensions were 120 inches wide, 39 inches high, and 24 inches thick. It was made of cast concrete with a steel bottom surface. After the testing was concluded, the concrete block was pulled horizontally. A load cell was used to measure the maximum and average force required. The coefficient of friction for the concrete block (steel bottom surface on concrete runway) was measured to be 0.40 dynamic and 0.46 static.

The two 1989 Ford Thunderbirds, donated by Texas Engineering Extension Service (TEEX), were complete but had minor damage to the front bumper. The vehicles weighed 3590 pounds (maroon vehicle) and 3576 (silver vehicle). The vehicle weight given by Expert Autostats was 3559 pounds with a weight distribution of 57% (front) to 43 % (rear). The measured weight distribution was identical to the figures listed above. The maroon vehicle front axle weight was 2052 pounds, while the rear axle weight was 1538 pounds. The silver vehicle front axle weight was 2031 pounds, while the rear axle weight was 1545 pounds. The wheelbase was 113 inches and the overall length was 199 inches.

Figure 3: Motorcycle tow system

The motorcycle tow system pictured in Figure 3 consisted of a 2 X 2 steel tubing boom protruding from a fixture welded to a trailer. A wooden 2 X 2 square bar, which extended out approximately 3 feet, was inserted into the steel tubing. Adjustable straps connected the wooden 2 X 2 bar to the top of the motorcycle's fork. The straps were adjusted to provide enough tension to support the motorcycle upright and extend the forks about 2 inches more than normal. The boom height was sufficient to clear the tops of the car and the target block. The boom arm extended about 7 feet beyond the side of the trailer. The trailer was towed by a Ford Expedition. None of the motorcycles carried rider ballast and none were braked, so the fork extension was greater than would be found in most highway collisions. The motorcycles were manually stabilized until they were moving at a speed sufficient to remain upright while in motion. The motorcycles were towed to the target and released by fracture of the wooden 2 X 2 bar during the early phase of the collision. The impact occasionally broke the tow straps. The tow arrangement permitted impacts extending up to about 8.5 feet from the near end of the target vehicle. Two of the motorcycles could not be used because they had stiff head bearings and were unstable when towed. The driver controlled the tow vehicle speed and impact speed was measured with a hand-held Stalker model radar gun.

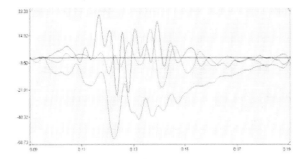

Figure 4. Typical motorcycle accelerometer trace. Vertical axis acceleration values are in units of g. Horizontal axis time values are in seconds. The lowest curve is the x axis response, above which are the z and y axis responses.

The acceleration data acquisition system consisted of four units provided by Instrument Sensor Technology, Inc. (IST). These units were sequentially installed on all of the test vehicles. The IST system contained identical X, Y, and Z-axis piezoelectric accelerometers. Full-scale acceleration was 60 g's. Output from the accelerometers was digitized in a 12-bit signal sampled 3200 times per second, and stored in memory. A 3 g, 10-millisecond (ms) trigger level was required to initiate data storage on the motorcycle. Data was stored starting 0.3 seconds before the trigger event. The trigger for the automobiles was set at ½ to 1 g for 10 ms. For all vehicles, +X was forward, +Y to the left, and +Z upward. After each collision the accelerometer system memory was copied to a hard drive file and the memory was cleared. Subsequently, the data was passed through a 180Hz filter. A typical set of X, Y, and Z impact pulses are shown in Figure 4. The impact acceleration pulse was defined as the interval from when the X-axis acceleration started to rise until it returned to the X axis. The pulse integrals and pulse duration were taken over that interval. Ringing effects observed following the acceleration pulse were not included in the integration.

On each automobile an IST three-axis data acquisition package was fastened to the steel tunnel between the front seats centered approximately 50 inches forward of the rear axle and 20 inches above the pavement. On each motorcycle, the seat was removed and an IST three-axis data acquisition package was attached to a steel plate and guard that was fastened to the motorcycle frame with U-bolts. Figure 5 shows a typical installation.

Figure 5: IST three axis accelerometer.

A crash test plan was designed to provide impact speeds from 10 to 49 mph. The impact into the barrier was to be as close to the center as possible. Impact locations on the cars were planned at the location of the bumper, door, fenders, and wheels.

TEST RESULTS

Seven motorcycles were impacted into the 11,080 lb rigid concrete target. Ten motorcycles were impacted at various locations into the two 1989 Ford Thunderbirds. The tests occurred over two days. The ambient temperature ranged from 42 to 72 degrees F. No rainfall occurred immediately before or during the tests.

For each test the motorcycle was strapped to the tow bar and towed along a nominally straight path to the impact area. Multiple video and still cameras recorded the collision from the sides and rear. Immediately after impact the overall scene was photographed, one team measured post-impact positions of the target and motorcycle while a second team measured the deformation of the vehicles. Rest positions were measured relative to the original position of the target with a steel tape, plumb bob, and chalk marks. Impact speed was based on a radar gun measurement of the tow vehicle speed. The collision dynamics were recorded by the on-board three-axis accelerometer system. Both vehicles were equipped with accelerometer systems during the motorcycle to car impacts, but no accelerometer was mounted on the concrete barrier. The post collision vehicle positions were determined by inspection and dimensional measurement, and recorded with still and video imagery. Drawings showing the measurements of pre and post impact scene data are presented in Appendix 1.

CRASH TEST OVERVIEW

Table 1 shows the crash test conditions and resulting motorcycle wheelbase dimension changes. The change in wheelbase with impact speed is plotted in Figure 6.

Table 1. Crash Test Conditions & Motorcycle Wheelbase Change

M/C No.	Target	Impact Location and Comments	Speed mph	WB Change inches	MC Weight pounds
1	Block	Vertical face	42	11.75	606
2	Block	Vertical face (M/C leaning left 30 deg at impact)	10	1.13	590
3	Block	Vertical face	31	10.25	599
4	Block	Vertical face	20	5.25	620
5	Block	Vertical face	24	8.25	602
6	Block	Vertical face	21	8.0	616
7	Block	Vertical face	35	13.0	618
8	Car (M)	Body between B-post and LR wheel well	46	10.75	615
9	Car (M)	Body LR, between wheel well and bumper	39	7.63	620
10	Car (M)	Rear bumper, 17 inches left of right end	34	8.25	608
11	Car (M)	Right side, between front wheel well and door	25	5.63	611
12	Car (M)	Right front wheel	30	3.25	631
13	Car (S)	Right door, center	42	6.81	625
14	Car (M)	Front bumper, 6 inches right of centerline	30	5.75	633
15	Car (S)	No target impact	--		602
16	Car (S)	Right front fender between wheel well and bumper	41	7.5	595
17	Car (S)	No target impact	--		548
18	Car (S)	Front bumper, right of center	45	8.81	610
19	Car (S)	Body left rear fender, between wheel well and bumper	49	7.25	611
		Targets			
	Concrete block, (M) Maroon Thunderbird, (S) Silver Thunderbird				

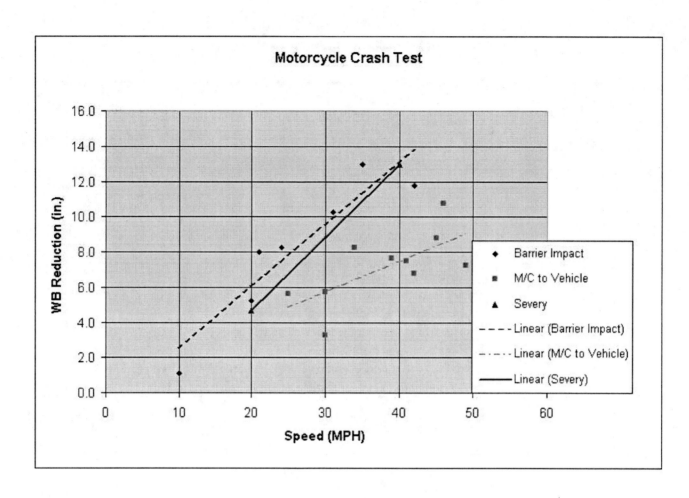

Figure 6: Motorcycle Wheelbase Change vs. Impact Speed

BARRIER IMPACT TEST RESULTS

In Tests 1 through 7, the motorcycle path was normal to the barrier face. The motorcycle impact displaced the barrier, rebounded rearward in an upright position, and fell over near the end of its trajectory. The motorcycle in Test 2 was leaning approximately 30 degrees left of vertical at impact.

High-speed video was recorded and analyzed on tests 3 and 4. Analysis of the video was conducted using the software program "Phantom" provided by Photosonics Inc. High-speed video was also recorded on tests 10 and 12 but was not compatible with the Phantom format.

For Test 3, video analysis indicated the impact speed to be 31.1 MPH compared to 31 MPH measured by the radar gun. The motorcycle velocity dropped to zero 40 ms after barrier contact. The dynamic motorcycle wheelbase change was 15 inches and the static wheelbase change was 10.25 inches. The motorcycle rebounded approximately 29 inches from the barrier face. The motorcycle post-impact speed was 5.2 MPH rearward compared to 5.8 MPH derived from analysis of the acceleration pulse. During the collision, the motorcycle front suspension compressed to maximum before the forks began to bend. The front wheel did not rotate after contact with the barrier.

In Test 4, video analysis indicated the impact speed to be 21.6 MPH compared to 20 MPH using the radar gun. The motorcycle velocity dropped to zero 55 ms after barrier contact. The dynamic motorcycle wheelbase change was 7.9 inches and the static wheelbase change was 5.7. Motorcycle rebound distance was not determined as the video ended before motorcycle motion stopped. The motorcycle post-impact speed was 6.1 mph compared to 6.1 MPH from analysis of the acceleration pulse. The motorcycle front suspension compressed to maximum before the forks began to bend, and the front wheel did not rotate after contact with the barrier.

In both Tests 3 and 4 the kick stand/center stand dropped down. In Test 5 the motorcycle's final rest position was upright on the kickstand. In Test 12, the right front wheel of the car was contacted forward of the centerline of the axle, the wheel rotated (i.e., Steer angle changed) counter clockwise (as viewed from above), the steering components were damaged, and the steering angle was altered. Test 11 was performed out of order just after Test 12. Test 11 impacted the same car on the right front, between the wheel and the A-pillar location. The steering angle of the car wheels was a factor in the post impact movement of the car. In test 16, the right front wheel was contacted and the wheel rotated counter clockwise. The front wheels of the cars freewheeled during the tests, since the steering was not restrained and front brakes were not applied.

VEHICLE IMPACT TEST RESULTS

In Tests 8 through 19, the motorcycle path was normal to the car centerline. Tests 15 and 17 resulted in no impacts. The impact locations are indicated in Table 1. Post-impact points of rest were measured and are recorded in Appendix 1.

Tests 10, 11, 13, 14 and 18 resulted in post-impact translational motion of the target cars. Cars in Tests 8, 9, 12, 16 and 19 displayed significant post-impact rotation. Table 2 presents a comparison of the car speed change as determined from (a) momentum conservation, (b) time integral of the car accelerometer pulses, and (c) slide-to-stop calculations. The three values for each test are in relatively good agreement.

Table 2. Car Speed Change (mph) for Translational Post-Impact Motion Crashes

Test No.	From Momentum Conservations, (Restitution, ε = 0, Minimum Possible ΔV for car)	From Acceleration pulse time integral	From slide-to-stop of post-impact movement
10	5.0	4.8	6.2
11	3.7	2.8	3.6
13	6.3	9.1	6.7
14	4.5	4.3	3.3
18	6.6	7.7	5.2

The speed change (post-impact speed) of car calculations assumed:
- Collisions assumed to be inelastic
- For acceleration pulse time integrals, $\Delta V = [(\Delta V_X)^2 + (\Delta V_Y)^2]^{1/2}$
- Four-wheel slides for lateral post-impact motion (Tests 11 and 14)
- Only rear wheels locked for longitudinal post-impact motion (Tests 10, 14, 18)
- Car weights, weight distributions and motorcycle weights as listed earlier
- Coefficient of friction of the car tires on the concrete surface to be 0.72, as measured by the Texas Transportation Institute.
- For slide-to-stop calculation, sum kinetic energy for each sliding tire (for the measured static wheel load), determine equivalent speed for car

MOTORCYCLE – BARRIER CRASH

Acceleration data recorded during the barrier crash tests were used to determine the motorcycle speed change for each crash, including the effect of rebound speed. These results are presented in Table 3. In Table 3 the impulse duration and speed change were computed from the integral of the X-axis accelerometer collision pulse. The impact speed in Test #4 was changed from 20 mph as indicated by the radar gun to 21.6 mph as indicated in the high-speed film analysis, because the high-speed video analysis was felt to be more accurate. The barrier center of mass displacement is also indicated in Table 3.

Table 3. Data for Barrier Crashes.

M/C TEST No.	Target	Impact Speed (m.p.h.)	Collision Impulse Duration (sec.)	Speed Change Delta-V (mph)	Barrier CG Displacement (ft)
1	Concrete Block	42	0.054	47.7*	0.83
2	Concrete Block	10	0.125	15.5	0.0
3	Concrete Block	31	0.073	36.3*	0.23
4	Concrete Block	21.6	No data	27.7**	0.11
5	Concrete Block	24	0.065	30.6*	0.08
6	Concrete Block	21	0.080	24.7	0.46
7	Concrete Block	35	0.063	38.5*	0.42

* The accelerometer briefly saturated above 60 G's. The values shown were determined from impact speed of the radar gun (except as noted) and the integration of the post-impact acceleration pulse.
** Pre & post impact, rebound speed, was determined through high-speed video analysis.

MOTORCYCLE – AUTOMOBILE CRASHES

Acceleration data recorded during the automobile crash tests was used to determine the motorcycle speed change for each crash, including the effect of rebound speed. These results are presented in Table 4. The three-axis accelerometer data was integrated to determine the speed change of the vehicle. Y axis movement during impact was taken into account by computing the acceleration as the square root of the sum of the squares of the X and Y axis accelerations. A detailed analysis of the variance in the different methods for computing speed changes will be the subject of a later paper.

Table 4. Data for Automobile Crashes

M/C TEST No.	Target	Radar Impact Speed (m.p.h.)	Collision Impulse Duration (sec.)	M/C Speed Change Delta V (mph)	Automobile Speed Change Delta V (mph)
8	Automobile	46	0.094	38*	8.9***
9	Automobile	39	0.098	28.3	8.3
10	Automobile	34	0.098	35*	4.7
11	Automobile	25	0.118	23.1	2.8
12	Automobile	30	0.171	27.5	3.6
13	Automobile	42	0.12	41*	9.1
14	Automobile	30	0.12	33.3	4.3
15	Automobile	No Data**			**
16	Automobile	41	0.090	40*	4.3
17	Automobile	No Data**			**
18	Automobile	45	0.121	42*	7.7
19	Automobile	49	0.063	39	8.9

* The motorcycle accelerometer briefly saturated at 60 G's. The actual speed change was computed from the addition of the impact speed and the post impact speed. Post impact speed was computed from post impact distance traveled and slide to stop energy method.
** Motorcycle accelerometer data was not recorded.
*** Vehicle acceleration data was not recorded. Delta V was calculated by vehicle displacement energy method.

VEHICLE DAMAGE

Damage to the 1989 Ford Thunderbirds, for selected collisions, is depicted in the following photographs. Appendix 2 presents additional damage data.

Figure 7: 46 mph collision – Test 8

Figure 8: 30 mph collision – Test 14

Figure 9: 42 mph collision – Test 13

Figure 11: Rear impact on bumper at 34 mph – Test 10

Figure 10: 49 mph collision – Test 19

The maximum car crush versus motorcycle impact speed is shown in Figure 12. Tests 8,9,13 and 19 involved impacts to various points on the bodies of the cars.

A linear regression of these points (circles) yields the solid curve with a slope of 0.319 inches/m.p.h., intercept of −1.60 inches, and correlation coefficient of 0.932. Tests 10,12,14,16 and 18 involved impacts to bumpers or wheels. A linear regression of these points (squares) yields the dashed curve with a slope of 0.259 inches/m.p.h., an intercept of −3.15 inches, and a correlation coefficient of 0.655. As would be expected the bumpers and wheels appear to exhibit somewhat more rigidity than the body structure.

Figure 12. Maximum Car Crush vs. Impact Speed

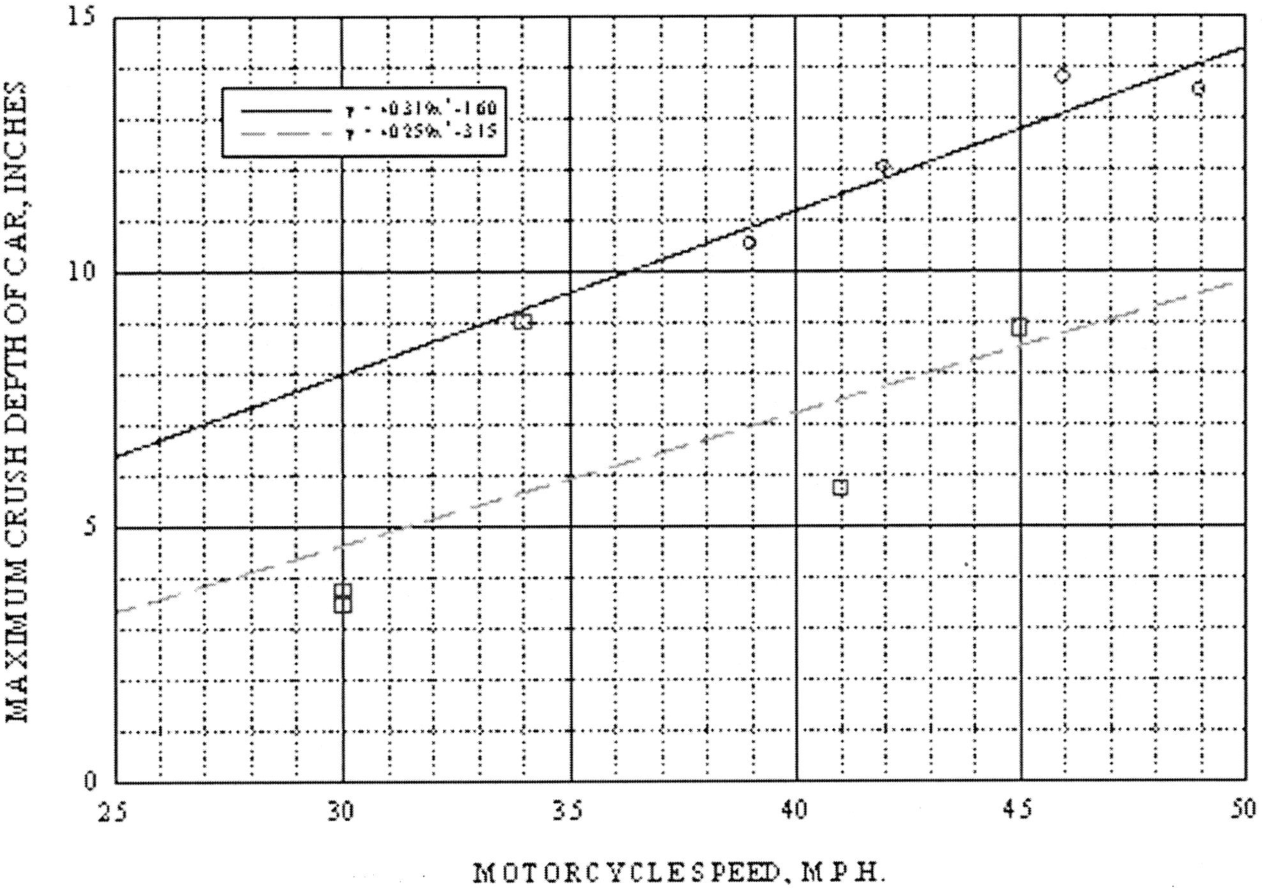

DISCUSSION

Severy's Results: Comparison and Contrast

In the 1970's, Severy [1] crashed seven motorcycles with wire spoke wheels into the sides of automobiles and published a relationship between the impact speed and the change in motorcycle wheelbase. His motorcycles were all Hondas and ranged in size from 90 to 750 cc and in weight from 200 to 480 lbs. Impact speeds ranged from 20 to 40 mph and all impacts were to the front door or front fender of a 1964 Plymouth 4 door sedan (Expert Autostats curb weight: 3900 pounds). His motorcycles were nominally upright, aligned perpendicular to the side of the car at impact, and carried a helmeted dummy. Severy concluded that his results were not dependent on motorcycle weight or size and indicated that the wheelbase change was proportional to the impact speed. The wheelbase change trend line from Severy's tests is shown in Figure 6. Delta V values were not reported for motorcycles or cars. Not surprisingly, the stiffness, or resistance to wheelbase shortening during impact, is significantly different for the current tests using larger Kawasaki motorcycles and for Severy's tests using the smaller, lighter Honda motorcycles. The motorcycle models were very different involving cast aluminum alloy seven-spoke wheels vs. wire spoke wheels. These tests employed no load vs. the dummy rider used in Severy's tests. The test cars were of different make, four- door vs.

two-door cars and more than 25 years different in model year. The greater extension of the unloaded forks in the current test series would not be expected to affect the stiffness because of the compression of the front suspension at impact prior to bending of the forks.

Current Tests

The current results show much more data scatter than Severy's data, which displayed only a 1- inch total spread between 5 tests at 30 mph. Some factors that may contribute to the data point scatter in our tests include:

- Variations of the steering angle at impact and change of steering angle during collision

- How quickly and how much the fork and front wheel turned to one side upon impact

- Front wheel displacement sufficient to allow contact with the engine for some higher speed tests

- The cast aluminum front wheel fracture in some tests

- The point of impact alignment with one of the spokes or between spokes

- The tire remained inflated (e.g., after one of the higher speed impacts, the tire remained inflated and the wheelbase was not shortened as much as in lower speed impacts)

- Motorcycle weight differences of 15%

The maximum wheel base reduction is limited by the deflection of the forks and the rim. Speeds higher than those tested would be expected to result in less wheel base reduction than a linear extrapolation of present results due to direct contact of the front wheel with the engine.

A linear regression of motorcycle/barrier wheelbase reduction data yields a slope (m) of 0.354 inches/m.p.h., an intercept (b) of –1.03 inches and a correlation coefficient (r) of 0.933. A linear regression of all motorcycle/car data yields m = 0.174 inches/m.p.h., b = 0.550 inches and r = 0.684. Linear regression of wheel/bumper impacts only yields m = 0.258 inches/m.p.h., b = -2.55 inches and r = 0.766 while for body impacts only m = 0.126 inches/m.p.h., b = 2.55 inches and r = 0.615.

CONCLUSIONS

Post impact motorcycle trajectories were measured for a variety of impact speeds and were presented in Appendix 1. They show possible rest locations for motorcycles in collision with rigid massive objects and cars.

Motorcycle to car crush profiles were measured at a variety of impact speeds and appear in Appendix 2. They are indicative of crush that might be observed in real accidental collision between motorcycles and cars. The car damage photographs, motorcycle crush measurements, and car crush measurements will provide a useful database for reconstruction of motorcycle collisions.

For a variety of reasons, explained above, the results from these tests show less wheelbase change than did the tests of Severy.

Data points from our tests have significant scatter. The best linear fits involve data depicting wheelbase reduction for barrier crashes and maximum crush depth for body impacts (excluding wheel/bumper impacts). The latter is unexpected since such varied points of the car body were impacted.

ACKNOWLEDGEMENTS

WREX2000 participants sincerely thank Mike Sunseri for procurement of the motorcycles and the New Orleans Police Department for making the test motorcycles available. Thanks also to TEEX and Texas A & M University for furnishing the test automobiles and the test facilities. The test series was organized and directed by Kelley Adamson and executed, at the exposition, by the authors plus other participants:

- Mr. Albert G. Fonda, P.E.
- Mr. Mike Winborn,
- Mr. Philip V. Hight, P.E,
- Mr. John P. Smith P.E.,
- Mr. Donald F. Rudny
- Mr. James R. Lock
- Mr. Marc Decote
- Mr. Paul Schubert

- Mr. Mike Sunseri

Instrumented Sensor Technology provided the accelerometers and the DynaMax software program. Mr Dan Burk of IST was gracious enough to allow us to use four units.

Photosonics & Instrumentation Marketing Corporation provided high-speed video cameras and video analysis.

REFERENCES

1. Severy, Brink, and Blaisdell, Motorcycle Collision Experiments, 14th Stapp Car Crash Conference, 1970, SAE 700897

CONTACT

Mr. Kelley Adamson, M.E., P.E.
Unified Building Sciences & Engineering, Inc.
2700 Earl Rudder Freeway South, Suite 2400
College Station, Texas 77845
(979) 696-6681
kadamson@cox-internet.com

Dr. Peter Alexander, Ph.D.
Raymond P. Smith & Associates
43766 Buckskin Rd.
Parker, CO 80138
(303) 840-0549
mt.man4@attbi.com

Dr. Edward L. Robinson, Ph.D.
Mr. Gary Johnson
Robinson & Associates, LLC
233 Oakmont
Birmingham, AL 35244-3264
(205) 408-1692
alrobinson@charter.net
gary@elrobinson.ws

Mr. Claude I. Burkhead, III, P.E.
Advanced Engineering Resources, PA
PO Box 1505
Apex, NC 27502
(919) 387-8811
cburk@deltaforce.net

Mr. John McManus, P.E.
Consulting Engineering Services
3990 Purchase Street
Purchase, New York 10577
(914) 422-0521
jcm3990@aol.com

Mr. Greg Anderson, P.E.
Scalia Safety Engineering
521 E. Washington Ave.
Madison, WI. 53703
(608) 256-8010
gerganderson522@cs.com

Mr. Ralph Aronberg, P.E.
Aronberg and Associates Consulting Engineers, Inc.
1304 S.W. 160 Avenue, Suite 220
Fort Lauderdale, FL 33326
(954) 236-6605
aronconeng@prodigy.net

Dr. J Rollly Kinney, Sc.D., P.E.
Kinney Engineering, Inc.
2107 NW Fillmore
Corvallis OR 97330
(541) 757-0349
rolly@afiassociates.com

Mr. David W. Sallmann, P.E.
Rudny & Sallmann Engineering Ltd.
3575 Grand Ave.
Gurnee, IL 60031
(847) 244-8868
dave@rseng.com

APPENDIX 1

Impact and Post Impact Diagrams

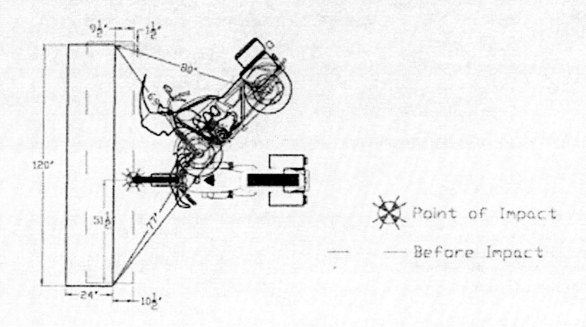

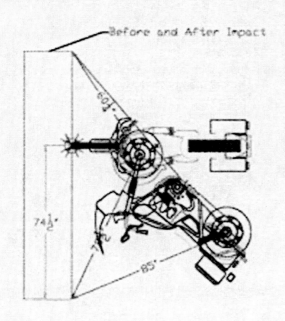

Motorcycle Crash Tests
Test # 3
Impact Speed 31 MPH

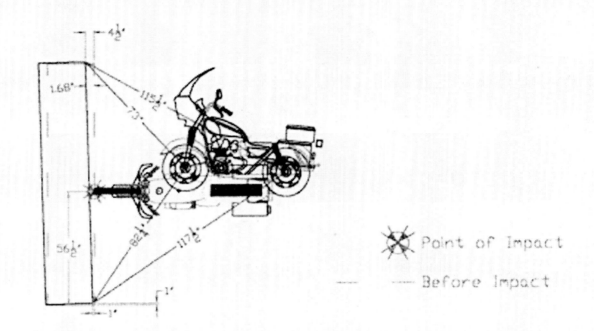

Motorcycle Crash Tests
Test # 4
Impact Speed 20 MPH

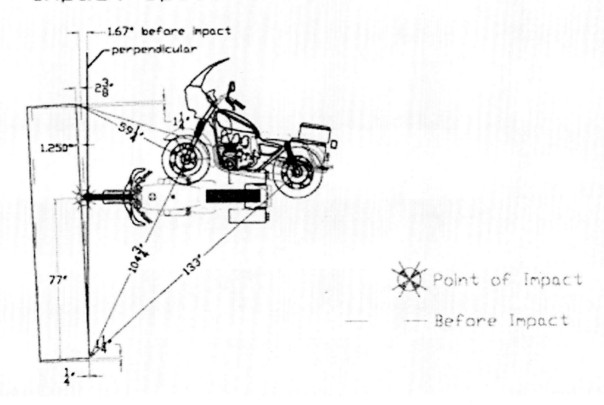

Motorcycle Crash Tests
Test # 5
Impact Speed 24 MPH

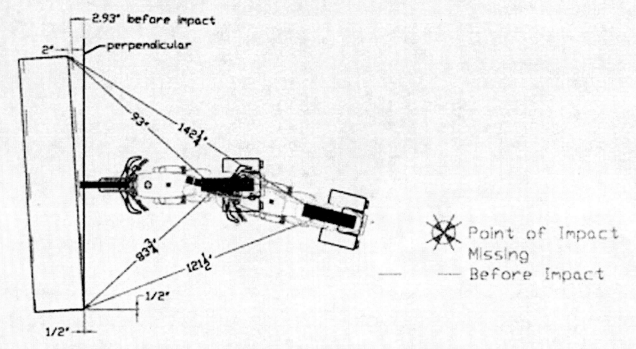

Motorcycle Crash Tests
Test # 6
Impact Speed 21 MPH

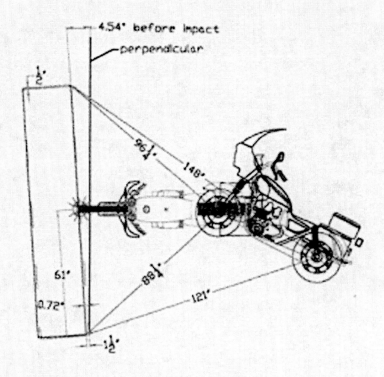

Motorcycle Crash Tests
Test # 7
Impact Speed 35 MPH

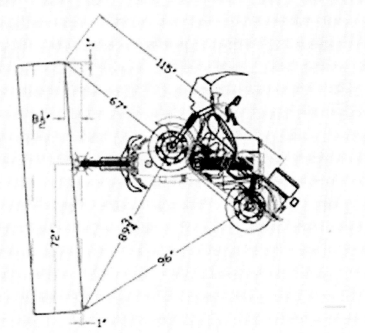

Motorcycle Crash Tests
Test # 8
Impact Speed 46 MPH

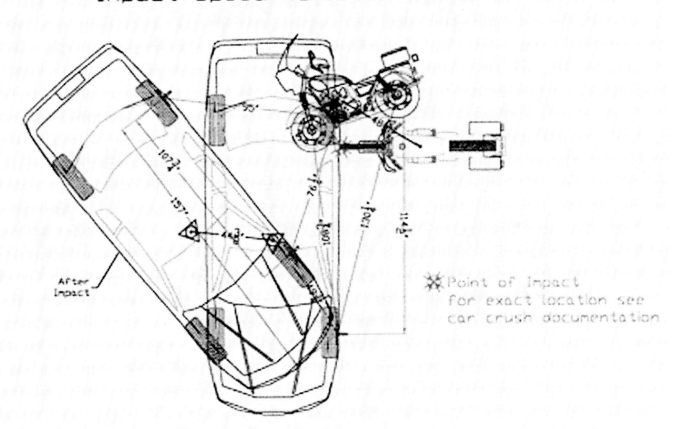

Motorcycle Crash Tests
Test # 9
Impact Speed 39 MPH

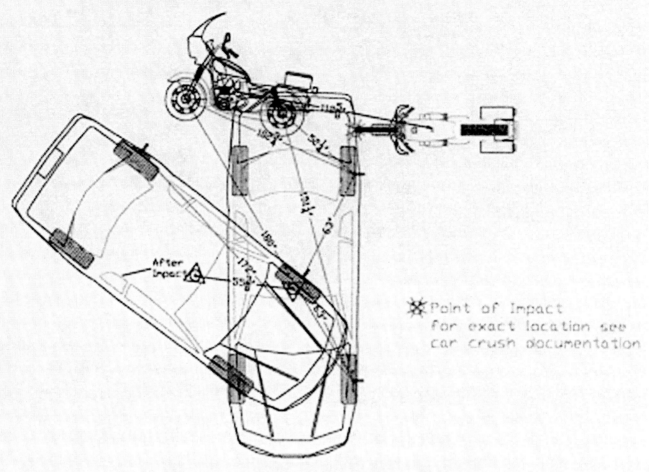

✵ Point of Impact
for exact location see
car crush documentation

Motorcycle Crash Tests
Test # 10
Impact Speed 34 MPH

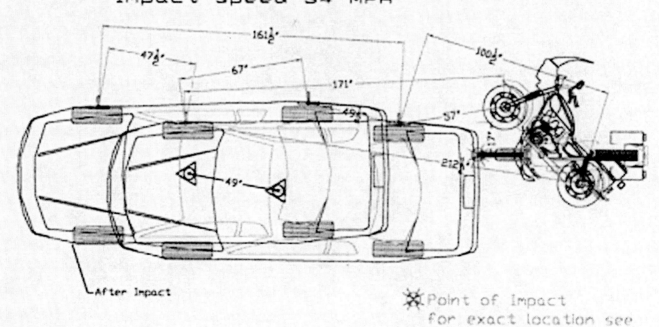

✵ Point of Impact
for exact location see
car crush documentation

Motorcycle Crash Tests
Test # 11
Impact Speed 25

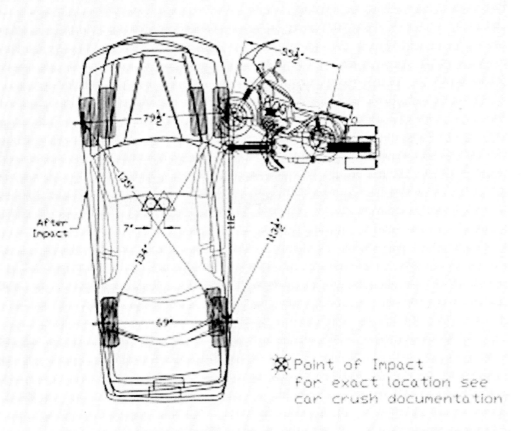

✷ Point of Impact
for exact location see
car crush documentation

Motorcycle Crash Tests
Test # 12
Impact Speed 30 MPH

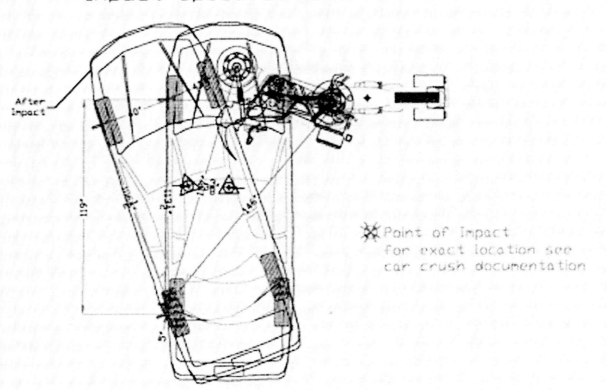

✷ Point of Impact
for exact location see
car crush documentation

Motorcycle Crash Tests
Test # 13
Impact Speed 42 MPH

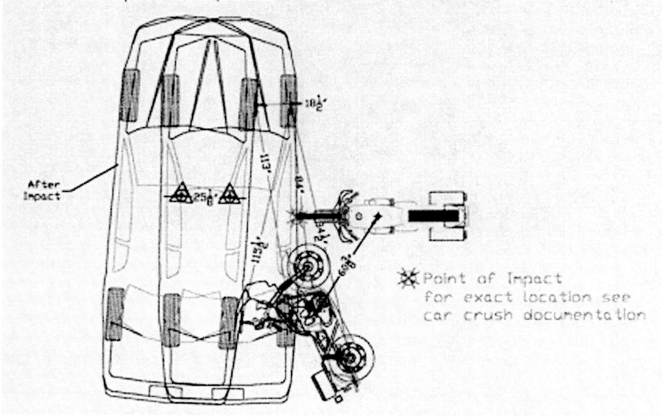

Motorcycle Crash Tests
Test # 14
Impact Speed 30

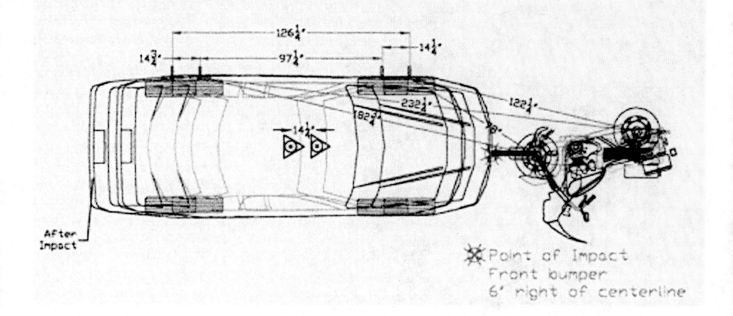

Motorcycle Crash Tests
Test # 16
Impact Speed 41 MPH

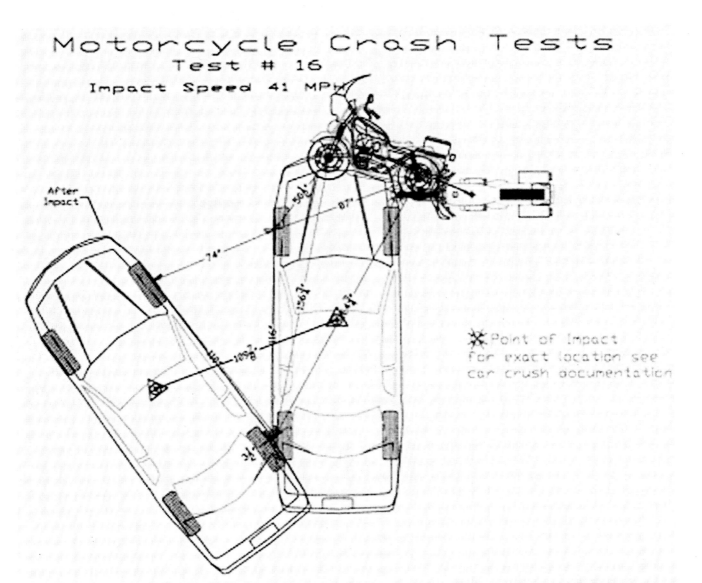

✳ Point of Impact
for exact location see
car crush documentation

Motorcycle Crash Tests
Test # 18
45 MPH

✳ Point of Impact
for exact location see
car crush documentation

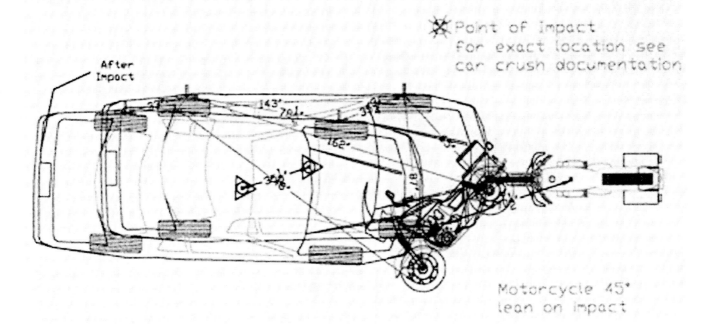

Motorcycle 45°
lean on impact

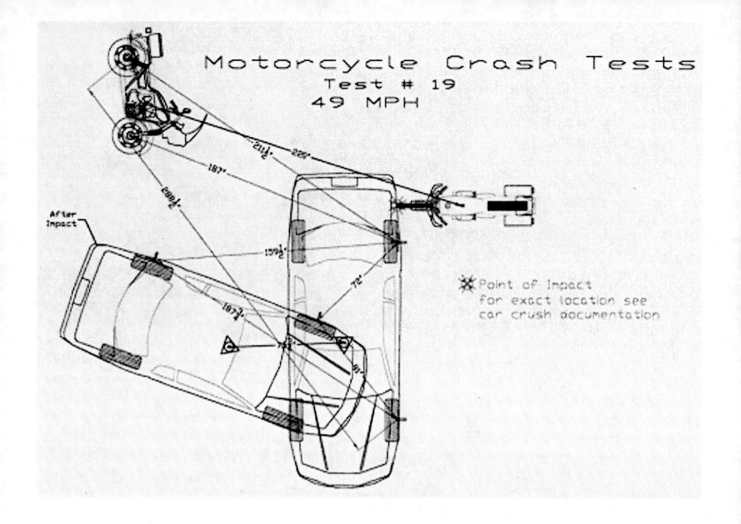

APPENDIX 2

Measured Vehicle Crush Values

Crush depth measurements were made inward from the original (convex) surface positions of the data points and were made at the vertical height of the maximum crush depth except in diagrams noted otherwise.

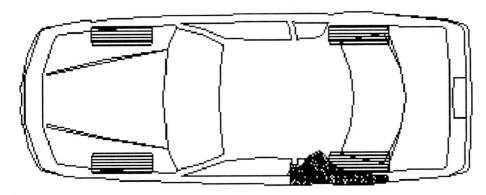

TEST #08 46 MPH, LEFT REAR, FRONT OF WHEEL WELL

DISTANCE FROM C/L "B" PILLAR	CRUSH DEPTH (INCHES)	
	H = 18"	H = 30"
0.0	9	11
4.3	7.5	13.5
8.6	12.25	11.75
13	13.75	12.75
21.6	WHEEL WELL	6.5
26	"	5.75
30	"	6
34.5	"	4.5
38.8	"	3.75
43	"	3

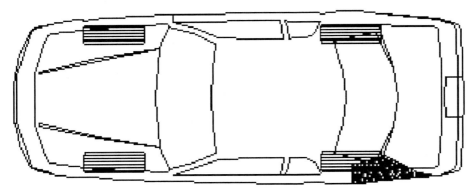

TEST # 9 39 MPH, LEFT REAR, BEHIND WHEEL WELL

DISTANCE FROM C/L REAR WHEEL	CRUSH DEPTH (INCHES)
0	8
7.2	9.25
14.4	10.5
21.6	6.5
28.8	1.75
36	1

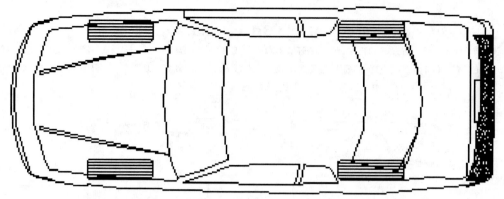

TEST #10 34 MPH, REAR, RIGHT OF C/L

DISTANCE FROM LEFT SIDE	CRUSH DEPTH (INCHES)
2.4	9
8	7
20	7
32.1	5.25
44	5.25
56	5.5
60	7.5

TEST #11 25 MPH, NO CRUSH DATA

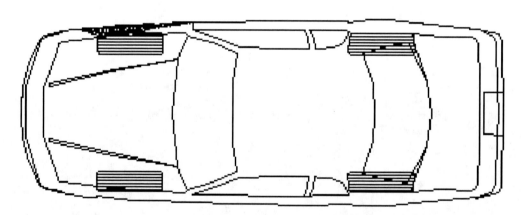

TEST #12 30 MPH, RIGHT SIDE, FRONT WHEEL

DISTANCE FROM "A" PILLAR BASE	CRUSH DEPTH (INCHES)
0	0
14	0
18	1.75
22	3
26	3.25
30	3.5
34	3.5
38	3.25
42	2.25
44	.75
48	0

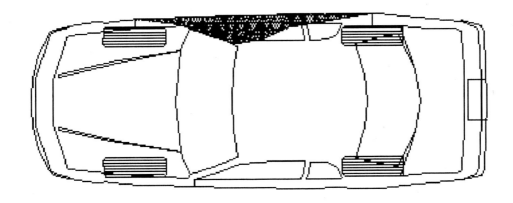

TEST #13 42 MPH, RIGHT SIDE

DISTANCE FROM C/L REAR WHEEL	CRUSH DEPTH (INCHES)
0	0
12	0
24	3.5
36	5
48	10
60	12
72	8
84	3

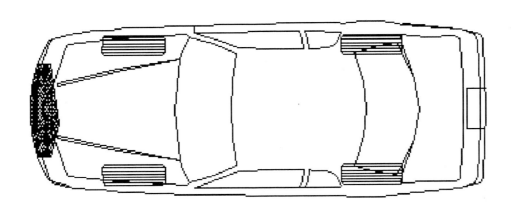

TEST #14 30 MPH, FRONTAL
FRONT BUMPER FELL OFF, MEASUREMENTS ARE DAMAGE TO FRAME
0 IS OUTSIDE EDGE OF RIGHT FRAME RAIL

DISTANCE FROM RAIL	CRUSH DEPTH (INCHES)
0	0
25	3.75
39	0

TEST #15 NO TARGET IMPACT

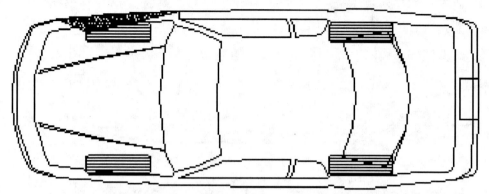

TEST #16 41 MPH, RIGHT FRONT WHEEL

DISTANCE FROM "A" PILLAR BASE	CRUSH DEPTH (INCHES)
0	0
28	5
29	5.75
39	6
52	0

TEST #17 NO TARGET IMPACT

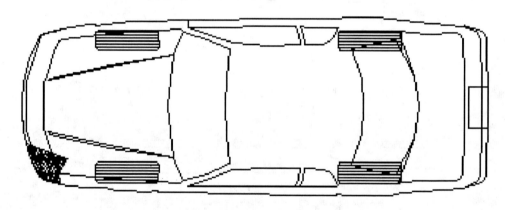

TEST #18 45 MPH, FRONT LEFT

DISTANCE FROM FRONT RIGHT CORNER	CRUSH DEPTH (INCHES)
49.2	0
55.2	15
61.2	8.875
64.8	7

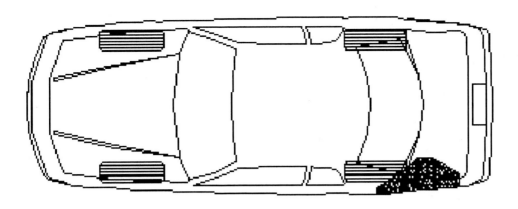

TEST #19 49 MPH, LEFT REAR

DISTANCE FROM C/L REAR WHEEL	CRUSH DEPTH (INCHES)
0	2.5
12	8.5
21.5	13.5
24	13
36	5.5

Paper Title: Seventeen Motorcycle Crash Tests into Vehicles and a Barrier

Paper Number: 2002-01-0551

Reviewer's Discussion
By John F. Kerkhoff, KEVA Engineering

As the authors stated in their ABSTRACT: "These crash tests will expand the existing motorcycle crash test database." Ten (10) equal weight motorcycle to the Thunderbird car impacts at nine (9) different locations on the same type car at essentially seven (7) different speeds plus seven (7) motorcycle to barrier tests certainly does that. The challenge will come in the future interpretations of the results. Due to the confounding responses caused by the disparate impact locations and speeds (see-attached sketches) it is difficult to check the consistency of the researchers' methodologies. As an example, it can be seen that there are only two motorcycle to car tests (#9 and #19) that are directly comparable in impact locations and structure and thus resulting vehicle dynamic response. Even then, the results from those two tests contain a conflict that was not explained or data provided to allow clarification by the reader. Test # 19, while impacting the same type structure at 49 MPH sustained less wheelbase collapse than the lower speed test #9 at 39 MPH.

It may not be appropriate to compare these multi-variable newer series tests with the earlier more restrictive variable series by Severy at UCLA, until both series are analyzed more thoroughly. The differences in parameters under study and variables controlled make even preliminary implied comparative conclusions premature at this stage

More motorcycle crash test research is certainly needed and this research is a valuable contribution. We should all look forward to a detailed analysis in future related writings. The authors should be commended for their efforts in bringing this valuable data rich material into the accident reconstruction community.

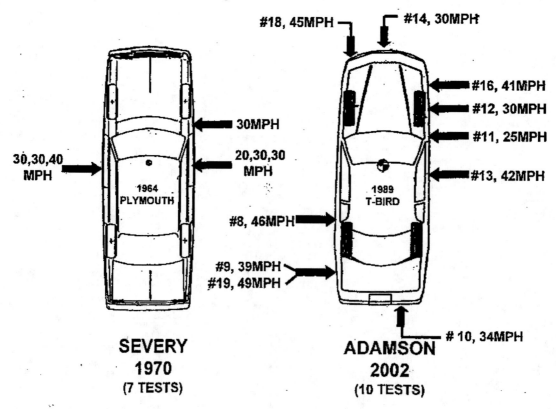

2002-01-0552

Narrow Object Impact Analysis and Comparison with Flat Barrier Impacts

Alan F. Asay, Dagmar B. Jewkes and Ronald L. Woolley
Woolley Engineering Research Corp.

Copyright © 2002 Society of Automotive Engineers, Inc.

ABSTRACT

Crash behavior in narrow object impacts was examined for the perimeter of a 4-door full size sedan. Additional test data was obtained for this vehicle by impacting four sedans with a rigid pole mounted to a massive moving barrier (MMB) in the front, right front oblique, right side, and rear. The vehicles were stationary when impacted by the MMB. Two of the four cars were repeatedly impacted with increasing closing speeds in the front and side, respectively. Each test was documented and the resulting deformation accurately measured. The stiffness characteristics were calculated for the perimeter of car and were presented using the power law damage analysis model. The vehicle's crash performance in these pole tests was compared to that of NHTSA's flat fixed barrier tests (deformable and non-deformable) for the front, side, and rear of this vehicle.

INTRODUCTION

In accident reconstruction, the damage energy method is often applied to quantify the severity of an accident. The accuracy of this procedure depends on the validity of the stiffness parameters used in the damage model and accurate representation of the crush-profile. The National Highway Traffic Safety Administration (NHTSA) and most vehicle manufacturers have published crash-test results since the 1970's, in accordance with the Federal Motor Vehicle Safety Standards (FMVSS) 208, 214 and 301 for frontal, side and rear collision performance, respectively. Furthermore, New Car Assessment Program (NCAP) tests have been performed in frontal and rear modes at higher speeds than required by the FMVSS. Additionally, NHTSA performed a limited number of repeated frontal impacts, under-ride guard tests and frontal pole impacts. The majority of publicly available crash-tests consists of frontal and rear impacts with full width barriers at 30 to 35 mph, followed by side tests with a full width deformable moving barrier into the side occupant compartment at 33 mph (FMVSS 214). Finally, the Insurance Institute for Highway Safety has conducted frontal offset deformable barrier tests and side pole impact tests, as well as other institutes throughout the world (EuroNCAP, ADAC). However, the latter data are not readily available for thorough analyses as required for application in accident reconstruction.

In real-world accidents, the vehicle's damaged area is not always well represented by the publicly available crash-test data. It is more difficult to estimate an appropriate stiffness for accident-vehicles damaged in the wheel/axle area, for those impacted at excessive speeds or for those that impacted narrow objects, since vehicle stiffness (expressed in stiffness per unit width) varies with the area impacted, depth of crush (Strother et al. 1986, Wood and Mooney 1996, Wood 1997), and impact width (Nilsson-Ehle et al. 1982, Buzeman-Jewkes 1999). A greater variation of test-conditions is needed to estimate stiffness data for reconstruction of narrow object collisions or of impacts into non-homogeneous structures. A method needs to be developed to compare the new test data with the publicly available crash-tests.

OBJECTIVE

The objective of this paper was to analyze the stiffness characteristics of a 4-door full size sedan resulting from narrow object perimeter impacts. The data was then compared to those publicly available for frontal, side, and rear impact tests of 1986 through 1991 Ford Taurus'.

TEST METHODOLOGY

GENERAL- A versatile crash-test method was used in which a Massive Moving Barrier (MMB) was driven into a stationary test vehicle (Woolley et al. 2000). A rigid pole was attached to the front of the MMB. The test method allowed the MMB to impact the vehicles repeatedly in the same area, in various other locations, and with varying speeds. The reinforced MMB is typically 10 times heavier than the target vehicle, therefore it experiences a delta-V of about 10% of the impact speed during a collision with the target vehicle. Because of the relatively low delta-V, the MMB can be driven into a stationary test vehicle with any orientation.

TEST PROCEDURE- Four 4-door Ford Taurus vehicles were impacted in seven tests by a 12 inch rigid pole mounted to the front of the MMB. Each of the four Taurus' were weighed along with the MMB prior to testing. One Taurus was repeatedly impacted in the front center. A second Taurus was subjected to a single front oblique impact directed in the area of the right front wheel and fender. A third was repeatedly impacted in the right front passenger door near the handle. The fourth Taurus was struck once in the center rear.

TEST DOCUMENTATION- The speed of the MMB was measured by a laser speed trap just prior to each impact. Each of the four Taurus' tested were photographed and measured. Three-dimensional, point-to-point residual crush was documented for each test. Several video cameras were also used to document each test from various vantage points.

Table 1: Test Results- Summary of Test Parameters for all Seven Tests

Test Number	Ford Taurus Year	Narrow Object Impact Location	Mass Vehicle (lbs)	Mass MMB (lbs)	MMB Impact Speed (mph)
1	1987	Front Center	2940	26,690	12.1
2	1st Repeated	Front Center	2940	26,630	22.8
3	2nd Repeated	Front Center	2940	26,630	30.7
4	1988	Right Front Oblique	3020	26,630	30.6
5	1988	Passenger Side Front Door	3020	26,680	12.2
6	1st Repeated	Passenger Side Front Door	3020	26,710	22.8
7	1987	Rear Center	2900	26,640	30.8

TEST RESULTS

Table 1 is a summary of all seven tests and the respective test parameters. Figures 1-7 each summarize the damage measured post test 1-7.

FORD TAURUS FRONTAL REPEATED TESTS- The MMB with a rigid 12-inch diameter steel pole mounted to the front impacted the stationary 1987 Ford Taurus head-on, centered on the front bumper. This Taurus was subjected to a total of three impacts at closing speeds of 12.1 mph, 22.8 mph and 30.7 mph.

Two-dimensional residual crush was documented for each test. Various points were permanently placed across the front bumper, front edge of the hood, and at various other locations of the vehicle. The location of the points was measured prior to the first test and then after each test to document the point-to-point displacement of the front structure.

Frontal Test 1- The first test at 12.1 mph caused damage to the front bumper and hood. (Figure 1) The wrap around plastic bumper detached as a result of induced buckling about the frame mounts (see crush profile). Post-collision, the front bumper structure did not conform to the shape of the pole. Because of the material properties of the plastic, the front bumper creased locally at the center of the pole impact and restored partially after the pole was removed.

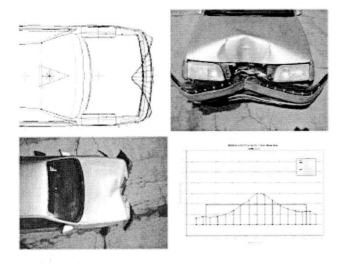

Figure 1. Ford Taurus Frontal Test 1

Frontal Test 2- The second test deformed the hood profile in a similar way as the bumper-profile, and involved the vehicle's engine more significantly. The bumper mounts were pushed deeper into the vehicle structure (Figure 2). However, the detached bumper corners further wrapped around the pole, causing 'negative crush'.

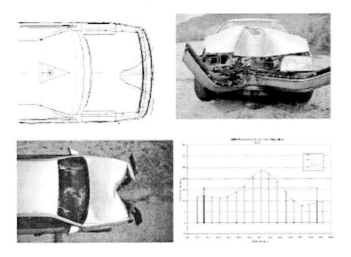

Figure 2. Ford Taurus Frontal Test 2.

Frontal Test 3- In the third test the engine was deflected further rearward and the front bumper corners were pulled even more inward toward the center of the vehicle (Figure 3). Following this test, as the pole was removed, the front bumper along the driver side fell away leaving a small section in the middle and another on the passenger side. It was noted, however, that the bumper in the crash sequence was forced against the radiator, so the profile of the radiator was taken to represent the bumper in that area. The depth of the pole penetration was sufficient to force the bumper corners rearward such that no 'negative crush' occurred.

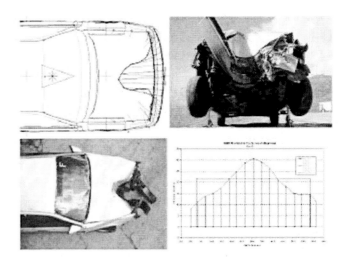

Figure 3. Ford Taurus Frontal Test 3

FORD TAURUS OBLIQUE POLE IMPACT TEST- The MMB impacted a 1988 Ford Taurus obliquely at a speed of 30.6 mph into the right front wheel. The Taurus was pre-positioned such that the impact force was directed at the vehicle center of mass, at 30 degrees clockwise from head-on. The impact pushed the wheel in and caused the axle to buckle. The right front of the vehicle was pulled towards the point of impact by induced bowing (Figure 4).

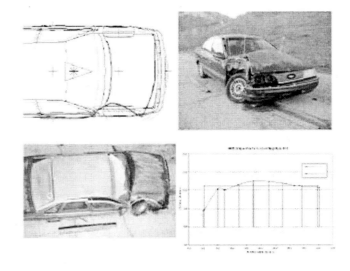

Figure 4. Ford Taurus Frontal Oblique Test

FORD TAURUS REPEATED SIDE IMPACT TESTS- The MMB with the 12" pole mounted to the front, was impacted into the side of a 1988 Ford Taurus at the front passenger door near the door handle. The Taurus was positioned at 90 degrees to the approaching MMB. Two impacts were conducted at 12.2 mph and at 22.7 mph, respectively.

Side Impact Test 1- The first side impact damaged both the rocker panel and the roof rail. There was no sign of bowing in the vehicle after the first impact (Figure 5).

Side Impact Test 2- The second test caused extensive damage to the roof and rocker structure resulting in significant bowing of the vehicle structure about the area of impact (Figure 6).

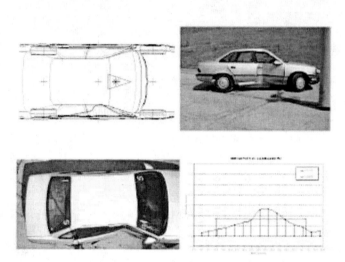

Figure 5. Ford Taurus Passenger Side Test 1

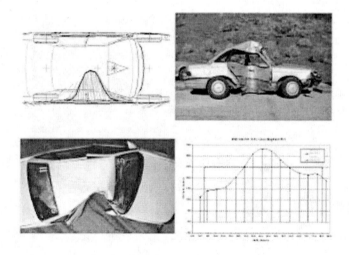

Figure 6. Ford Taurus Passenger Side Test 2

FORD TAURUS REAR IMPACT TEST- A single high-speed pole-impact was conducted by driving the MMB at a speed of 30.8 mph into the center rear of a 1987 Ford Taurus. The vehicle had been involved in a rollover prior to testing, and several small pieces of sheet metal had been removed. It was believed that the rear structural integrity had not been affected by the rollover or removal. The Taurus experienced a delta-V that was approximately twice as high as the delta-V experienced in a FMVSS 301 test. There was significant displacement, causing the rear structure to wrap around the pole. The pole penetrated through the rear trunk, into the rear window, and nearly reached the roof of the vehicle (Figure 7).

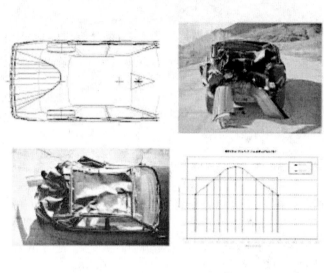

Figure 7. Ford Taurus Rear Test

DAMAGE ANALYSIS METHOD

CRUSH PROFILE MEASUREMENT- Irregular residual crush profiles were obtained by measuring a set of pre-determined points on the vehicle's exterior before and after each of the impacts. This data was collected through use of a surveying instrument. Three common non-displaced points were used to align the post-test vehicle(s) measurements with the undamaged pre-test vehicle measurements. The two-dimensional displacement of the points or damage vectors were obtained from the aligned data sets. The damage vectors had a longitudinal and a lateral component. The longitudinal component of the crush was omitted when negative, which was observed for the bumper corners in frontal and rear tests. Other methods were evaluated and were discussed in the section 'Discussion of Damage Measurement'.

The profile width dimension, w, was taken to be the undamaged width perpendicular to the principle direction of penetration. In general, the width comprises the direct damage caused by the narrow object impact, plus additional width from induced damage.

In the oblique crash, the width was measured perpendicular to the crush vectors and included the induced crush print along the side and front. However, damage produced solely by induced bowing was excluded.

CRUSH PROFILE REPRESENTATION

The area defined by the irregular crush profile can be represented by an equivalent rectangular crush profile having the same area (Woolley 2001).

$$A = \int dA = \int_0^{w_o} x\,dw \quad (1)$$

The depth of the rectangular crush profile is represented by the area based average crush, X_{ave} (Equation 2).

$$X_{Ave} = \frac{1}{A}\int x\,dA = \frac{\int_0^{w_o} x^2\,dw}{\int_0^{w_o} x\,dw} \quad (2)$$

The width of the equivalent rectangular crush profile, w_r, is calculated by taking the total area of the damage profile and dividing it by X_{ave} (Equation 3).

$$w_r = \frac{A}{X_{Ave}} = \frac{\left(\int_0^{w_o} x\,dw\right)^2}{\int_0^{w_o} x^2\,dw} \quad (3)$$

NOTATION

Vehicle or test characteristics pertaining to the target or subject vehicle are represented by subscript 'a' while the other vehicle is represented by subscript 'b', which becomes in these tests, the Massive Moving Barrier, 'MMB'.

DAMAGE MODEL

The damage energy method is based upon the observation that energy absorbed by vehicle crush can be related to the closing speed (Equation 4). The amount of crush is a measure for the energy absorbed in the accident, and therefore of the closing speed. The relationship between the crush and the absorbed energy can be established by crash tests. The relationship is generally found by plotting a derivative of the absorbed energy (calculated from known impact speeds) against the vehicle's average crush (given by Equation 2).

ENERGY ESTIMATION

In the crash-tests, the closing speeds were known, and the energy absorbed by the vehicle deformation was calculated from the closing speed:

$$E_{abs,a} + E_{abs,b} = \frac{1}{2}\frac{m_a m_b}{(m_a + m_b)}\left(V_{Close}^2 - V_{Sep}^2\right) \quad (4)$$

For the narrow object testing, the target vehicle absorbed all of the energy in each test since the MMB and pole did not deform. In all tests, the closing speed was the same as the MMB impact speed since the target car was stationary. In addition, the pole-impact force was directed at or near the vehicle center of mass to minimize post-impact vehicle rotation. This allowed the separation speed to be directly related to the restitution, ε:

$$E_{abs,a} = \frac{1}{2}\frac{m_a m_{MMB}}{(m_a + m_{MMB})} V_{imp}^2 (1-\varepsilon^2) \quad (5)$$

REPEATED TEST ENERGY

The technique of repeated testing was utilized for the side and front orientations to achieve stiffness data for various energy levels without having to test more vehicles. Warner et al. (1986), Prasad (1990) and Buzeman-Jewkes (1999) showed that the energy absorbed in repeated tests can be super-imposed and related to the observed vehicle crush to obtain the vehicle stiffness at higher speeds. The total energy absorbed after n test repetitions, $E_{abs,n}$, and the equivalent barrier energy (EBE) was calculated. EBE was defined as the summation of energy absorbed in all test repetitions (Equation 5) plus the restitution energy of the last test (Warner et al. 1986) and are given in Equation (6). The restitution coefficients were estimated from general passenger car crash-tests, and it was assumed that the restitution coefficient was higher for collisions at lower speeds.

$$EBE = \sum_{i=1}^{n-1} \frac{1}{2}\frac{m_a m_{MMB}}{(m_a + m_{MMB})} V_{imp,i}^2 (1-\varepsilon_i^2) + \frac{1}{2}\frac{m_a m_{MMB}}{(m_a + m_{MMB})} V_{imp,n}^2 \quad (6)$$

CRASH PLOT

The square root of the energy per unit width absorbed in the collision was plotted against the area averaged residual crush in the so-called crash-plot to estimate the vehicle stiffness for the front, side, and rear structures (Woolley 1983, Strother 1986). In lower speed collisions, the vehicle structure often shows a linear relationship between force per unit width and damage (Campbell 1974). A vehicle with constant stiffness behavior can be represented by a straight line in the crash plot (Equation 8), where K represents the stiffness per unit width and X_o the crush offset.

$$E = \frac{1}{2}\int_{w=0}^{w=w_0} K(X + X_0)^2\,dw \quad (7)$$

$$\sqrt{2E/w_r} = \sqrt{K}(X_{Ave} + X_0) \quad (8)$$

However, the vehicle stiffness often decreases with higher accident severities or crush. A force saturation model (Strother et al. 1986) and a power law model (Woolley 2001) have been developed to account for the energy absorbing capabilities of the structure at higher crush levels.

The power law model offers the ability to model a vehicle's structural behavior from constant force (n=0) to constant stiffness (n=1) (Equations 9, 10). This model was selected to represent the crash behavior of the Ford Taurus. Stiffness parameters are defined graphically using the crash plot: k_o the reference stiffness, X_N, the reference crush and X_o, the crush offset. The value of the power, n, in the model may be found from a log-log plot of the absorbed energy and crush data (Equation 11).

$$F = \int_0^{w_o} k_o X_N \left(\frac{x+X_o}{X_N}\right)^n dw \qquad (9)$$

$$E = \int_0^{w_o} \frac{k_o X_N^2}{n+1} \left(\frac{x+X_o}{X_N}\right)^{n+1} dw \qquad (10)$$

$$Ln(E/w) = (n+1) Ln(x+X_o) + Const. \qquad (11)$$

The crash tests performed in this study provided additional data points to evaluate the relationship between absorbed energy and vehicle crush in pole-impacts to the vehicle front, oblique, side, and rear.

RESULTS OF DAMAGE ANALYSIS

TEST RESULTS OF MMB NARROW OBJECT-

The conditions of the narrow object tests, and the resulting absorbed energy and crush representation are presented in Table 2. The crash-plot of the tests conducted in this study are shown in Figure 8, and the data illustrates the substantial stiffness difference for the various impact areas of the Taurus.

Table 2. MMB Damage Analysis Results

Test#	Impact Location	Impact Speed (mph)	Restitution Coefficient	Delta-V (mph)	Total Absorbed Energy (ft-lbf)	EBS (mph)	X_{ave} (inch)	W_{rect} (inch)
1	Front (1)	12.1	0.25	13.6	12961.7	11.5	9.464	50.975
2	Front (2)	22.8	0.20	24.6	58162.9	23.9	15.741	54.731
3	Front (3)	30.7	0.15	31.8	139742.8	37.7	21.781	55.339
4	Oblique	30.6	0.15	31.6	84903.2	29.0	11.289	39.375
5	Side (1)	12.2	0.25	13.7	13498.4	11.6	8.366	62.946
6	Side (2)	22.8	0.20	24.6	59805.0	24.3	24.694	77.335
7	Rear	30.8	0.15	31.9	82937.5	29.3	35.907	52.495

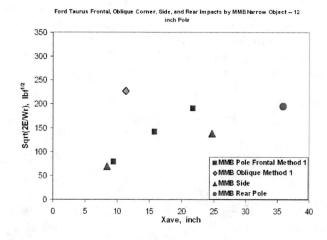

Figure 8. MMB Crash Plot of All Tests

FORD TAURUS FRONT/OBLIQUE IMPACTS

The square root of the energy per unit width versus the area averaged crush from the MMB pole impacts was included in the crash-plot for NHTSA's database of Taurus frontal tests (Figure 9). The oblique data point reflects a stiffness that is relatively high compared to the average stiffness behavior of the Taurus in full width frontal barrier tests.

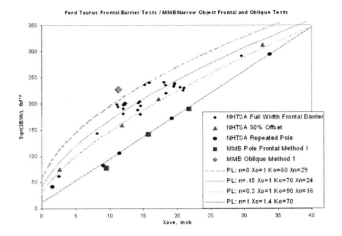

Figure 9. Crash Plot for the front structure of the Ford Taurus

The pole impact results agree well with previously conducted repeated pole impacts by NHTSA. The pole data shows that narrow-object impacts sustain a deeper average crush than the full width barrier impacts for equal energy input. Furthermore, the pole impacts revealed that the stiffness remained constant with increasing crush (n=1 in power-law), as opposed to the decreasing stiffness observed in the full width barrier tests (n<1, see Figure 10). However, as a generalization, 50% offset and narrow object tests demonstrated a reduction in stiffness from that of full width tests (Figures 10 & 11).

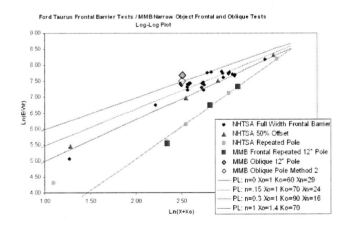

Figure 10. Log-Log Plot for the front structure of the Ford Taurus

FORD TAURUS SIDE IMPACTS

Two side pole-impact tests were added to the crash-plot of the Taurus side impacts with a deformable moving barrier conducted by NHTSA (Figure 11). The maximum damage of the MMB side-tests exceeded that of the NHTSA tests.

The deformable moving barrier and narrow-object side tests show a similar power-law coefficient, as reflected by the parallel lines in the log-log plot of the energy versus crush (Figure 12). However, the narrow object tests require lower energy for the same deformation.

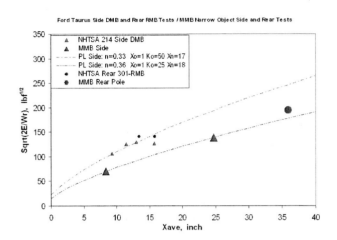

Figure 11. Crash Plot for the side and rear structure of the Ford Taurus

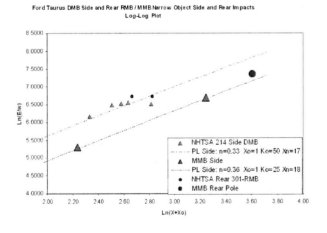

Figure 12. Log-Log Plot for the side and rear structure of the Ford Taurus

FORD TAURUS REAR IMPACTS

A single high-speed pole-impact was conducted with the MMB to the center rear of a 1987 Ford Taurus at 30.8 mph (Figure 10). The Taurus experienced a delta-V of approximately 31.9 mph in this test, which is approximately twice as high as the delta-V experienced in a FMVSS 301 test. The narrow-object tests to the center rear of the Taurus revealed a greatly reduced stiffness compared to that found in FMVSS 301 tests with a full width barrier, since the center rear of the vehicle contains virtually no load-bearing structures.

DISCUSSION OF DAMAGE MEASUREMENT

DAMAGE ANALYSIS METHODOLOGY- Three-dimensional vehicle crush vectors were measured at bumper-level in frontal and rear impacts and at trim-level in the side tests. The frontal and rear impacts caused the bumper corners to detach from the vehicle structure, which resulted in a negative longitudinal crush component.

FRONTAL METHOD- Four methods were applied to quantify the vehicle's frontal crush profile.

Method 1: Planar, two-dimensional crush vectors were used. The undamaged width was applied to calculate the area averaged crush and equivalent width. The longitudinal component of the crush was omitted when negative.

Method 2: Planar, two-dimensional crush vectors were used. Wherever the longitudinal crush component was negative, that portion of the profile was excluded. This resulted in exclusion of the edges of the undamaged width where the corners bent forward thus reducing the equivalent width and producing a variation with each test repetition.

Method 3: The two-dimensional hood damage better represented the vehicle's structural deformation than the detached bumper in tests 2 and 3. The change in hood deformation from tests 2 and 3 was added to the bumper crush profile of test 1 to attain the total crush profile of all test repetitions. The profile width was taken to be the undamaged bumper width.

Method 4: One-dimensional longitudinal crush was measured and included in the crush profile when reflecting a positive crush. The lateral deformation of the vehicle structure was not included in the crush estimate. The distorted bumper width as measured post impact was used for the crush profile width. The width decreased greatly with increasing crush amount.

Figure 13 shows the differences between the methods in terms of the crash-plot results.

In this paper, method 1 was applied to determine the vehicle crush. The same method was applied to determine the two-dimensional point-to-point crush vectors of the repeated frontal pole crash-tests previously published by NHTSA, which allowed a direct comparison of the new pole-test data with the NHTSA data.

Alternative method 2 caused a wide variation in crush width between the test repetitions and thus in the factor $\sqrt{(2E/w)}$, which would greatly complicate the analysis.

Method 3 was the only method that used vehicle structure other than the bumper to define the two-dimensional crush. These measurements were not available in the NHTSA tests even though similar bumper detachment problems were observed in the test photos (Method 1 was used for NHTSA pole data). Therefore, MMB test dataset could not be compared to NHTSA's dataset using Method 3 even though it better represented the structural damage. Other structural measurements could also be considered.

Method 4 is most often used in damage analysis for accident reconstruction purposes. However, comparison of the four methods revealed that the last method dramatically overestimated the stiffness for two reasons:

a) The method did not contribute any of the crash-test energy to the lateral deformation of the vehicle structure.

b) The method overestimated the factor $\sqrt{(2E/w)}$, since the post-crash width is much smaller than the undamaged width in narrow-object collisions. This caused an upward shift of the data-points in the crash-plot (Figure 13).

The authors of this paper believe that method 4 is unreliable.

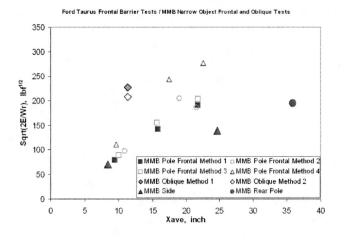

Figure 13. Comparison of crush profile measurement methods.

OBLIQUE METHOD- The oblique two-dimensional crush depth was measured at the level of the trim-line. The pre-impact width was defined perpendicular to the principle direction of penetration, 30 degrees clockwise from head-on. The width included all the direct and induced damage from pole-contact. An alternative method defined the crush width along the side of the vehicle from the front corner to the base of the A-pillar. The first method better represents the actual work generated by the impact force ($E=\int F \cdot dS$), and was applied in this paper.

SIDE DAMAGE ANALYSIS- The two-dimensional side crush was measured at the level of the trim-line. The additional crush vectors produced in the second test were added to those of the first test to obtain the combined, path dependant, displacement. The width was measured from the a-pillar structure to the c-pillar structure.

REAR DAMAGE ANALYSIS- Two-dimensional crush was measured at bumper level, and the undamaged bumper width of the vehicle was used for damage analysis.

STRUCTURAL STIFFNESS COMPARISONS- The Ford's frontal structure was stiffer than the side and rear structures, while the vehicle's stiffness was the highest when the wheel and axle structures were involved (oblique impact). The side and rear structures had a similar stiffness. The Taurus generally showed a lower stiffness for crush below approximately 10 inches for narrow object impacts than for full width collisions. For higher crush, the stiffness remained constant for narrow object impacts, while a reduction in stiffness or saturation was observed for collisions with broad objects.

The frontal structure is often stiffer than the side or rear structures, because of the location of the energy absorbing rails and the engine. Furthermore, frontal collisions are on average more severe (in terms of impact speed and delta-V) than side or rear impacts due to the greater speed difference between two vehicle at the time of impact. Most vehicles have perimeter stiffness values that have been optimized for absorbing the energy of the test requirements (like FMVSS 208, 214 and 301 for front, side and rear, respectively, Wood 1997). The severity of the test requirements reflects the severity of the majority of real world accidents (Evans 1994).

The narrow object impacts showed a reduced stiffness in the front, side and rear collisions versus that in full width collisions, at least up to a certain amount of crush. The pole was aimed at the vehicle center in front and rear crashes, and the door structure in the side impact. These structural areas have reduced load-bearing capability. In contrast, a higher stiffness was found in the oblique pole-impact into the wheel. This illustrates that the vehicle structures are not homogeneous, which is generally assumed for reconstruction purposes (CRASHIII Algorithm). The non-homogeneous behavior of vehicle structure was previously demonstrated by Buzeman-Jewkes et al. (1999) and Digges and Eigen (2000), who conducted crash-tests with a load-cell barrier.

The structural stiffness of the vehicle remained constant for greater crush in crash-tests with narrow objects, as opposed to the force saturation behavior seen in vehicle structures impacted by broad objects.

Nilsson-Ehle et al. (1982) and Buzeman-Jewkes (1999) addressed the stiffness increase in their underride and offset crash-tests, and explained it by the shear-stiffness of the vehicle. The shear stiffness is reflected in indirect damage to the vehicle.

CONCLUSIONS

The MMB proved to be a convenient tool to obtain crush data in less standard conditions.

Vehicle stiffness varied greatly between the different pole impact locations. In general, front stiffness was higher than rear and side stiffness, which were similar. The highest stiffness was found in the oblique test into the front wheel.

Stiffness values determined in the pole impacts depended on load-bearing ability of the structures in the struck area.

In accident reconstruction, the crash-tests evaluated for vehicle stiffness should closely represent the location, direction and local structure impacted in the accident.

Methods used for damage measurements of a subject vehicle should be consistent with the method used in evaluating crash-tests to correctly quantify the severity of the accident compared to that of the crash-test.

It is preferable to measure the actual vehicle structure's damage (like method 3) versus that of a detached or distorted bumper. However, a reliable estimate of the accident severity may not be obtained if the structural damage of the subject vehicle is compared to the damage of the detached bumper of the crash-test vehicle evaluated for stiffness estimates.

The irregular crush profile analysis allows narrow object crash-test data to be compared with flat barrier data. The data presented in this paper (Figures 9 and 11) showed that full width barrier data can be used as a first approximation of the vehicle behavior in narrow object collisions.

The results indicate the usefulness of having a wide variety of crash-test data in order to more accurately estimate stiffness coefficients.

ACKNOWLEDGMENTS

The authors would like to acknowledge and thank the employees of Woolley Engineering Research Corporation along with GMH Engineering and Wilderness Design for their assistance in testing and research.

REFERENCES

ADAC, Allgemeiner Deutscher Automobil-Club e V (ADAC), www.adac.de

Buzeman-Jewkes, D. G.; Lövsund, P.; Viano, D.C. (1999) Use of Repeated Crash-Tests to Determine Local Longitudinal and Shear Stiffness of the Vehicle Front with Crush. Proc. SAE Conf. SP-1432. Paper No. 1999-01-0637.

Campbell, K. L. (1974) Energy Basis for Collision Severity. Proc. SAE Conf. Paper No. 740565.

Digges, K.; Eigen, A. (2000) Analysis of Load Cell Barrier Data to Assess Vehicle Compatibility. Proc. SAE Conf. Sp-1516. Paper No. 2000-01-0051.

EuroNCAP, www.euroncap.com

Evans, L. (1994) Driver Injury and Fatality Risk in two-car Crashes versus Mass Ratio Iinferred Using Newtonian Mechanics. Accident Analysis and Prevention 26:609-616.

IIHS, www.iihs.org

Nilsson-Ehle, A.; Norin, H.; Gustafsson, C. (1982) Evaluation of a Method for Determining the Velocity Change in Traffic Accidents. Proc. 9th Int. Conf. On ESV. Pp 741-750.

Prasad, A.K. (1990) Energy Dissipated in Vehicle Crush – A Study Using the Repeated Test Technique. Proc. SAE Conf. Paper No. 900412.

Strother, C. E.; Woolley, R.L.; James, M. B.; Warner, C.Y. (1986) Crush Energy in Accident Reconstruction. Proc. SAE Conf., Paper No. 860371. Warrendale, PA.

Warner, C.Y.; Allsop, D.L.; Germane, G.J. (1986) A Repeated Crash Test Technique for Assessment of Structural Impact Behavior. Proc. SAE Conf. Paper No. 860208.

Wood, D. P. (1997) Safety and Car Size Effect: A Fundamental Explanation. Accident Analysis and Prevention 29:139-151

Wood, D.P.; Mooney, S.J. (1996) Car size and Relative Safety: Fundamental Theory and Real Life Experience Compared. Proc. Int. IRCOBI Conf. Dublin: 223-229

Woolley, R.L. (1983) Method for Determining Crush Energy Coefficients from Crash Test Data. Collision Safety Engineering. Library # 2502.

Woolley, R.L. (2001) Non-Linear Damage Analysis in Accident Reconstruction. Proc. Of SAE Conf. SP-1572. Paper No. 2001-01-0504.

Woolley, R.L.; Asay, A.F.; Jewkes, D.B. (2000) Crash Testing with a Massive Moving Barrier as an Accident Reconstruction Tool. Proc. SAE Conf. Sp-1516. Paper No. 2000-01-0604

CONTACT

Alan F. Asay M.S., P.E.
Woolley Engineering Research Corporation
5314 North 250 West Suite 330
Provo, Utah 84604   (801) 431-0220

APPENDIX A. ENLARGEMENTS OF TEST DATA GRAPHICS

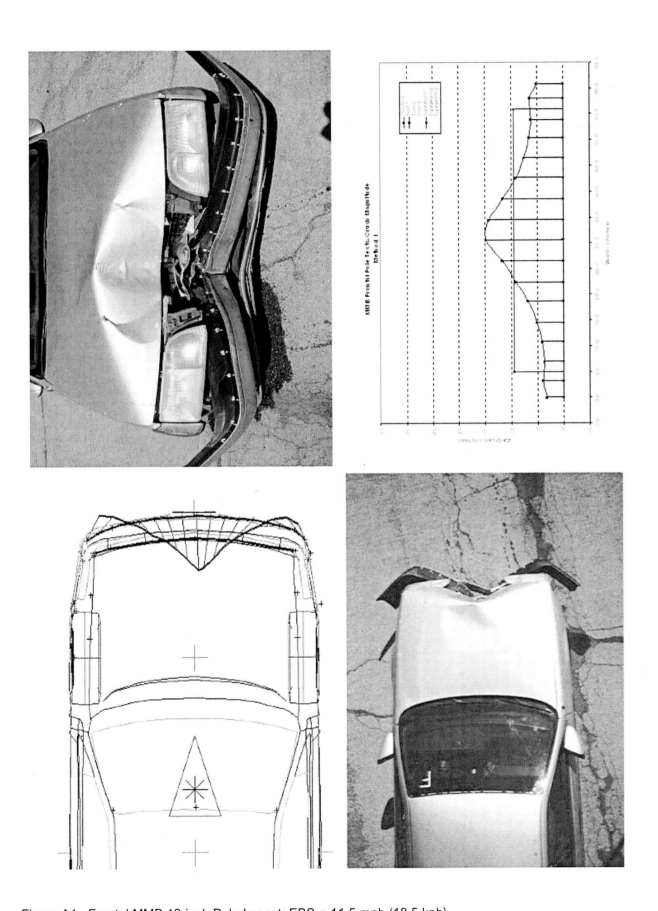

Figure A1. Frontal MMB 12-inch Pole Impact, EBS = 11.5 mph (18.5 kph).

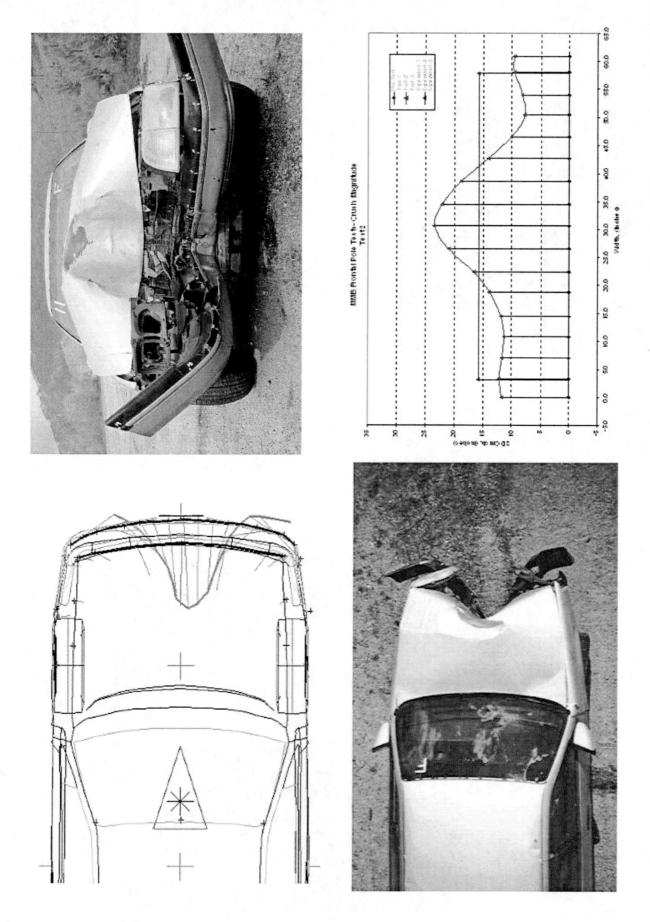

Figure A2. Frontal MMB 12-inch Pole Impact, EBS = 23.9 mph (38.5 kph).

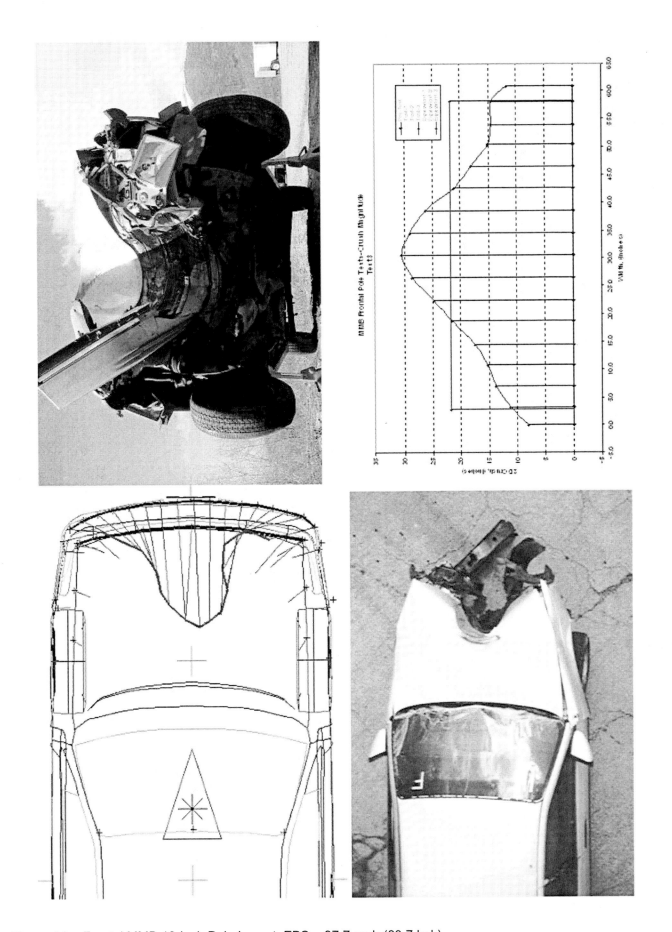

Figure A3. Frontal MMB 12-inch Pole Impact, EBS = 37.7 mph (60.7 kph).

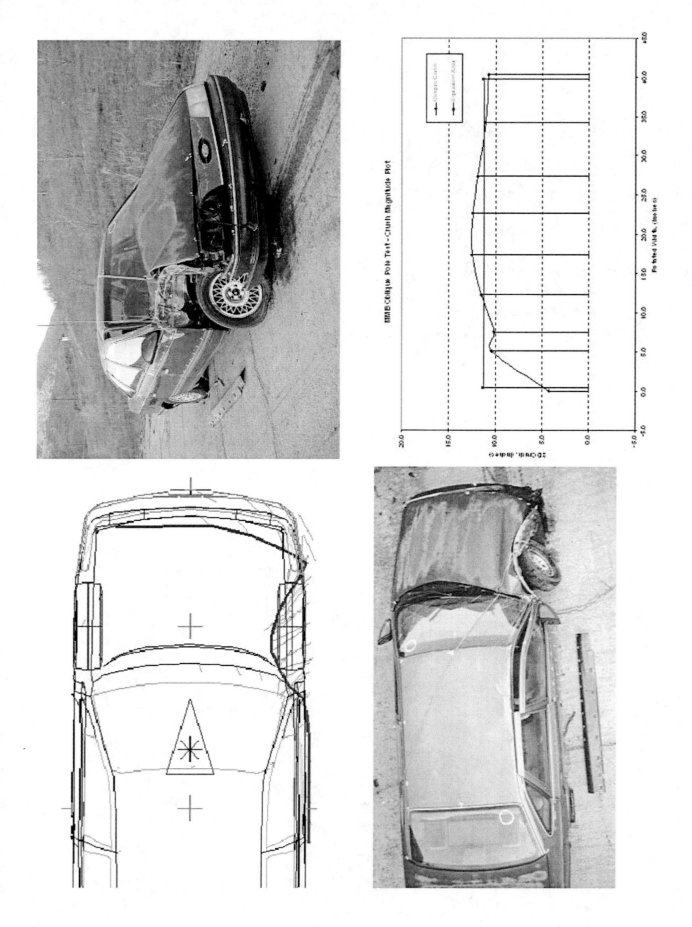

Figure A4. Oblique MMB 12-inch Pole Impact, EBS = 29.0 mph (46.7 kph).

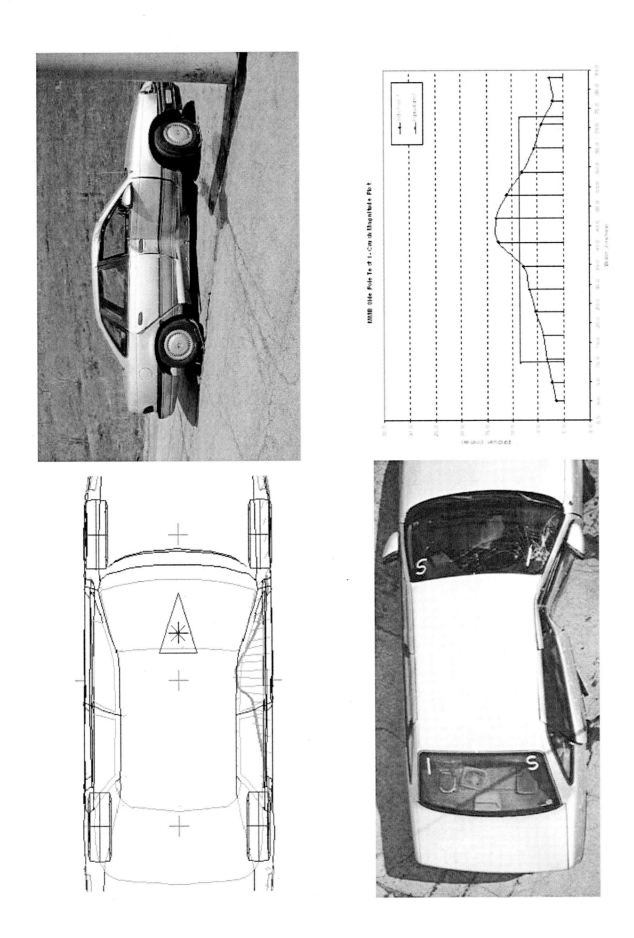

Figure A5. Side MMB 12-inch Pole Impact, EBS = 11.6 mph (18.7 kph).

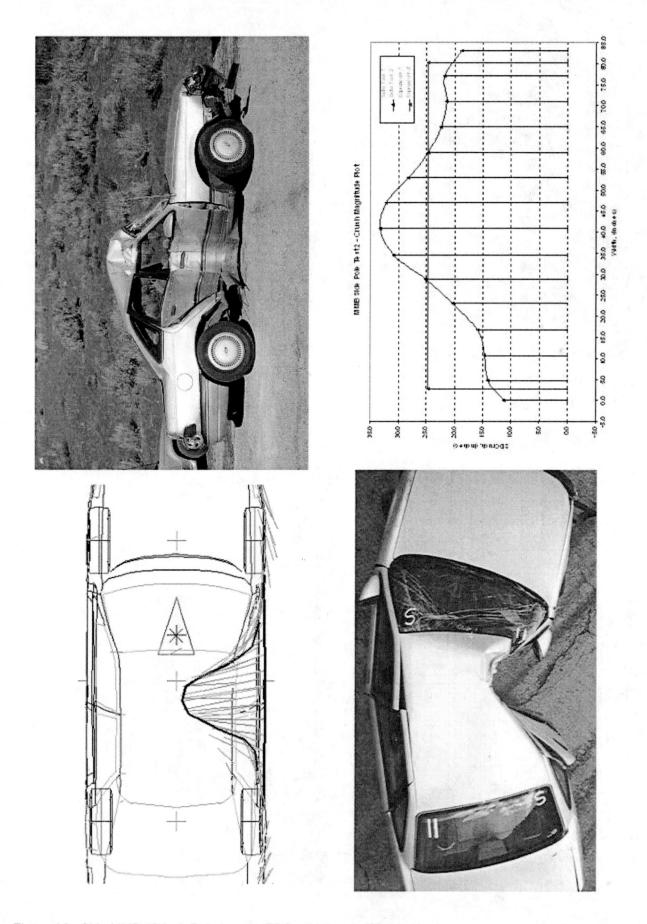

Figure A6. Side MMB 12-inch Pole Impact, EBS = 24.3 mph (39.1 kph).

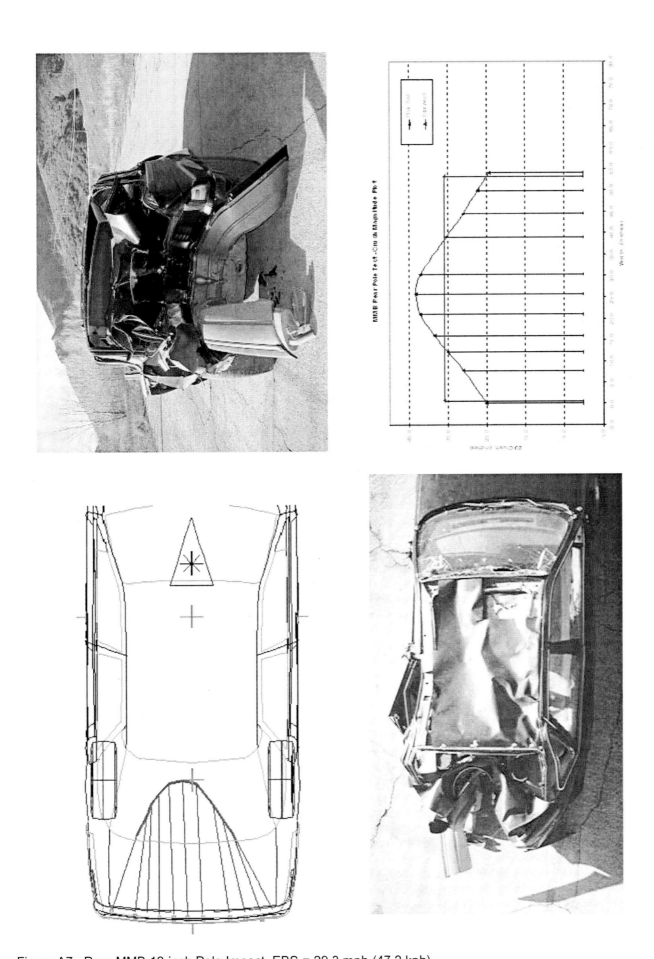

Figure A7. Rear MMB 12-inch Pole Impact, EBS = 29.3 mph (47.2 kph).

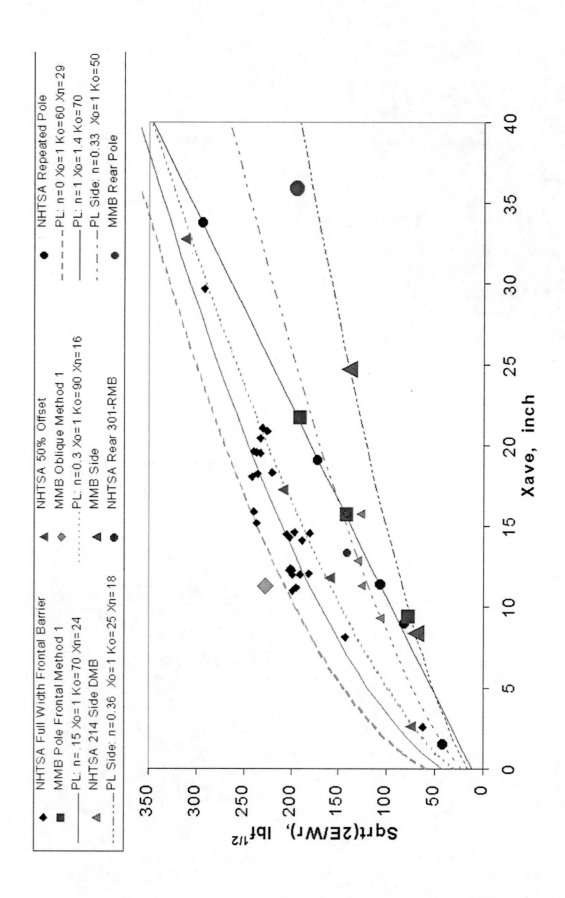

Figure A8. Crash Plot Comparison of MMB Pole Tests with All NHTSA Test Data For 86-91 Ford Taurus.

Reviewer Discussion

Reviewers Name: Michael S. Varat
Paper Title: Narrow Object Impact Analysis and Comparison with Flat Barrier Impacts
Paper Number: 2002-01-0552

The authors have performed impressive research into the structural performance of vehicles in impacts with narrow objects. Several important conclusions should be noted form this recent addition to the engineering literature.
1. Vehicles are not always well represented by the homogeneous, isotropic assumptions inherent in the popular CRASH III algorithm.
2. Consistency must be maintained between measurement methodologies used in the determination of stiffness and the measurement methodology used in an accident reconstruction under study.
3. The application of the power law formulation to narrow object impacts provides a general model that is well suited to represent crash test data.
4. Vehicle structural stiffness varies between full width flat barrier impacts and narrow object impacts.

I would like to applaud the authors on an interesting technical paper that is a valuable long term reference to the accident reconstruction community.

2002-01-0553

Single Vehicle Wet Road Loss of Control; Effects of Tire Tread Depth and Placement

William Blythe
Consulting Engineer

Terry D. Day
Engineering Dynamics Corp.

Copyright © 2002 Society of Automotive Engineers, Inc.

ABSTRACT

When an automobile is driven on wet roads, its tires must remove water from between the tread and road surfaces. It is well known that the ability of a tire to remove water depends heavily on tread depth, water depth and speed, as well as other factors, such as tire load, air pressure and tread design. It is less well known that tire tread depth combined with placement can have an adverse effect on vehicle handling on wet roads.

This paper investigates passenger car handling on wet roads. Flat bed tire testing, three-dimensional computer simulation and skid pad experimental testing are used to determine how handling is affected by tire tread depth and front/rear position of low-tread-depth tires on the vehicle.

Some skid pad test results are given, along with corresponding simulations. A literature review also is presented.

Significant changes in tire-road longitudinal and lateral friction are shown to occur as speed, tread depth and water depth vary, even before hydroplaning occurs.

Three-dimensional computer simulation confirms and illustrates the effect of worn-tire placement on vehicle handling characteristics during wet road maneuvers.

INTRODUCTION

A misunderstanding exists in the after-market tire industry regarding the significance of tire placement when two new tires are being installed and two rather worn tires remain on the vehicle. Personal case studies by one of the authors have shown that, when a customer requests the installation of only two new tires, it is not uncommon for tire installers to place the two new tires on the front wheels while leaving worn tires of considerably less tread depth on the rear.

Reasons for this choice often are stated as: (1) steering occurs at the front wheels, (2) most of the braking occurs at the front wheels, and (3) it is easier to control a blowout at a rear wheel as compared to a front wheel.

The first two observations above are, of course, correct, while the third is not [38][1]. Reduction in lateral (cornering) force at a rear tire due to loss of air potentially will turn a normally understeering vehicle into an oversteering vehicle. Similarly, on a wet road a reduction of lateral friction capability at the rear tires due to low tread depth can lead to oversteer. For the average driving population, oversteer must be avoided in order to achieve proper and safe vehicle handling.

The second finding (above) suggests that if tires of significantly different tread depth are placed on the front and on the rear of a passenger car, handling characteristics on a wet road can be affected seriously. For example, if new tires are placed on the front wheels and worn tires remain on the rear, a normally understeering car can oversteer[2] on a wet road, potentially resulting in loss of control. In some cases, brake or throttle application can contribute further to rear wheel slip and subsequent oversteer.

This paper quantifies the variation in tire/road friction capability as a function of tread depth, water depth and speed. The results then are used to study the vehicle

---

[1] Numbers in brackets designate references found at the end of the paper, preceding the Appendices.
[2] No distinction is made herein between understeer and oversteer at the "limit" of control and those same handling characteristics below the limit state, which is included in the normal technical definition. See [35], pp. 361-364.

handling effects on wet roads produced by well-treaded tires and worn tires at various wheel positions.

LITERATURE REVIEW

Except for attempts to quantify the conditions for full hydroplaning[3], no quantification has been presented, in recent literature, of variation in tire/surface friction on wet surfaces, for various tread depths, water depths and speeds.

In 1965 Maycock [1] examined the effects of tread pattern, tread material and tire casing construction on the skidding resistance of passenger car tires on wet surfaces. Tread depths were nominally 9/32 inches and water depths ranged from 0.005 to 0.080 inches above the asperities. These were on-vehicle tests. The effect of tread depth variations was not studied.

Kelley [2], in 1968, studied a number of factors affecting tire traction on wet roads, but tread depth and water depth were not among them.

Staughton [3], in 1970, investigated the effects of tread depth on peak and slide braking coefficients, on a variety of wet surfaces, by vehicle testing. Tread depth varied from zero to approximately 10.5/32 inches (a new tire). Water depth above the asperities varied from approximately 0.016 inches to 0.046 inches. Tests were conducted at 50, 80 and 130 km/h (nominally 30, 50 and 80 mph). Reductions in peak and slide braking coefficients, comparing tires with approximately 2/32 inches tread to new tires, ranged from 44 to 60 percent for the greater water depth, at 50 and 80 mph.

Staughton and Williams [4] published an extensive report in 1970, investigating the effect of water depth on free-rolling wheel spin down and locked wheel braking force coefficients. Parameters studied included tire type, loads, inflation pressures and speeds. Tire tread depth was not a focus of these studies. The authors conclude that because "there can be a progressive loss in braking force coefficient" before full hydroplaning occurs "it is… not so much the fully aquaplaning (hydroplaning) condition that is likely to produce increased stopping distances and loss of adhesion but the losses that occur long before this condition is apparent."

Sabey, et al. [5] in 1970 studied the effect of water depth on surfaces of different texture, using fully treaded and smooth tires, measuring braking force coefficients and vehicle speed to produce spin down. They concluded that "tread pattern cannot compensate for lack of road texture" and, consistent with Staughton and Williams, "the main problem on roads is the lubricating effect of relatively thin water films, not aquaplaning with thick films."

Also in 1970, Harvey and Brenner [6], in an extensive report issued by the National Bureau of Standards, concluded "there is strong evidence that it is safest to have one's least worn tires on the rear axle." This is the earliest reference found on the significance of tire placement.

Bergman et al. [7] in 1971 compared tire friction measured in the laboratory with field measurements. The laboratory tests were conducted on an internal drum machine (internal diameter 12.5 feet) lined with a 3M manufactured 80 grit abrasive surface, the same surface used in the present tests. This is described in [7] as a "traction surface simulating a modern highway constructed with high friction surface materials." Following a correction of the laboratory measurements to a base road coefficient, excellent agreement was found between laboratory and road test measurements.

In an exhaustive Monograph published in 1971 [8], the National Bureau of Standards included a graph (p. 470) of braking force coefficient versus tread depth as a function of speed, in water depths of 0.04 to 0.06 inches, which showed that a 50% worn radial tire could lose 50% or more of its braking ability at 60 to 80 mph, as compared with a new tire.

The earliest direct industry notification found, of the significance of new tire placement, is dated February 22, 1974 from Sears Roebuck and states, in part, "When mounting two new tires, regardless of the type of tires on the vehicle or the type of new tires, the new tires must always go on the rear axle" [9][4].

Vieth and Pottinger, in 1974 [10], investigated tire cornering forces on wet surfaces, with trailer tests. Water depths varied from 0.005 inches to 0.25 inches and the bias ply tires were new and half worn. Speeds ranged from 20 to 60 mph. Significant decreases in cornering force were noted with decreased tread depth and increased water depth and speed. Vehicle handling was not addressed directly.

Lippmann and Oblizajek [11] studied the effects of tire wear on cornering stiffness and aligning torque stiffness of several passenger car tire constructions, in 1974. No wet surfaces were considered.

Also in 1974, Dijks [12] examined the effects of tread depth on the wet skid resistance of passenger car tires. Preliminary tests utilized water depths of approximately 0.012 and 0.024 inches, and for these values the effects of water depth were found to be small.

---

[3] The term "full hydroplaning" or "hydroplaning" is used herein to indicate complete loss of tire contact with the road surface, whether the cause is dynamic water pressure (dynamic or inertial hydroplaning) or water viscosity (viscous hydroplaning). See [35], pp. 133, 134.

[4] Several references listed are not in the "open literature" and thus not readily available. Interested persons may contact the first author.

Browne [13] provided an early (1975) mathematical treatment of hydroplaning and stated, "...the general wet traction problem could rightly be termed partial hydroplaning."

Agrawal and Henry (1977)[14] studied the hydroplaning potential of pavements, including the change in brake force coefficient with speed and water depth, but used only the ASTM tires (E-249 and E-524).

In 1978 Sakai et al. [15] studied tire braking and cornering forces under wet conditions on a drum-type tire testing machine. No variations in tread depth were included. They state, "...the problem in the actual use of tires is not simply the hydroplaning phenomenon but the entire slip phenomenon in the presence of a thin water film, the so-called wet skid...".

The Gallaway et al. report of December, 1979 [16], contains extensive results from hydroplaning testing and presents empirical equations for determining hydroplaning speed as a function of tread depth, surface texture depth, water depth above the asperities, and tire inflation pressure. The study does not address reduced friction prior to full hydroplaning.

A series of SAE papers in 1980 [17, 18, 19] addressed techniques for testing on ice and snow, but not wet surfaces.

In 1982 the Tire Industry Safety Council stated "If selecting only a pair of replacement tires... put the two new tires on the rear wheels for better handling" [20].

Williams and Evans [21], in 1983, reported test results from both laboratory and field work, for radial tire braking and cornering coefficients over a range of tread and water depths, and speeds, similar to those investigated in the present paper. They also considered vehicle stability as a function of tire condition and placement, by calculating an understeer/oversteer factor based upon the usual equation for a linear suspensionless vehicle:

$$U = \frac{F_{z1}}{C_{\alpha 1}} - \frac{F_{z2}}{C_{\alpha 2}} \qquad (1)$$

where U = understeer/oversteer gradient, degrees,

F_{z1}, F_{z2} = front and rear axle load, lbs.,

and $C_{\alpha 1}, C_{\alpha 2}$ = total cornering stiffness at the front and rear axles, lbs./deg.

Negative U implies oversteer and the potential for a spin mode loss of control. They also considered the effects of uneven tread wear on tire cornering ability, something not considered in the present investigation. They state: "The study has shown that vehicle stability under wet conditions... does depend on the state and type of tire wear, with tread depths greater than the generally accepted legal limit likely to create stability problems if the fitment is not favorable. The fitting of new front tires can make a vehicle unstable when tires with low wet cornering power are fitted to the rear." Their results for evenly-worn tire braking and cornering are compared to results from the present investigation.

Hayes et al. [22] investigated ASTM E 274 skid numbers as well as spin down, over speeds from 20 to 60 mph, on three surfaces, in 1983. The E 274 tests employed a fully-treaded ASTM E501 tire. Comparison tests were run with an ASTM E501 tire at 2/32 inches tread depth, but on standing water[5] 0.25 inches deep. Remarkable decreases in "skid numbers"[6] are reported with the decrease in tread depth on standing water as compared to the standard ASTM procedure. Tests also were run with a commercial radial tire at 2/32 inches tread depth on standing water 0.25 inches deep. For the tread and water depths used, the "skid numbers" were similar for the two tires. Unfortunately, no comparisons of the effect of the water application procedure were made using the same tire at the same tread and water depth.

In 1986 Sears Roebuck stated in a training manual [23] "The construction type with the best traction must be on the rear axle, regardless of whether the vehicle is front or rear wheel drive."

French [24] in 1988 reported tire laboratory data produced by the Dunlop Company "in the 1960s" showing graphs of "relative braking adhesion" as a function of speed (from 20 to 90 mph) and tread depth (from 8 mm to "smooth", in 2 mm steps) for water depths of 1 mm and 2.5 mm (approximately 0.04 to 0.10 inches). The tires tested are not identified as to model. These results are compared to data developed in the present paper.

Michelin, in a tire "owners manual" published in 1988 [25] stated, "Whenever only two tires are replaced, the new tires should be put on the rear axle." The same advice is given for specific tires [34].

Allen et al. [26] in 1991 discussed in detail the importance of understeer/oversteer considerations on lateral stability. Tire and road surface conditions were not a part of these investigations.

---

[5] ASTM E 274 specifies that water be sprayed onto the pavement surface by a nozzle mounted on the test trailer in front of the test tire.
[6] Technically these numbers from tests not following the ASTM E 274 procedure are normalized longitudinal tire forces (or brake force coefficients), since *skid numbers* are defined only under the conditions of ASTM E 274.

The 21st Edition of the Pirelli Cinturado Guide (1992)[27] gave the following reasons "the two best tires should generally be put in the rear... : a. better grip when starting uphill or on slippery ground (with rear-wheel drive), b. better and safer roadholding, as understeer is less dangerous than oversteer, c. more reliable braking...the tires with the best grip should be where there is the least weight... ."

Dunlop, in 1994, issued a Technical Service Bulletin [28] containing a paragraph headed "Fit newest tires on rear axle", which discussed the importance of avoiding oversteer. A Dunlop 1995 Fitment Guide [30] states "If replacing two tires put the best or new tires on the rear of the vehicle."

The 1995 Michelin America Small Tire (MAST) Fitment Guide [29] states plainly: "When replacing only two tires they are to be installed on the rear axle."

A review and discussion of various methods of estimating the speed at which hydroplaning may be expected to occur is presented by Nevin [31]. The issue of pre-hydroplaning loss of friction is not discussed.

In 1996 Wohanka and Essers [32] conducted road tests on a dry road and on a water film thickness of approximately 0.12 inches, comparing results for lateral and longitudinal tire force coefficients. Tread depth was approximately 8/32 inches and no variation in tread depth was studied. Decreases of approximately 20 percent in lateral force coefficients and approximately 40 percent in longitudinal force coefficients were noted under wet conditions.

Mooney and Wood [36] studied tire-road contact on wet surfaces in some detail and concluded that tire-to-road friction levels are determined at low speeds primarily by the road surface condition and at high speeds primarily by the tire tread condition.

The question of tire placement, when differences in tire tread are to be considered, still seems to be a subject of confusion outside the research community, as evidenced by the popular car magazines, where discussions on the merits of front versus rear placement of new tires continue [33,39].

TIRE LABORATORY TESTS

Laboratory tests on a number of tires were conducted at Veridian Engineering/Calspan Operations in Buffalo, New York, at the Tire Research Facility (TIRF), in December, 2000 and February, 2001. The tests were run on their flat bed tire test machine (see Figure 1) on a surface of 80 grit polycut material manufactured by 3M Corporation. The capabilities of this tire test equipment are detailed in Figure 2. A tire running under wet conditions is shown in Figure 3.

Figure 1

Tire tread surfaces were prepared by shaving to desired tread depths, from a nominal 2/32 inches to 10/32 inches, using an Amermac Tire Truer. The tires used in the tests are listed, with their tread depths, in Appendix 1.

Tests included straight-ahead braking and free-rolling cornering at water depths of 0.05 inches, 0.10 inches and 0.15 inches. Test speeds varied from 20 to 80 mph, in 20 mph increments. Inclination angles were zero in all tests. Vertical load was maintained at 800 pounds and tire pressure was held at 30 psi during testing for most tires. Two tires were tested at 35 psi.

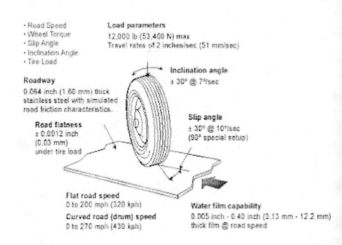

Figure 2

Figure 3

Tire No.	DOT Number	Nominal Tread Depth, inches	Inflation, psi
1	HNBBELTX099	New-10/32	30
2	HNBBELTX099	6/32	30
3	HNBBELTX199	8/32	30
4	HNBBELTX199	2/32	30
5	HNBBELTX209	4/32	30
6	B3BBAFHX235	3/32	35
7	B9BBAFHX317	9/32	35
8	ULXOEKLR068	10/32	30
9	PJXOKHKR225	1/32 (-)	30

Table 2

Longitudinal and lateral forces were measured as a function of speed, tread depth and water depth, for the full range of longitudinal slip and slip angles. Tire forces were normalized with respect to vertical tire load, which was kept constant during testing. All primary tests were conducted with the same tire design. A few tests were conducted on tires of other designs, for comparison.

BRAKING TESTS

The tires subjected to braking tests are described in Tables 1 and 2. A total of nine tires was tested. Examples of raw data from the braking tests are shown in Appendix 2. The friction values were interpreted from the raw data visually; no algorithm or curve fitting techniques were employed.

0.05 Inch Water Depth - Values of peak and slide friction for tires number 1 through 5 are plotted against speed for a water depth of 0.05 inches in Figures 4 and 5.

Tire No.	Brand	Model	Size
1	Michelin	Symmetry	P215/75R15
2	Michelin	Symmetry	P215/75R15
3	Michelin	Symmetry	P215/75R15
4	Michelin	Symmetry	P215/75R15
5	Michelin	Symmetry	P215/75R15
6	Michelin	Radial-X	P215/75R15
7	Michelin	Radial-X	P215/75R15
8	Grand-Am	Multi-Mile G/T	P225/60R16
9	Grand Spirit	Aqua Flow GTX	P225/60R16

Table 1

Test results for peak friction were well behaved, as seen in Figure 4. Test friction values for 100 percent slip (slide) had considerable scatter at speeds below 60 mph, as indicated in Figure 5. Results at 60 mph and above had the expected relationships.

Generally, both peak and slide friction forces at highway speeds are reduced to half or less of the new tire value if the tire wear exceeds about 50 percent.

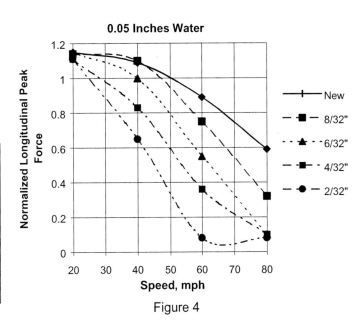

Figure 4

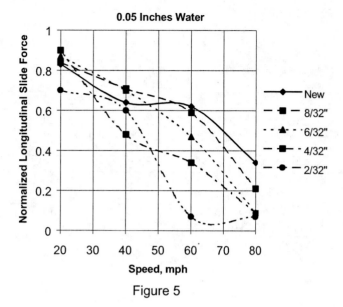

Figure 5

If hydroplaning is defined as the presence of friction values less than 0.10, then tires with 2/32 inches of tread depth or less can be expected to hydroplane at highway speeds, under the conditions of load, tire pressure and water depth present in these tests.

0.10 Inch Water Depth - Values of peak and slide friction for tires number 1 through 5 are plotted against speed for a water depth of 0.10 inches in Figures 6 and 7.

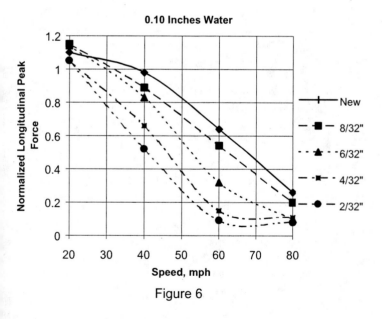

Figure 6

Peak friction values for 0.10 inches of water follow an expected pattern, with values generally less than those found for 0.05 inches of water (compare Figures 4 and 6). Data scatter again is significant for speeds under 60 mph for slide friction, but values at 60 mph and above compare well with those for 0.05 inches of water (Figures 5 and 7).

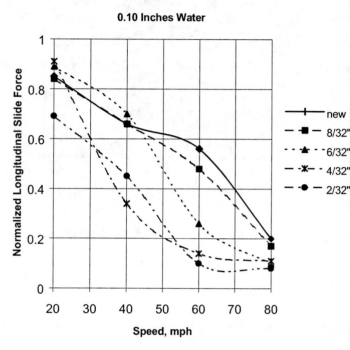

Figure 7

0.15 Inch Water Depth - Values of peak and slide friction for tires number 1 through 5 are plotted against speed for a water depth of 0.15 inches in Figures 8 and 9.

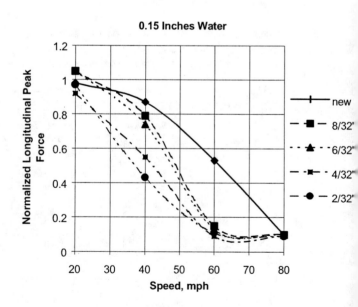

Figure 8

Peak friction values indicate tires of all tread depths, except the new tires, are close to or at hydroplaning (friction values of 0.10 or less) at 60 mph (Figure 8). Slide friction results indicate the same behavior (Figure 9). The two least-treaded tires produced no readable data below 40 mph at 100 percent slip.

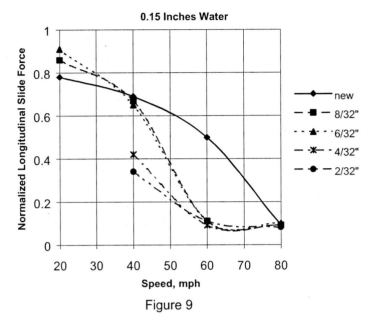

Figure 9

These test results suggest the need for data, at least for the new tires, at 70 mph, since the plot may have an inflection point in that region. Therefore, caution should be used when interpolating near 70 mph.

Figures 4 through 9 provide a clear indication that, at low speeds, some tire wear actually provides a slight increase in friction capability. The benefit is small, and generally has disappeared at speeds of 40 mph and above.[7]

Friction vs. Water Depth - Although the focus of these tire tests was on the effect of tread depth on wet road friction as a function of speed, the effect of water depth on a given tire also is of interest. Those relationships can, of course, be deduced from Figures 4 through 9, however Figure 10 shows some of those results directly. (Refer to Tables 1 and 2 for the identification of the tires.)

Figure 10 shows that a well-treaded tire in 0.15 inches of water behaves similarly to a worn tire in 0.05 inches of water.

Differences Between Tire Models - All tires in the tests thus far discussed were of exactly the same model. Figures 11, 12 and 13 show results for two different tire models (one of the same brand as the other tires and run at a slightly higher pressure). These tests are compared to the previous tests of tires with the closest tread depths.

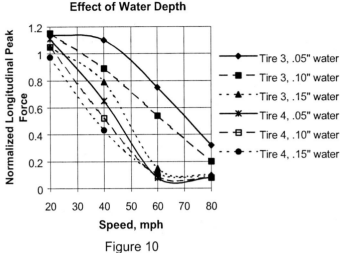

Figure 10

In the upper four plots of Figure 11, tire 7 (at nominally 9/32 inches tread depth) is compared to tires 1 and 3 (with nominally 10/32 and 8/32 inches of tread depth respectively) Tire 8 (nominal tread depth 10/32 inches) also is included. In the lower four plots, tire 6 (at nominally 3/32 inches tread depth) is compared to tires 4 and 5 (with nominally 2/32 and 4/32 inches of tread depth respectively). Tire 9 (an essentially bald tire) also is included.

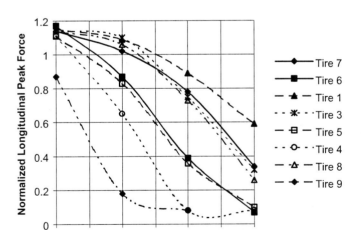

Figure 11

Tire 1 (a new tire) performed best, and considerably better than tire 8 (an equally well-treaded tire of a different brand, but at the same load and tire pressure)

---

[7] Although the phenomenon remains unexplained, it has been observed by other researchers. See [21], and compare their Figures 2 and 6.

at this water depth (0.05 inches). Tire 8 performed no better than tires 3 and 7, each of which had slightly less tread depth.

The lower four plots of Figure 11 follow an expected pattern, with the performance difference between tires 5 and 6 essentially indistinquishable. Tire 6 has slightly less tread depth (3/32 inches) than tire 5 (4/32 inches) but it has slightly higher pressure and a different tread design.

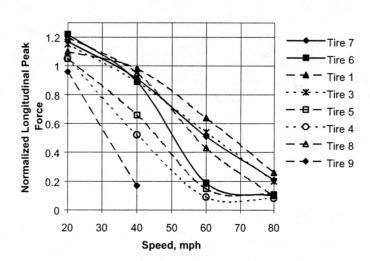

Figure 12

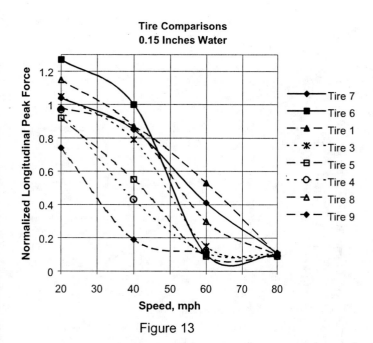

Figure 13

In Figure 12, similar comparisons are made to those shown in Figure 11, except for a water depth of 0.10 inches. Tire 1 again performed better than tire 8, and tire 8 performed slightly less well than tires 3 and 7. Tire 6 performed better than tire 5. Tire 9 (the bald tire) did not produce readable data above 40 mph; it essentially is hydroplaning at speeds above 40 mph (even at 0.05 inches of water).

Figure 13 shows comparisons at 0.15 inches of water. Tire 1 performed better than tire 8 at speeds above 40 mph. Tire 8 performed less well than tire 7 at speeds above 40 mph but better than tire 3 through the entire speed range. Tire 6 again performed better than tire 5.

These comparisons suggest possible better wet road performance of higher pressured tires (compare tires 6 and 7 at 35 psi with simlilarly-treaded tires at 30 psi), although these data are not extensive enough to be conclusive. That issue is not the focus of the present research, however, from other research, such a conclusion probably is to be expected. In any case, the same tire pressure was used in most of the present tests. Tire pressure is recognized as an important parameter in tire/wet road friction performance.

The data also suggest possible differences in performance due to tire size and/or tire model, but again that was not the focus of this work.

Comparisons of the present test results with other published data are not easily made, due to differences in test methods, tire models and sizes, and tire pressures and normal loads, among other test variables. The two references that have presented similar data are Williams and Evans [21] and French [24].

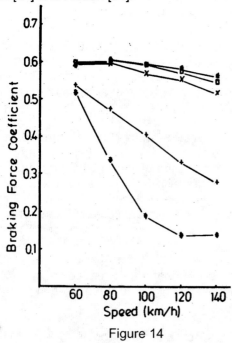

Figure 14

Figure 14 is from Figure 2 of [21]. These results are for an evenly worn tire of approximately 9/32 inches tread depth. The tire size was 155SR13; the brand was not given; neither was the tire pressure nor the normal load. The tests were conducted on the University of Karlsruhe internal drum tire testing machine. The test surface was not described. The "braking force coefficients" are for peak longitudinal braking. The plots are for water depths of 0.2, 0.5, 1.0, 2.0 and 3.0 mm, with the top plot corresponding to a water depth of 0.2 mm (approximately .008 inches) and the bottom plot corresponding to a water depth of 3.0 mm (approximately 0.12 inches), with the others ranged in between.

The two lower plots, for 2.0 mm (0.08 inches) and 3.0 mm (0.12 inches) of water depth, display the same general shape as those in Figure 10 for tire 3 (8/32 inches of tread depth), although the absolute values of peak braking coefficients are quite different.

Figure 15 is from Figure 4.22 of [24]. Laboratory test results are given as "relative braking adhesion", without a clear definition of the meaning of the scale. The results are described as "produced by the Dunlop Company in the 1960s", with no information on tire size,

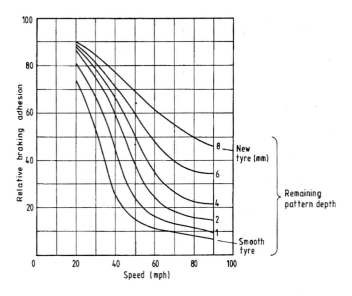

Figure 15

pressure, vertical load, or other details of the test procedure. The water depth is given as 2.5 mm (0.10 inches). Regardless of the inability for a direct quantitative comparison, it is valuable to note that the shapes of the plots of Figure 15 are very similar to those produced in the current research (compare with Figures 4 and 6).

Based upon the above comparisons, the results of other researchers and of tests on tires of different tread design reported herein, it seems reasonable to expect the general behavior to apply widely. That is, the large reduction of braking friction with tread wear, at highway speeds on wet roads, can be expected to occur well before full dynamic hydroplaning develops.

CORNERING TESTS

Six tires of the same model were subjected to cornering tests. The tires were Michelin Symmetry, P215/75R15. They are described in detail in Table 3. Tire 10 had been tested as a new tire in the braking tests.

Tire No.	DOT Number	Nominal Tread Depth, inches	Inflation, psi
10	HNBBELTX099	10/32	30
11	HNBBELTX099	6/32	30
12	HNBBELTX199	8/32	30
13	HNBBELTX199	2/32	30
14	HNBBELTX209	4/32	30
15	HNBBELTX129	New-10/32	30

Table 3

Examples of raw data from the cornering tests are shown in Appendix 2. The tires were free rolling, with slip angles ranging from approximately 15 degrees right and left. Values of normalized lateral peak force are plotted against speed, for the three different water depths, in Figures 16, 17 and 18. The normalized force values were interpreted from the raw data visually; no algorithm or curve fitting techniques were employed.

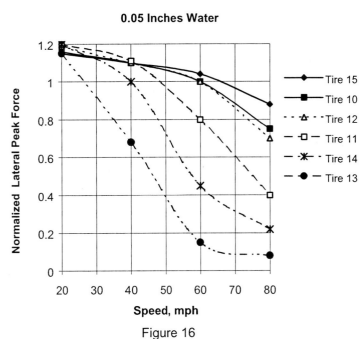

Figure 16

0.05 Inch Water Depth - Results for lateral friction at a 0.05 inch water depth are shown in Figure 16. The results follow an expected pattern, and are similar to those for longitudinal friction shown in Figure 4, although lateral friction values are higher than longitudinal friction values, as may be expected. Again, however, for a tire approximately 50% worn, lateral friction forces at highway speeds are reduced to about half of the new tire value for this water depth.

0.10 and 0.15 Inch Water Depth - Figure 17, at a water depth of 0.10 inches, also shows expected results, as does Figure 18 at 0.15 inch water depth. Both figures show the "reversal" at speeds below 40 mph, where lower tread depths produce better performance; this phenomenon was discussed earlier for the braking tests. It is not present in the results of Figure 17.

At 0.10 inches of water, and above, the tires with tread depths of 6/32 inches or less have lost essentially all lateral friction and either are hydroplaning, or very close to hydroplaning, at 60 mph and higher.

Perhaps more important is the remarkable decrease in lateral friction at highway speeds, well before hydroplaning is expected to occur, even at the lowest water depth considered.

Figure 19 is from Figure 2 of reference 21. As in Figure 14, these results are for an evenly worn 155SR13 tire of approximately 9/32 inches tread depth. The tire brand, pressure, normal force and test surface were not given in reference 21. Therefore, quantitative comparisons are not possible, but it is useful to note that the shapes of the plots for tires 10 and 12 in Figures 16, 17 and 18 are similar to the plots in Figure 19, when similar water depths are compared.

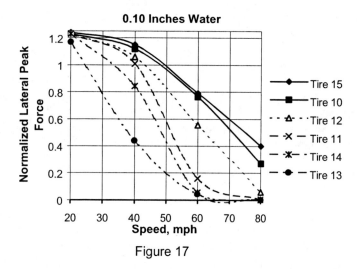

Figure 17

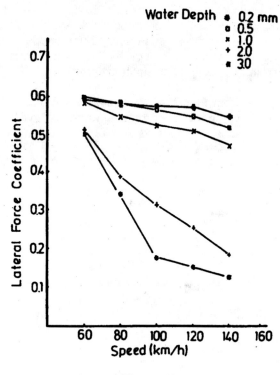

Figure 19

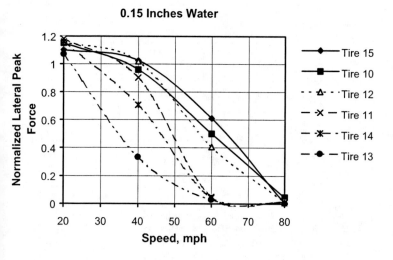

Figure 18

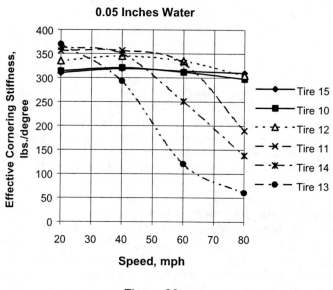

Figure 20

It generally is understood that cornering stiffness increases as tread depth decreases. This is correct for

-170-

tires on dry surfaces, however, with water present effective cornering stiffness[8] decreases with increased speed and water depth, and with decreasing tread depth. Figures 20, 21 and 22 show the change in effective cornering stiffness for the range of tread depths, water depths and speeds considered.

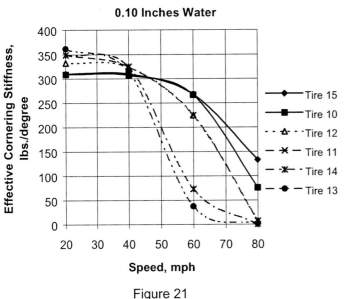

Figure 21

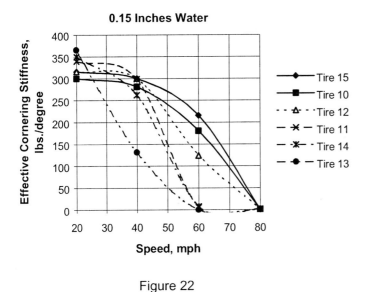

Figure 22

At 0.05 inches of water (Figure 20) there essentially is no loss of effective cornering stiffness for tires with at least 8/32 inches of tread. For tires with tread depths of 6/32 inches or less, the decrease in effective cornering stiffness is rapid and significant, especially above 60 mph.

At larger water depths (Figures 21 and 22), even the best-treaded tires show significant decreases in effective cornering stiffness. The poorest tires essentially loose effective cornering stiffness at speeds above 60 mph.

SKID PAD TESTS

A series of circle tests at a 100 foot radius was conducted on a skid pad with a 1990 Ford Thunderbird Super Coupe (see Appendix 3 for the vehicle specifications). Tires 8 and 9 were used (see Tables 1 and 2) and were matched, at each axle, with an essentially identical tire.

In one series of tests, the best-treaded tires (tire 8 and a matched tire) were fitted on the front (case 1). In a second series, the tire positions were reversed (case 2).

In case 1, speed on the circle was increased gradually until the vehicle left the circle due to oversteer. In case 2 the car was run at the maximum speed of case 1 to demonstrate that it would remain stable. Data from both tests were recorded at essentially steady-state conditions and, therefore, calculation of understeer/oversteer gradient is not possible by the standard SAE method (see equation 3).

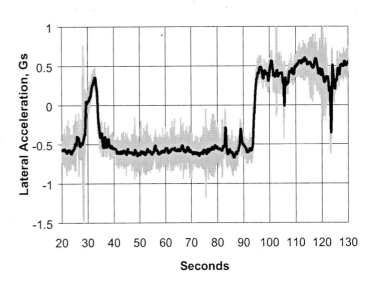

Figure 23

Figure 23 shows the lateral acceleration from case 1.

---

[8] Effective cornering stiffness is defined as the slope of the lateral force vs. slip angle plot, measured between plus and minus one degree slip angle.

Skid Pad - Case 1

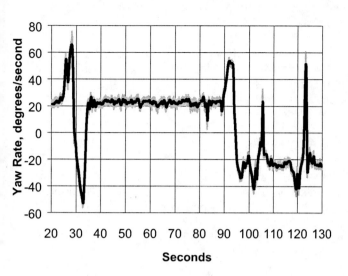

Figure 24

Skid Pad - Case 2

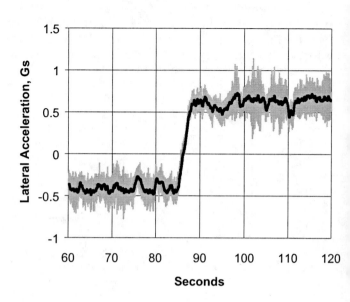

Figure 25

Skid Pad - Case 2

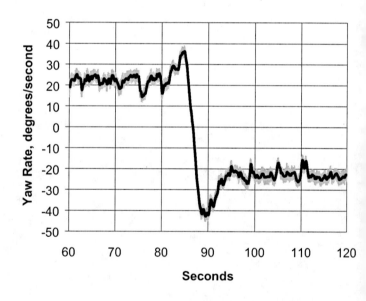

Figure 26

Figure 24 gives yaw rate. This test (case 1) consisted of first a clockwise run and then a counterclockwise run. The dark line is the data filtered at 1.25 hertz.

Yaw instability (limit oversteer) can be observed to occur at approximately 30 seconds into the clockwise run. The lateral acceleration is approximately 0.6 G. Elementary point mass kinematics suggest a speed of approximately 30 mph (speed was not recorded) and a yaw velocity of approximately 25 degrees/second, reasonably consistent with Figure 24. In the counterclockwise portion of the test, the lateral acceleration is approximately 0.45 G, suggesting a speed of approximately 25 mph. Tire 9 (a nearly-bald tire) was mounted on the right-rear and matched on the left-rear with a tire of slightly deeper tread, likely accounting for at least some of the difference in performance counterclockwise versus clockwise. However, some lack of symmetry in performance, even with perfectly-matched tires, is not unexpected. Two instances of yaw instability are shown in the counterclockwise run.

In case 2, where the best treaded tires (tire 8 and a matched tire) were on the rear, lateral accelerations in the counterclockwise direction exceeded 0.6 G (figure 25) and no yaw instability developed (Figure 26). The dark line is the data filtered at 2.5 hertz.

No direct measures of tire/skid pad surface friction were made.

Tires 8 and 9 were not subjected to cornering laboratory tests. However, recognizing the similarity of the longitudinal friction to cornering friction plots, insight can be gained from Figure 11. At 25 mph, the ratio of peak longitudinal friction for tire 9 to tire 8 is approximately 60 percent. The weight distribution for this vehicle was 57 percent to the front wheels. Therefore, equation (1) becomes, for case 1,

$$U = \frac{F_{z1}}{C_{\alpha 1}} - \frac{F_{z2}}{C_{\alpha 2}} = (W/K)[0.57/1 - 0.43/0.6]$$

$$= (W/K)[0.57 - 0.72] = -0.15(W/K) \quad (2)$$

where W = total vehicle weight and K = a common factor in the cornering stiffness. This elementary formula predicts oversteer for case 1 and understeer for case 2.

SIMULATIONS

The Engineering Dynamics Corporation Vehicle Simulation Model (EDVSM), part of the HVE simulation environment, was used to investigate vehicle handling during circle tests. This is a three-dimensional single-vehicle simulator [37]. Simulations of circle tests provide quantitative data on understeer/oversteer gradients.

Simulations were run that corresponded to the skid pad tests, with a range of tire/road fundamental friction vaues. Individual tire/road friction values were set according to the tire test results discussed previously. The simulations produced plots of steering wheel angle versus lateral acceleration, so that the SAE defined understeer gradient could be computed [35]:

$$k_U = \frac{d\delta_U}{dA} \qquad (3)$$

where k_U = understeer gradient,

δ_U = steering wheel angle, degrees,

and A = lateral acceleration, G.

A positive value of k_U indicates understeer, a negative value indicates oversteer, and a zero value indicates a neutral steering vehicle.

Vehicle specifications are given in Appendix 3. The circle radius was 100 feet and a reference peak lateral friction of 1.15 for the front tires (case 1) was used. The friction at the worn tires was estimated from Figure 11. The values used are shown in Table 4 for the two skid pad cases.

Skid Pad Case	Speed, mph	Peak Lateral Friction	
		Front	Rear
1	15	1.15	1.0
	20	1.15	0.86
	25	1.15	0.70
2	15	1.0	1.15
	20	0.86	1.15
	25	0.70	1.15

Table 4

For Case 1, with the worn tires at the rear, the results shown in Figure 27 (solid line) were obtained, indicating oversteer for lateral accelerations above approximately .28 G.

Yaw instability developed at approximately 25 mph. This corresponds to the counterclockwise skid pad test, since both rear tires in the simulation were based upon test results for tire 9.

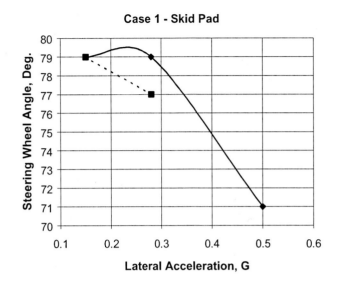

Figure 27

Case 2 simulation results are presented (solid line) in Figure 28. This case simulated the fully-treaded tires (tire 8 and a matched tire) at the rear. At approximately 25 mph the car left the circle in understeer in the simulation.

Both of these simulations agree well with the results of the skid pad tests. Simulation runs were in 3rd gear, with sufficient throttle to maintain the target speed.

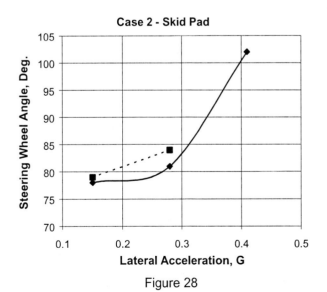

Figure 28

The skid pad Case 1 and Case 2 simulations were repeated with a lower reference friction, to determine if only the ratio of friction capability front to rear, or the absolute value of friction, was the determining factor for understeer/oversteer behavior. A reference value of 0.70 was used, with the same ratio of friction at the worn tires as shown in Table 4. The results are shown by the dashed lines in Figures 27 and 28. The basic understeer/oversteer characteristics essentially were unchanged (case 1 still was oversteer, case 2 still was understeer), but in each case loss of control occurred at lateral accelerations above 0.28 G (about 20 mph). When compared to the results of the skid pad physical tests, it appears the higher values of friction (Table 4) more closely represent the test surface.

At highway speeds (50 mph and above) tires with approximately 50 percent wear begin to show significant decreases in cornering friction capability, even with the least amount of water for which tests were conducted (see Figure 16). Simulations at 60 mph and above (not detailed in this paper), with rear tires at 3/32 and 2/32 inches tread depth and front tires at 8/32 to 10/32 inches tread depth, show the potential for yaw instability and loss of control during normal lane change maneuvers, with even the smallest water depth considered. When the tire positions are reversed, the simulations demonstrate stability in the same maneuvers.

CONCLUSIONS

Care must be exercised when applying the results of the present research, since important parameters such as tire model, air pressure and tire load generally were held constant. These factors may have significant effects on tire performance on wet surfaces.

Test results presented by the National Bureau of Standards in 1971 [8] relative to the loss of tire/road friction on wet surfaces are consistent with data developed in the present research.

The basic conclusion drawn is that the best rubber should be on the rear of passenger cars, regardless of vehicle configuration, in order to preserve proper vehicle handling on wet surfaces and avoid a dangerous oversteer condition. A review of the literature shows that this conclusion has been recognized by many engineers for many years, but perhaps without the quantitative basis provided herein, and still is not recognized as a serious consideration by others.

The importance of tire placement is not widely appreciated in the tire replacement and installing industry. This paper has presented quantitative test and analytical data on current tires to demonstrate further the significance of tire placement when differences in tread depth exist.

The work by Williams and Evans [21] was the closest found to the present investigation. The present work confirmed (on essential points) and extended their tire test data and also extended their conclusions regarding the vehicle handling implications of tire condition and placement, on wet roads.

Longitudinal and cornering friction forces can be reduced, at highway speeds, to less than half of new tire values on a wet road if the tread wear is 50 percent or more.

The effects of tread depth and water depth on lateral peak friction forces are similar to the effects on longitudinal peak friction forces. Therefore, in the absence of tests of one type, the results of the other type can be used with reasonable accuracy.

Significant loss of tire/roadway friction occurs with tread wear, at highway speeds on wet roads, well before full hydroplaning. This probably is a more significant safety issue than full hydroplaning, which is a relatively rare event.

A well-treaded tire, running in 0.15 inches of water above the asperities, behaves similarly to a worn tire in 0.05 inches of water.

Tires of different brand (which of course means different construction and tread design), while producing different levels of friction, show the same type of decrease in friction with decreasing tread depth and increasing water depth and speed. This does suggest, however, that it is not desirable to mix tire brands.

While cornering stiffness increases with tread wear when a tire is run on a dry surface, effective cornering stiffness decreases for worn tires running on wet surfaces. Even with well-treaded tires, with water depths at 0.10 inches or greater above the asperities, significant decreases in effective cornering stiffness will occur at highway speeds.

Normal lane change maneuvers can lead to loss of control on a wet road if sufficient difference in tread depth exists front to rear, with the better treaded tires on the front axle of a passenger car.

The Engineering Dynamics Corporation three-dimensional vehicle simulator, EDVSM, modeled well the understeer and yaw stability performance of the car tested on the skid pad in these investigations, and is expected to perform similarly for other passenger cars.

The difference in friction capability, front to rear, is the significant determinant in establishing understeer or oversteer behavior, rather than the absolute value of friction. Actual friction values will determine the lateral acceleration needed to cause yaw instability.

ACKNOWLEDGMENTS

Special mention is due Richard Hille of Syson/Hille and Associates, Goleta, CA and Anil V. Khadilkar, Ph.D., of Biodynamics Engineering, Inc., Pacific Palisades, CA for designing, conducting and providing the data from the skid pad tests discussed herein.

The authors wish to thank Terry O'Reilly and Jim Collins of San Francisco, California for their support and encouragement and also Tammy Van Sant and the Rhyne, Olson and Gibbins families.

Thanks also to Charles P. Dickerson of Collision Engineering Associates, Inc. of Mesa, Arizona, for arranging for the shaving of the tires.

CONTACT

William Blythe, P. E., Ph.D.
William Blythe, Inc.
1545 University Avenue
Palo Alto, CA 94301-3140
650-323-2062
wbinc8831@aol.com

Terry D. Day
Engineering Dynamics Corp
Beaverton, Orgeon 97008-7100
day@edccorp.com

REFERENCES

1. Maycock, G., "Studies on the Skidding Resistance of Passenger-car Tyres on Wet Surfaces," Proceedings of the Institution of Mechanical Engineers, Vol. 180, Pt. 2A, No. 4, 1965-66.

2. Kelley, J. D. Jr., "Factors Affecting Passenger Tire Traction on the Wet Road," SAE Paper No. 680138, 1968.

3. Staughton, G. C., "The Effect of Tread Pattern Depth on Skidding Resistance," Road Research Laboratory, Report LR 323, Ministry of Transport, U. K. 1970.

4. Staughton, G. C. and Williams, T., "Tyre Performance in Wet Surface Conditions," Transport and Road Research Laboratory, Laboratory Report 355, Ministry of Transport, U. K., 1970.

5. Sabey, B. E., Williams, T. and Lupton, G. N., "Factors Affecting the Friction of Tires on Wet Roads," SAE Paper No. 700376, 1970.

6. Harvey, J. L. and Brenner, F. C., "Tire Use Survey: The Physical Condition, Use, and Performance of Passenger Car Tires in the United States of America," Technical Note 528, National Bureau of Standards, Washington, D. C. May, 1970.

7. Bergman, W., Clemett, H. R., and Sheth, N. J., "Tire Traction Measurement on the Road and in the Laboratory," SAE Paper No. 710630, 1971.

8. Clark, S. K., (Editor), Mechanics of Pneumatic Tires, National Bureau of Standards (U.S.) Monograph 122, November, 1971.

9. "Tech-Let," for Division 93-95, Sears Roebuck and Company, Chicago, Illinois. February 22, 1974.

10. Veith, A. G., and Pottinger, M. G., "Tire Wet Traction: Operational Severity and its Influence on Performance," in The Physics of Tire Traction, D. F. Hays and A. L. Browne, Editors, Plenum Press, New York-London, 1974, pp. 5-20.

11. Lippman, S. A. and Oblizajek, K. L., "The Influence of Tire Wear on Steering Properties and the Corresponding Stresses at the Tread-Road Interference," SAE Paper No. 741102, October, 1974.

12. Dijks, I. A., "A Multifactor Examination of Wet Skid Resistance of Car Tires," SAE Paper No. 741106, October, 1974.

13. Browne, A. L., "Mathematical Analysis for Pneumatic Tire Hydroplaning," in "Surface Texture versus Skidding: Measurements, Frictional Aspects, and Safety Features of Tire-Pavement Interactions, " ASTM STP 583, American Society for Testing and Materials, 1975, pp. 75-94.

14. Agrawal, S. K. and Henry, J. J., "A Technique for Evaluating the Hydroplaning Potential of Pavements," Transportation Research Board, 56th Annual Meeting, Washington, D. C., January, 1977.

15. Sakai, H., Kanaya, O. and Okayama, T., "The Effect of Hydroplaning on the Dynamic Characteristics of Car, Truck and Bus Tires," SAE Paper No. 780195, February, 1978.

16. Gallaway, B. M., Ivey, D. L., Hayes, G., Ledbetter, W. B., Olson, R. M., Woods, D. L. and Schiller, R. F., Jr., "Pavement and Geometric Design Criteria for Minimizing Hydroplaning," Final Report, Contract DOT-FH-11-8269, Report No. FHWA-RD-79-31, Federal Highway Administration, Washington, D. C., December, 1979.

17. Davis, J. B., Wild, J. R. and St. John, N. W., "Winter Tire Testing," SAE Paper No. 800838, June, 1980.

18. Kneip, J. V., Derooy, A. F. and Kind, G. J., "European Winter Tire Testing," SAE Paper No. 800837, June, 1980.

19. Janowski, W. R., "Tire Traction Testing in the Winter Environment," SAE Paper No. 800839, June, 1980.

20. "New Consumer Tire Guide," Tire Industry Safety Council, Washington, D.C. 1982.

21. Williams, A. R. and Evans, M. S., "Influence of Tread Wear Irregularity on Wet Friction Performance of

Tires," in "Frictional Interaction of Tire and Pavement", ASTM STP 793, W. E. Meyer and J. D. Walter, Eds., American Society for Testing and Materials, pp. 41-64, 1983.

22. Hayes, G. G., Ivey, D. L., and Gallaway, B. M., "Hydroplaning, Hydrodynamic Drag, and Vehicle Stability," in "Frictional Interaction of Tire and Pavement", ASTM STP 793, W. E. Meyer and J. D. Walter, Eds., American Society for Testing and Materials, pp. 151-166, 1983.

23. "Basic Training," Sears Tires, Division 95, Sears Roebuck and Company, Chicago, Illinois. 1986.

24. French, T., Tyre Technology, Adam Hilger, Publisher, Bristol and New York. ISBN 0-85274-360-2. 1988.

25. Michelin Passenger & Light Truck Tire Owners Manual, Publication No. 2620-69. February, 1988.

26. Allen, R. W., Szostak, H. T., Rosenthal, T. J., Klyde, D. H. and Owens, K. J., "Characteristics Influencing Ground Vehicle Lateral/Directional Dynamic Stability," SAE Paper No. 910234, February, 1991.

27. Pirelli Cinturado Guide, 21st Edition, Pirelli Armstrong Tire Corporation, 500 Sargent Drive, New Haven, CT 06536-0201, 1992.

28. Dunlop Technical Service Bulletin, "Tire Replacement Update," Buffalo, New York. 1994.

29. Michelin America Small Tires (MAST) 1995 Fitment Guide.

30. Dunlop Passenger, Performance, Light Truck, Van and Sport Utility Vehicle Tire Fitment Guide, Dunlop Tire Corporation, Buffalo, New York. 1995.

31. Nevin, F., "Hydroplaning and Accident Reconstruction," SAE Paper No. 950138, 1995.

32. Wohanka, U. and Essers, U., "Influence of Waterfilm Thickness on Tyre Force and Moment Characteristics, Measured on Public Roads," European Automobile Engineers Cooperation, 5th International Congress, Strasbourg, June, 1995.

33. Winfield, B., "Stretching the Rubber Envelope Isn't Easy," p. 36, Car and Driver, July 1995.

34. Michelin Technical Bulletin PE-95-18, "RainForce MX4 Tire Mixing/Compatibility," Michelin North America, Inc., Greenville, South Carolina. December 11, 1995.

35. Dixon, J. C., Tires, Suspension and Handling, 2nd Edition, published by The Society of Automotive Engineers, Warrendale, PA, 1996.

36. Mooney, S. and Wood, D., "Locked Wheel Car Braking in Shallow Water," SAE Paper No. 960653, February, 1996.

37. Day, T. D., "Validation of the EDVSM 3-Dimensional Vehicle Simulator," SAE Paper No. 970958, February, 1997.

38. Blythe, W., Day, T. D., and Grimes, W. D., "3-Dimensional Simulation of Vehicle Response to Tire Blow-outs," SAE Paper No. 980221, February, 1998.

39. "Technical Correspondence", Letter: "Best Place for Trouble," p. 130, Road & Track, February, 2001.

APPENDIX 1

TIRES TESTED AND PRE-TEST TREAD DEPTH MEASUREMENTS

Note: "Depth 1" is the measurement of the groove nearest the side with the DOT number and "Depth 4" (or "Depth 6") is the measurement of the groove farthest from the side with the DOT number. Some tread depths were measured in decimal form and some in 32nds of inches.

Braking Tests								
Tire: Micheln Symmetry P215/75R15								
DOT Number	HNBBELTX099				HNBBELTX199			
Nominal Depth	New				8/32 (0.25) inches			
Radial Position	12:00	3:00	6:00	9:00	12:00	3:00	6:00	9:00

Depth 1	Nominal 10/32 inches; Refer to the measurements for this tire under "Cornering Tests" in this Appendix	0.253"	0.254"	0.257"	0.254"
Depth 2		0.275"	0.275"	0.275"	0.273"
Depth 3		0.261"	0.259"	0.266"	0.256"
Depth 4		0.258"	0.254"	0.257"	0.255"

Braking Tests								
Tire: Michelin Symmetry P215/75R15								
DOT Number	HNBBELTX099				HNBBELTX209			
Nominal Depth	6/32 (0.188) inches				4/32 (0.125) inches			
Radial Position	12:00	3:00	6:00	9:00	12:00	3:00	6:00	9:00
Depth 1	0.183"	0.19"	0.195"	0.186"	0.129"	0.124"	0.125"	0.129"
Depth 2	0.206"	0.216"	0.2"	0.208"	0.126"	0.134"	0.14"	0.153"
Depth 3	0.196"	0.207"	0.196"	0.193"	0.133"	0.131"	0.131"	0.151"
Depth 4	0.182"	0.182"	0.179"	0.167"	0.12"	0.123"	0.128"	0.129"

Braking Tests						
Tire	Michelin Symmetry P215/75R15				Michelin Radial-X P215/75R15	
DOT Number	HNBBELTX199				B9BBAFHX317	
Nominal Depth	2/32 (0.063) inches				9/32"	
Radial Position	12:00	3:00	6:00	9:00	12:00	6:00
Depth 1	0.101"	0.079"	0.088"	0.094"	9/32"	9/32"
Depth 2	0.096"	0.088"	0.079"	0.092"	10-/32"	9+/32"
Depth 3	0.076"	0.063"	0.070"	0.074"	10-/32"	10-/32"
Depth 4	0.077"	0.062"	0.074"	0.083"	9/32"	9/32"

Braking Tests			
Tire	Michelin Radial-X P215/75R15	Grand Am Radial G/T P225/60R16	Grand Spirit Aqua Flow GTX P225/60R16
DOT Number	B3BBAFHX235	UXLOEKLR068	PJXOKHKR225

Nominal Depth	3/32"		10/32"		1-/32"	
Radial Position	12:00	6:00	12:00	6:00	12:00	6:00
Depth 1	3/32"	3+/32"	10/32"	10/32"	1/32"	1/32"
Depth 2	3/32"	4-/32"	10-/32"	10-/32"	0	1-/32"
Depth 3	3/32"	3+/32"	10-/32"	10-/32"	0	0
Depth 4	2+/32"	2/32"	10/32"	10/32"	0	0
Depth 5					0	0
Depth 6					1-/32"	1-/32"

Cornering Tests								
Tire: Michelin Symmetry P215/75R15								
DOT Number	HNBBELTX129				HNBBELTX099* *This tire was tested as "new" in the braking tests			
Nominal Depth	New				10/32 (0.312) inches			
Radial Position	12:00	3:00	6:00	9:00	12:00	3:00	6:00	9:00
Depth 1	0.292"	0.297"	0.295"	0.298"	0.300"	0.300"	0.298"	0.301"
Depth 2	0.300"	0.302"	0.299"	0.303"	0.311"	0.307"	0.304"	0.308"
Depth 3	0.298"	0.301"	0.300"	0.300"	0.311"	0.308"	0.302"	0.308"
Depth 4	0.296"	0.298"	0.295"	0.295"	0.305"	0.298"	0.297"	0.298"

Cornering Tests								
Tire: Michelin Symmetry P215/75R15								
DOT Number	HNBBELTX199				HNBBELTX099			
Nominal Depth	8/32 (0.25) inches				6/32 (0.188) inches			
Radial Position	12:00	3:00	6:00	9:00	12:00	3:00	6:00	9:00
Depth 1	0.254"	0.260"	0.254"	0.255"	0.181"	0.197"	0.186"	0.168"
Depth 2	0.264"	0.266"	0.265"	O.265"	0.185"	0.210"	0.197"	0.180"
Depth 3	0.276"	0.275"	0.270"	0.272"	0.196"	0.215"	0.207"	0.189"

| Depth 4 | 0.277" | 0.264" | 0.261" | 0.260" | 0.184" | 0.199" | 0.192" | 0.188" |

Cornering Tests								
Tire: Michelin Symmetry P215/75R15								
DOT Number	HNBBELTX209				HNBBELTX199			
Nominal Depth	4/32 (0.125) inches				2/32 (0.063) inches			
Radial Position	12:00	3:00	6:00	9:00	12:00	3:00	6:00	9:00
Depth 1	0.117"	0.120"	0.102"	0.126"	0.057"	0.034"	0.045"	0.046"
Depth 2	0.126"	0.127"	0.097"	0.131"	0.062"	0.035"	0.039"	0.032"
Depth 3	0.128"	0.131"	0.103"	0.134"	0.068"	0.059"	0.053"	0.046"
Depth 4	0.123"	0.122"	0.103"	0.133"	0.077"	0.060"	0.059"	0.055"

APPENDIX 2

RAW DATA FROM TIRE TESTS – EXAMPLES

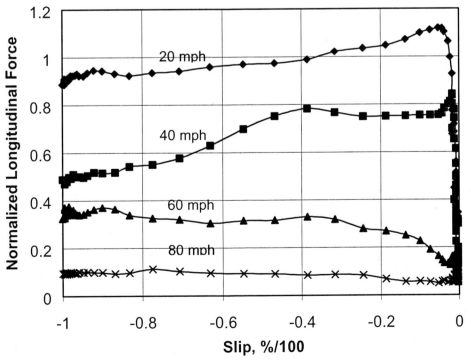

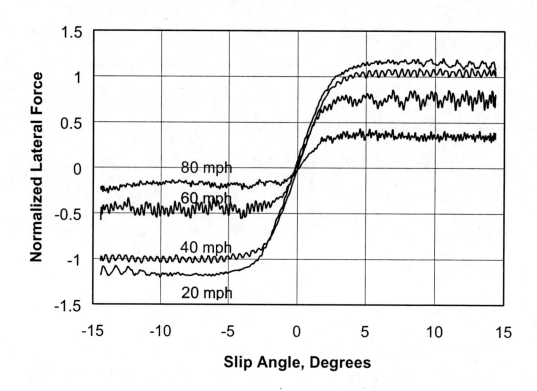

Cornering: Tire No. 14; 0.05 Inches Water

APPENDIX 3

TEST VEHICLE DATA

    General Vehicle Information

        Vehicle Name: Ford Thunderbird 2-Dr

Overall Length (in):	198.50
Overall Width (in):	71.80
CG to Front End (in):	89.50
CG to Rear End (in):	-109.00
Wheelbase (in):	113.00
Front Track Width (in):	61.60
Rear Track Width (in):	60.20
Front Overhang (in):	42.12
Rear Overhang (in):	-43.38
CG to Front Axle (in):	47.38
CG to Rear Axle (in):	-65.62
CG Height Above Ground (in):	20.40
Total Weight (lb):	4250.01
Effective Air Drag (in^2):	0.000058
Aerodynamic Drag, Cd:	0.3500
Frontal Area (in^2):	2940.20

    Sprung Mass Data

Mass (lb-sec^2/in):	10.48
Weight (lb):	4050.01

Rot Inertia (lb-sec^2-in), Ix: 5900.00
Iy: 34100.00
Iz: 36600.00
Ixz: 0.00

Suspension Data

	Front	Rear
Suspension Type:	Independent	Independent
Wheel Location (in), x:	47.38	-65.62
y:	30.80	30.10
z:	7.10	7.10
Unsprung Weight (lb):	100.00	00.00
Roll Steer - Const (deg):	0.00	0.00
Linear Rate (deg/in):	0.00	0.00
Quadratic Rate (deg/in^2):	0.00	0.00
Cubic Rate (deg/in^3):	0.00	0.00
Aux Roll Stiff (in-lb/deg):	429.00	0.00
Ride Rate @ Wheels (lb/in):	149.00	143.00
Damping @ Wheels (lb-sec/in):	7.26	6.40
Susp Friction Force (lb):	50.00	100.00
Min Vel for Friction (in/sec):	0.00	0.00
Jounce Stop (in):	-4.11	-4.76
Linear Rate (lb/in):	300.00	300.00
Cubic Rate (lb/in^3):	600.00	600.00
Rebound Stop (in):	3.96	4.41
Linear Rate (lb/in):	300.00	300.00
Cubic Rate (lb/in^3):	600.00	600.00
Energy Loss Ratio (%/100):	0.50	0.50

Camber and Half-track Tables

Front			Rear		
Susp Defl (in)	Camber (deg)	1/2-track Change (in)	Susp Defl (in)	Camber (deg)	1/2-track Change (in)
-4.00	-0.50	0.00	-4.00	-0.50	0.00
0.00	-0.50	0.00	0.00	-0.50	0.00
4.00	-0.50	0.00	4.00	-0.50	0.00

Anti-pitch Table

Front		Rear	
Susp Defl (in)	Anti-Pitch (lb/ft-lb)	Susp Defl (in)	Anti-Pitch (lb/ft-lb)
-4.00	0.00	-4.00	0.00
0.00	0.00	0.00	0.00
4.00	0.00	4.00	0.00

Drivetrain Data

Engine Description: 3.8L V-6, 4-speed automatic
Maximum Power (HP): 210
Maximum Torque (ft-lb): 300
Transmission Forward Speeds: 4
Differential Speeds: 1

Wide-open Throttle, Speed (RPM):	500	1200	2000	2400	3500	3800	4300	5000
Power (HP):	6	35	100	130	200	210	200	150
Torque (ft-lb):	60	153	263	284	300	290	244	158

Closed Throttle, Speed (RPM):	1000	2000	3000	4000	5000
Power (HP):	-3	-10	-23	-42	-65
Torque (ft-lb):	-14	-27	-41	-55	-68

Transmission Gear:	Reverse	1st	2nd	3rd	4th
Numerical Ratio:	-2.000	2.400	1.470	1.000	0.670

Differential Gear Ratio: 3.270

Steering System Data

    Steering Gear Ratio (deg/deg):   14.10

Brake System Data

    Pedal Ratio (psi/lb):   26.00

	Right Front	Left Front
Brake Type:	Disc Brake	Disc Brake
Torque Ratio (in-lb/psi):	13.89	13.89
Pushout Pressure (psi):	0.00	0.00
Lining/Drum T, Init. (farenheit):	68.00	68.00
Proportioning?	No	No

	Right Rear	Left Rear
Brake Type:	Disc Brake	Disc Brake
Torque Ratio (in-lb/psi):	10.48	10.48
Pushout Pressure (psi):	0.00	0.00
Lining/Drum T, Init. (farenheit):	68.00	68.00
Proportioning?	Yes	Yes
Proportioning, Pstart (psi):	200.00	200.00
P(wheel)/P(system) (%/100):	0.33	0.33

Tire Data – Data Base Values

	Right Front	Left Front
Tire Name:	Generic	Generic
Tire Type:	Passenger Car	Passenger Car
Tire Manufacturer:	Generic	Generic
Tire Model:	Generic	Generic
Tire Size:	P225/60R16	P225/60R16

Physical Data -

	Right Front	Left Front
Unloaded Radius (in):	13.30	13.30
Initial Ride Rate (lb/in):	1500.00	1500.00
2nd Ride Rate (lb/in):	15000.00	15000.00
Defl @ 2nd Rate (in):	4.24	4.24
Maximum Tire Defl (in):	7.00	7.00
Pneumatic Trail (in):	-1.61	-1.61
Weight, Tire+Rim (lb):	50.00	50.00
Spin Inertia (lb-sec^2-in):	7.60	7.60

A0 Coefficient:	-4360.21	-4360.21
A1 Coefficient:	26.71	26.71
A2 Coefficient:	3978.79	3978.79
A3 Coefficient:	1.71	1.71
A4 Coefficient:	4200.00	4200.00
Rolling Resistance Constant:	0.01	0.01

Friction Data -
    Location: R/F
    Number of Loads: 3
    Number of Speeds: 1
    In-use Factor: 1

Test Speed (in/sec):	528.00		
Test Load (lb):	650.00	1300.00	1950.00
Peak Longitudinal Mu:	0.92	0.88	0.84
Peak Lateral Mu:	0.92	0.88	0.84
Slide Mu:	0.72	0.68	0.64
Slip @ Peak Mu (%/100):	0.16	0.16	0.16
Long Stiffness (lb/slip):	13000.00	13000.00	13000.00

Friction Data -
    Location: L/F
    Number of Loads: 3
    Number of Speeds: 1
    In-use Factor: 1

Test Speed (in/sec):	528.00		
Test Load (lb):	650.00	1300.00	1950.00
Peak Longitudinal Mu:	0.92	0.88	0.84
Peak Lateral Mu:	0.92	0.88	0.84
Slide Mu:	0.72	0.68	0.64
Slip @ Peak Mu (%/100):	0.16	0.16	0.16
Long Stiffness (lb/slip):	13000.00	13000.00	13000.00

Cornering Stiffness Data -
    Location: R/F
    Number of Loads: 3
    Number of Speeds: 1
    In-use Factor: 1

Test Speed (in/sec):	528.00		
Test Load (lb):	650.00	1300.00	1950.00
Cornering Stfns (lb/deg):	177.40	331.90	387.40

Cornering Stiffness Data -
    Location: L/F
    Number of Loads: 3
    Number of Speeds: 1
    In-use Factor: 1

Test Speed (in/sec):	528.00		
Test Load (lb):	650.00	1300.00	1950.00
Cornering Stfns (lb/deg):	177.40	331.90	387.40

Camber Stiffness Data -
    Location: R/F
    Number of Loads: 3
    Number of Speeds: 1

In-use Factor: 1

Test Speed (in/sec): 528.00
Test Load (lb): 650.00 1300.00 1950.00
Camber Stfns (lb/deg): 17.70 33.19 38.74

Camber Stiffness Data -
   Location: L/F
   Number of Loads: 3
   Number of Speeds: 1
   In-use Factor: 1

   Test Speed (in/sec): 528.00
   Test Load (lb): 650.00 1300.00 1950.00
   Camber Stfns (lb/deg): 17.70 33.19 38.74

	Right Rear	Left Rear
Tire Name:	Generic	Generic
Tire Type:	Passenger Car	Passenger Car
Tire Manufacturer:	Generic	Generic
Tire Model:	Generic	Generic
Tire Size:	P225/60R16	P225/60R16

Physical Data -

	Right Rear	Left Rear
Unloaded Radius (in):	13.30	13.30
Initial Ride Rate (lb/in):	1500.00	1500.00
2nd Ride Rate (lb/in):	15000.00	15000.00
Defl @ 2nd Rate (in):	4.24	4.24
Maximum Tire Defl (in):	7.00	7.00
Pneumatic Trail (in):	-1.61	-1.61
Weight, Tire+Rim (lb):	50.00	50.00
Spin Inertia (lb-sec^2-in):	7.60	7.60
A0 Coefficient:	-4360.21	4360.21
A1 Coefficient:	26.71	26.71
A2 Coefficient:	3978.79	3978.79
A3 Coefficient:	1.71	1.71
A4 Coefficient:	4200.00	4200.00
Rolling Resistance Constant:	0.01	0.01

Friction Data -
   Location: R/R
   Number of Loads: 3
   Number of Speeds: 1
   In-use Factor: 1

   Test Speed (in/sec): 528.00
   Test Load (lb): 650.00 1300.00 1950.00
   Peak Longitudinal Mu: 0.92 0.88 0.84
   Peak Lateral Mu: 0.92 0.88 0.84
   Slide Mu: 0.72 0.68 0.64
   Slip @ Peak Mu (%/100): 0.16 0.16 0.16
   Long Stiffness (lb/slip): 13000.00 13000.00 13000.00

Friction Data -
   Location: L/R
   Number of Loads: 3
   Number of Speeds: 1
   In-use Factor: 1

```
Test Speed (in/sec):        528.00
Test Load (lb):             650.00    1300.00   1950.00
Peak Longitudinal Mu:         0.92       0.88      0.84
Peak Lateral Mu:              0.92       0.88      0.84
Slide Mu:                     0.72       0.68      0.64
Slip @ Peak Mu (%/100):       0.16       0.16      0.16
Long Stiffness (lb/slip): 13000.00   13000.00  13000.00
```

Cornering Stiffness Data -
   Location: R/R
   Number of Loads: 3
   Number of Speeds: 1
   In-use Factor: 1

```
Test Speed (in/sec):        528.00
Test Load (lb):             650.00   1300.00   1950.00
Cornering Stfns (lb/deg):   177.40    331.90    387.40
```

Cornering Stiffness Data -
   Location: L/R
   Number of Loads: 3
   Number of Speeds: 1
   In-use Factor: 1

```
Test Speed (in/sec):        528.00
Test Load (lb):             650.00   1300.00   1950.00
Cornering Stfns (lb/deg):   177.40    331.90    387.40
```

Camber Stiffness Data -
   Location: R/R
   Number of Loads: 3
   Number of Speeds: 1
   In-use Factor: 1

```
Test Speed (in/sec):        528.00
Test Load (lb):             650.00   1300.00   1950.00
Camber Stfns (lb/deg):       17.70     33.19     38.74
```

Camber Stiffness Data -
   Location: L/R
   Number of Loads: 3
   Number of Speeds: 1
   In-use Factor: 1

```
Test Speed (in/sec):        528.00
Test Load (lb):             650.00   1300.00   1950.00
Camber Stfns (lb/deg):       17.70     33.19     38.74
```

Reviewer Discussion

Reviewers Name: Ernest Z. Klein
Paper Title: *Single Vehicle Wet Road Loss of Control; Effects of Tire Tread Depth and Placement*
Paper Number: 2002-01-0553

This paper addresses the effect of water and other factors on a motor vehicle performance. It starts out stating that when only 2 tires are replaced on a motor vehicle, the new tires should be placed on the rear axle of the vehicle. This reviewer certainly agrees with that statement, a statement which has been previously discussed in scientific literature and accepted in the scientific community.

It follows then with a good literature review and summary. One may also wish to consider additional literature not summarized by the authors.

In this paper, valuable laboratory test data is being presented by the authors. The effect of water depth, tread depth, and speed on normalized longitudinal and lateral slide forces are provided. One must bear in mind, however that values presented were obtained on a flat bed test machine and do not take into account variations in roadway micro and macro textures, which are a significant factor in real life experiences. It also does not consider variations in tread patterns, tire construction, or tire pressure variations. These variations may have a significant effect in real life experiences of vehicle performance on wet roads.

A following statement by the authors may need to be clarified – "it generally is understood that cornering stiffness increases as tread depth decreases. This is correct for tires on dry surfaces, however, with water present effective cornering stiffness decreases…" Cornering stiffness is a tire parameter and does not depend on presence or absence of water on the road. The authors are correct, however when they state that "with water present, EFFECTIVE cornering stiffness, for tire road interaction decreases." Such lower values should be used in vehicle handling simulation.

Further work is recommended for pre-hydroplaning situations to quantify tire road interaction under wet conditions at various speeds, loads, tire pressure, etc.

2002-01-0554

Large School Bus Side Impact Stiffness Factors

Kristin Bolte, Shane Lack and Larry Jackson
National Transportation Safety Board

ABSTRACT

School bus travel is one of the safest forms of transportation on the road today. The passenger fatality rate in school buses is 0.2 fatalities per million vehicle miles traveled (VMT) as compared to 1.5 per million VMT for passenger cars and 1.3 per million VMT for light trucks. Each year on average, nine school bus passengers are fatally injured in school bus crashes while sixteen school-age pedestrians are fatally injured by the school bus. Although much has been done to improve the safety of school buses over the years, more research may reflect new ways to better protect school bus passengers.

The National Transportation Safety Board concluded in 1999 that current compartmentalization is incomplete in large school buses in that it does not protect passengers during lateral impacts. In order to better understand severe lateral impacts to school buses and the resulting passenger motion and injuries, the stiffness of the side of the school bus needed to be determined.

The purpose of this paper is to present the development of side impact stiffness factors for the large school bus based on previously investigated collisions where a train impacted the side of the school bus.

INTRODUCTION

School bus travel is one of the safest forms of transportation. The passenger fatality rate in school buses is 0.2 fatalities per million vehicle miles traveled (VMT) as compared to 1.5 per million VMT for passenger cars and 1.3 per million VMT for light trucks.[1] Each year on average, nine school bus passengers are fatally injured in school bus crashes while sixteen school-age pedestrians are fatally injured by the school bus.[2] Although much has been done to improve the safety of school buses over the years, more research may reflect new ways to better protect school bus passengers.

The National Transportation Safety Board concluded in 1999 that current compartmentalization is incomplete in large school buses in that it does not protect passengers during lateral impacts. In order to better understand severe lateral impacts to school buses and the resulting passenger motion and injuries, accurate models representing the side of the school bus are needed. These models must incorporate the material properties of the side of a large school bus so that the deformation and forces applied to the school bus are realistic resulting in accurate bus motion and predicted injuries to occupants. The stiffness of the side of the school bus needed to be determined to accurately reflect the forces applied to the side of a large school bus during a lateral impact collision.

Vehicle stiffness can be calculated based on the crush of the vehicle when impacted by a non-deformable barrier. The non-deformable barrier is essential in the calculation because the energy prior to the impact, the kinetic energy, is almost all distributed into the crush of the deformable vehicle and the acceleration of the vehicle. In addition, the barrier should be flat so that the crush distributed across the front of the barrier is uniform.

Data is currently not available for a side impact crash test where a non-deformable barrier impacts the large school bus but the Safety Board does have data on past investigations involving many different side impact collisions between large vehicles and large school buses. Typically these impacts involve large school buses and tractor-trailers where significant damage is seen to both vehicles but when a train impacts a school bus, the bus typically absorbs the majority of the impact energy.

Therefore, estimates were made for the stiffness factors based on several assumptions. The stiffness of the train was assumed to be almost infinite since little deformation occurred to the train during past investigations. In addition, the front of the train was assumed to be flat. This assumption enabled the assumption of a uniform crush profile on the bus as well. Although the train is not entirely flat across the front, the force distribution appeared uniform in many of the accidents investigated.

The purpose of this paper is to present the development of side impact stiffness factors for the large school bus based on previously investigated collisions where a train impacted the side of the school bus.

METHODS AND RESULTS

The collisions examined were Conasauga, Tennessee (3/28/00), Sinton, Texas (2/28/98), Buffalo, Montana (3/10/98), Fox River Grove, Illinois (10/25/95), Port St. Lucie, Florida (9/27/84)[i], and Carrsville, Virginia (4/12/84). Several other collisions between school buses and trains have been investigated by the Board but were not included in this study due to a lack of data on the crush profile of the vehicle or because the damage was so extensive that the vehicle 'disintegrated' in the area of impact. These collisions all involved pre-standard[ii] school buses and included the following collisions: Stratton, Nebraska (8/8/76), Aragon, Georgia (10/23/74), Congers, New York (3/24/72), and Waterloo, Nebraska (10/2/67).

For this collision, the crush coefficients were calculated based on McHenry Accident Reconstruction[3] crush coefficients calculations where W_1 is the weight of the bus and W_2 is the weight of the train engine. M_1 and M_2 represent the mass of the two vehicles. Q is a multiplier and delta V_c represents the change in velocity of the accident vehicle. L is the length of the contact damage and C_r is the residual crush. b_0 and b_1 are the intercept and slope of a linear curve fit of the Q*delta V_c versus C_r plot. A and B were calculated from b_0 and b_1 and represent the stiffness of the vehicle for the collision algorithm. These calculations are detailed in the following equations, where the data supplied is based on the Conasauga collision.

The data necessary to graph the crush coefficients for each of the six included collisions are listed in Table 1 and Table 2. The velocities listed in Table 1 refer to the forward velocity of the school bus and were not included when determining the change in lateral velocity during the impact. Detailed crush measurements were taken from the Conasauga, Buffalo, and the Fox River Grove buses but only the maximum residual crush was reported for the other collisions. Therefore, the maximum residual crush was used for the calculations of the crush coefficients.

The impact speed change at a point of common velocity[iii] (delta Vc) was calculated for each of the included collisions. The maximum crush was plotted versus the impact speed change and a linear fit was calculated through the data points representing the post-standard school buses.[iv] (See Figure 1.)

$$W_1 = \text{bus weight} = 17{,}846\, lbs$$
$$M_1 = \text{bus mass} = 46.2\, {lb\cdot sec^2}/{in}$$
$$W_2 = \text{train weight} = 4{,}930{,}000\, lbs$$
$$M_2 = \text{train mass} = 12758.8\, {lb\cdot sec^2}/{in}$$
$$Q = \sqrt{1 + M_1/M_2} = 1.0$$
$$\Delta V_c = \frac{M_1 V_1 + M_2 V_2}{M_1 + M_2} = 876.8\, in/sec$$
$$Q \cdot \Delta V_c = 878.4\, in/sec$$
$$L = \text{crush length} = 163.1"$$
$$C_r = \text{residual crush} = 39.4"$$
$$b_0 = \text{intercept}$$
$$b_1 = \frac{Q \cdot \Delta V_c - b_0}{C_r}$$
$$A = \frac{b_0 b_1 M_1}{L}$$
$$B = \frac{b_1^2 M_1}{L}$$

Table 1: This table details the weight and speed of the school bus at the moment of impact and the maximum crush on the school bus.

Collision	W_1 (lbs)	V_1 (mph)	$C_r$max (in)
Conasauga, TN	17,846	15	39.4
Sinton, TX	30,700[v]	~20	~5
Buffalo, MT	24,000[vi]	0	22.6
Fox River Grove, IL	23,390	0	40
Port St. Lucie, FL	15,450[vii]	0	50
Carrsville, VA	17,500[vii]	0	34

---

[i] The bus in this accident was manufactured in 1968 and therefore is a pre-standard school bus and does not conform to the current school bus standards for joint, roof, and floor strength.

[ii] A pre-standard school bus is a bus manufactured before 1977. After that date, buses were subject to standards for joint, roof, and floor strength.

[iii] A point of common velocity was assumed for this study.

[iv] The amount of crush for a similar impact speed-change was drastically different between the pre-standard school bus and the post-standard school buses. This most likely resulted because of the improvements in floor, roof, and body joint construction for the post-standard school buses.

[v] Gross Vehicle Weight Rating (GVWR)

[vi] This was the GVWR but the bus was a 48-passenger bus with only five students and one driver on board at the time of the accident. The weight was reduced by 4,000 lbs for this calculation.

[vii] Estimated based on the weight of the Easton, MD accident bus that was also a 66-passenger school bus. The passenger weights were also estimated and included.

Table 2: This table details the weight and speed of the train at the moment of impact.

Collision	W_2 (tons)	V_2 (mph)
Conasauga, TN	2465	51
Sinton, TX	6506	17
Buffalo, MT	4546	43
Fox River Grove, IL	570	60
Port St. Lucie, FL	2710	33
Carrsville, VA	5232	44

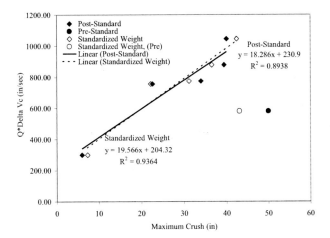

Figure 1: The graph of the residual crush versus the normalized impact speed change for both the post-standard school buses and the single pre-standard school bus. The post-standard school bus data was linearly fit with b_0 and b_1 as the intercept and slope of the fitted line.

Based on the linear fitted data for the post-standard school buses, the stiffness coefficients, A and B, were calculated as detailed in the equations above for the bus in the Conasauga, TN collision. Thus, A and B for this bus were calculated to be 1340 lb/in and 106 lb/in^2, respectively. For comparison, the A and B stiffness coefficients[3] for a front impact on an '84 Ford LTD were 344.94 lb/in and 44.51 lb/in^2, respectively. Also, the A and B stiffness coefficients[3] for a rear impact on an '86 Ford Taurus were 355.18 lb/in and 51.69 lb/in^2, respectively.

The above calculations assume a standardized weight of the impacted school bus and also assume that the entire weight of the school bus was involved in the collision. In reality, the weights of the school buses were not identical in each collision and the impacts were not at the center of gravity of the school bus. Therefore, a standardized vehicle weight and crush were calculated. The standardized weight was simply the average weight of all six school buses, 20,814 lbs. The standardized crush, C_S was calculated based on the measured crush, C_r, the measured bus weight, W_1, and the standardized weight, W_{ave}.[3]

$$C_S = C_r \sqrt{W_{ave}/W_1}$$

Figure 2 details the differences in the linearly fitted data for the standardized weight and crush and for the non-standardized weight and crush. The resulting differences in b_0 and b_1 were small. The calculated A and B stiffness coefficients for the standardized weight and crush were 1132 lb/in and 108 lb/in^2, respectively.

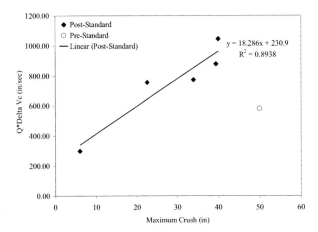

Figure 2: The graph of the residual crush versus the normalized impact speed change for school buses with standardized weight and crush. The post-standard school bus data was linearly fit for the standardized and non-standardized cases.

In addition, the effective mass[3] of the school bus in each collision was estimated. The effective mass accounts for the offset side impact and therefore only includes a percentage of the total weight of the impacted vehicle based on the amount the impact was offset from the center of gravity of the impacted vehicle. The multiplier for the effective mass, γ, was calculated based on the radius of gyration in yaw, $k^2 = \dfrac{l^2 + w^2}{12}$, and the moment arm of the resultant collision force, h, (the amount of offset from the center of gravity).[3]

$$\gamma = \frac{k^2}{k^2 + h^2}$$

Unfortunately, due to a lack of information concerning the axle weights of many of the buses and the exact impact location on the bus, the effective mass could only be calculated for the Conasauga bus and the Sinton bus. Table 3 details the calculation of the effective mass multiplier and the resulting $Q \cdot \Delta V_c$. Although the effective mass was only calculated for two of the six included collisions, the percent difference between the $Q \cdot \Delta V_c$ that accounted for the effective mass and the $Q \cdot \Delta V_c$ that did not was less than 0.25%. Therefore, it was assumed that the entire mass of the bus was representative of the effective mass of the bus for the calculation of the stiffness coefficients.

Table 3: The calculation of the effective mass multiplier and the resulting $Q \cdot \Delta V_c$ as compared to the $Q \cdot \Delta V_c$ calculated with the entire mass of the bus included.

	Conasauga, TN	Sinton, TX
L (in)	448.55	420.68
W (in)	94.5	~94
k^2 (in^2)	17,512	15,484
h (in)	107	62
γ	0.6	0.8
$Q \cdot \Delta V_c$	879.1	298.2
$Q \cdot \Delta V_c$ old	878.4	298.9

Since some of the values for each collision were estimates and variations in weight and crush affected the slope and intercept of the linearly fitted data, the sensitivity of the A and B stiffness coefficients was determined. In addition, the A and B values for the train were unknown but assumed to be large to approximate an extremely stiff vehicle, as demonstrated by the small amount of damage to the train despite the severity of the collisions.

SENSITIVITY

The stiffness of the train and the bus affected the amount of deformation predicted by the Human Vehicle Environment (HVE)[4] simulation program[viii], the maximum acceleration experienced by the bus and train, and the angular acceleration of the bus. The initial stiffness coefficients for the bus in the Conasauga collision were set based on the calculations accounting for the standardized weight and crush (A=1132 lb/in and B=108 lb/in^2). The stiffness coefficients for the locomotive were initially set as the maximum default values in HVE[4] (A=1000 lb/in and B=1000 lb/in^2).

The B stiffness coefficients for the bus and the train were varied by 50% in an effort to determine the sensitivity of the coefficients. The simulation was run for 0.15 seconds so that only the initial impact was included and the output frequency of the data was at 1 msec. In addition, the B stiffness values for the bus were increased by 100% and 300% while the stiffness of the train was maintained at the maximum default values in HVE. The peak acceleration of the bus and the train and the peak angular acceleration of the train are shown in Table 4.

Although the variations linear and angular accelerations are approximately linear in relation to the stiffness values of the bus and train, variations in the amount of crush were also noted. A visual comparison of the predicted crush to the actual crush revealed that the calculated bus stiffness values (A=1132 lb/in and B=108 lb/in^2) resulted in similar crush values to the actual bus while the lower bus stiffness values resulted in too much crush and the higher stiffness values resulted in too little crush. (Limitations in the model were present in that the model did not predict the bending of the bus body and frame around the front of the train body. Only the deformation due to the intrusion of the train body into the bus body was predicted.) In addition, variations in pulse duration were noted between the different stiffness values.

Table 4: The peak acceleration of the bus and the train and the peak angular acceleration of the train while the stiffness of the bus and the train were varied.

School Bus	Locomotive	Peak Acceleration (G)	Peak Acceleration (G)	Angular Acceleration (deg/sec)
		School Bus	Locomotive	School Bus
A=1132, B=108	A=1000, B=1000	-26.7	-1.2	2501.3
A=1132, B=162	A=1000, B=1000	-32.8	-1.4	3111.3
A=1132, B=54	A=1000, B=1000	-23.3	-1.0	1864.9
A=1132, B=108	A=1000, B=500	-24.4	-1.1	2417.0
A=1132, B=162	A=1000, B=500	-29.4	-1.3	2880.2
A=1132, B=54	A=1000, B=500	-20.2	-0.9	1889.0
A=1132, B=216	A=1000, B=1000	-33.2	-1.5	3136.8
A=1132, B=432	A=1000, B=1000	-40.9	-1.9	3588.4

DISCUSSION

The end goal for modeling the side impact stiffness factors of the large school bus was to transfer the collision pulse into an occupant simulation model. Then modeling could be used to better understand occupant motion in large school buses during side impact collisions and to investigate potential injury reducing mechanisms inside the large school bus.

Recently, the National Highway Traffic Safety Administration conducted two full-scale crash tests of large school buses.[5] The first was a frontal impact of the school bus into a non-deformable barrier at 30 mph. The second was a side impact crash test where the bus was stationary and was impacted in the side by a weighted tractor cab at a speed of 45 mph. The damage profile in the side impact crash test was similar to the damage seen in the Conasauga collision. In addition, the velocities were comparable. (The train was traveling at 51 mph and the truck was traveling at 45 mph. In the NHTSA crash test, the bus was stationary. In the Conasauga collision, the bus had a forward velocity of approximately 14.9 mph at impact and little to no lateral velocity.) Therefore, the acceleration values measured during the NHTSA side impact crash test were used as reference values or guidelines for the Conasauga collision. (The main difference was that the NHTSA side impact was behind the first axle while in this collision, the train impacted the rear axle.)

[viii] EDSMAC4 is a two-dimensional simulation analysis of vehicle collisions based on the smac model originally developed by Calspan for NHTSA. Multiple vehicles, trailers, and/or barriers may be included in the analysis.

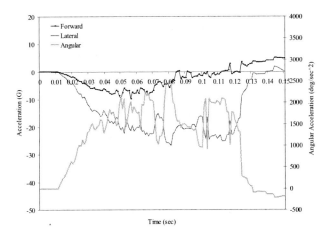

Figure 3: The linear and angular acceleration values for the original calculated bus stiffness values (A=1132 lb/in and B=108 lb/in^2) and the maximum default stiffness values for the train.

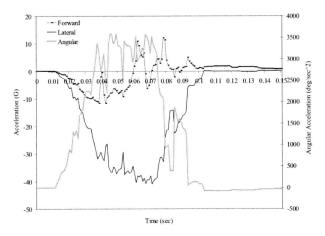

Figure 4: The linear and angular acceleration values for the bus stiffness values (A=1132 lb/in and B=432 lb/in^2) corresponding to the NHTSA side impact crash tests.

The preliminary results indicate that despite a peak side impact acceleration at the center of gravity of approximately 45 G[ix], significant injuries to the crash test dummies away from the area of impact were not predicted.[x] Unfortunately, the angular accelerations of the bus were not recorded during this crash test but the lateral accelerations away from the area of impact, especially at the rear of the bus, were low. In the Conasauga, TN collision, severe and fatal injuries were documented to passengers away from the area of impact in the rear of the vehicle. Thus, the bus design, bus motion or the occupant motion was significantly different between the two collisions.[xi]

Despite the match between the crush depth of the bus in the Conasauga, TN collision and the predicted crush

---

[ix] The impact occurred forward of the center of gravity.
[x] The school bus was a transit style bus with the engine at the rear of the vehicle. The instrumented dummy away from the area of impact was positioned near the rear of the vehicle. The increased weight of the bus at the rear may have reduced the accelerations experienced in that region.
[xi] The large school bus in the NHTSA crash tests was a 1995 Blue Bird transit style school bus (GVWR 36,200 lbs).

based on the calculated A and B stiffness factors, the actual accelerations experienced during the collision were unknown. To investigate the occupant motion and the predicted injuries to the occupants from the Conasauga, TN collision, the accelerations from both the original calculated stiffness values (A=1132 lb/in and B=108 lb/in^2) (see Figure 3) and the simulation comparable to the NHTSA side impact crash tests (A=1132 lb/in and B=432 lb/in^2) (Figure 4) were used to investigate the occupant kinematics in MADYMO. Preliminary results indicated that the predicted injuries based on the extremely stiff bus resulted in unrealtistically high injury values to the simulated occupants while the predicted injuries were more reasonable for the simulated occupants exposed to the bus with the original calculated stiffness values.

LIMITATIONS

The side impact crush coefficients were limited in several ways. To begin, the train was assumed to be a non-deformable impactor and the front was assumed to be flat so that the crush distribution across the side of the bus would be uniform. In addition, the train/school bus collisions were not separated into impacts into the axle and impacts into the bus body alone. Axle impacts will be stiffer than impacts into the body of the bus alone. Also, the bus body and chassis separated in several collisions. The crush measurements were based on the crush to the bus body. A point of common velocity between the bus and the train was assumed. This common velocity may not occur during impacts away from the bus's center of gravity.

In addition, the calculation assumes an engagement area covering the entire train width for the full height of the bus. Under-ride collisions or collisions where only a small vertical portion of the school bus side was impacted may not be well predicted with these stiffness factors. Furthermore, the stiffness of the impacting object must be known, such as the case with a train collision where the train is assumed to be non-deformable. These school bus side impact stiffness factors may be used to calculate the forces and accelerations during other accident scenarios within the limitations listed above.

CONCLUSION

The large school bus is one of the safest forms of transportation on the road today but improvements can be made to the occupant protection during side impact collisions. Side impact stiffness factors are essential to modeling the crush and accelerations during the collision sequence, thereby enabling more accurate occupant simulations.

Collisions between large school buses and trains were investigated to calculate side impact crush coefficients. These crush coefficients can be used for large school bus collisions when impacted on the side by another large vehicle or object of a known stiffness. Although all the included accidents were used to calculate the stiffness factors, these factors appear reliable based on the sensitivity analysis performed for the Conasauga accident.

Future work should focus on collecting accurate crush profiles, vehicle weights, and impact speeds so that the stiffness factors may be refined to better address axle impacts, impacts away from the center of gravity, and impacts to a portion of the school bus side.

REFERENCES

1. Federal Register Notice of October 26, 1998 (Volume 63, Number 206) on NHTSA's School Bus Research Plan.
2. School Buses Traffic Safety Facts, U.S. Department of Transportation, National Highway Traffic Safety Administration.
3. McHenry RR, McHenry BG, McHenry Accident Reconstruction, 1998 McHenry Seminar.
4. HVE Human Vehicle Environment, Version 3, Engineering Dynamics Corporation, Beaverton, OR 97008.
5. National Highway Traffic Safety Administration, Vehicle Research and Test Center, East Liberty, Ohio.

Reviewer Discussion

Reviewers Name: Ronald B. Heusser
Paper Title: Large School Bus Side Impact Stiffness Factors
Paper Number: 2002-01-0554

Reviewer's Discussion
By Ronald B. Heusser, Engineering Accident Analysis, EA[2]

SAE # 2002-01-0554
"Large School Bus Side Impact Stiffness Factors"
Kristin Bolte, Shane Lack, Larry Jackson, Authors.

The authors have discussed a method for calculating the side stiffness coefficients for large post standard school buses. They have then applied that methodology, using six accidents and one NHTSA crash test where most of the data was known, to arrive at A and B stiffness coefficients for the side of post standard school buses.

The data used is severely limited and would benefit from more accidents and / or tests added for further refinement. The methodology appears to be sound, but the limitations need to be kept in mind before using the actual calculated coefficients. Only collisions involving trains and heavy trucks into the side of post standard school buses would seem appropriate for use with these coefficients. Impacts with passenger cars or pickups would not fit the limitations since under-ride would most likely be involved. The more useful aspects of the paper would be the methodolgy and example of calculating stiffness coefficients for specific vehicles and types of collisions.

2002-01-0556

Crush Energy Considerations in Override/Underride Impacts

Micky C. Marine, Jeffrey L. Wirth and Terry M. Thomas
Thomas Engineering, Inc.

Copyright © 2002 Society of Automotive Engineers, Inc.

ABSTRACT

In automobile accident reconstruction it is often necessary to quantify the energy dissipated through plastic deformation of vehicle structures. For collisions involving the front structures of accident vehicles, data from Federal Motor Vehicle Safety Standard (FMVSS) 208 and New Car Assessment Program (NCAP) frontal barrier impact tests have been used to derive stiffness coefficients for use in crush energy calculations. These coefficients are commonly applied to the residual crush profile of the front bumper in real-world traffic accidents. This has been accepted as a reasonable approach, especially if there has been significant involvement of the front bumper and its supporting structures. For impacts where the structures above the bumper level are deformed more than the bumper itself, this approach may not be so readily applied. These types of impacts are called override/underride, and are encountered quite often in truck-to-car accidents where there is a vertical difference in bumper heights and also in accidents where bumper height mismatches are created through vehicle brake dive. In this paper we examine the crush-energy considerations of override/underride impacts. The limited available literature and test data are reviewed. Two test programs that involved impacts over a wide range of severities are analyzed in detail.

INTRODUCTION

In the reconstruction of automobile accidents it is often necessary to quantify the energy that has been dissipated through plastic structural deformation of the involved vehicles. For frontal collisions, the classical approach has involved measuring the residual bumper crush profile and comparing this information to available crush data from full-frontal barrier crash tests of a like or similar vehicle. The barrier test data are usually generated through 30 mph FMVSS 208 compliance testing or the 35 mph NCAP testing. In both of these test protocols a vehicle impacts a rigid, non-moving flat-surfaced barrier. From this test data crush stiffness coefficients are determined and subsequently used in calculations based on the subject vehicle bumper profile. These full-frontal barrier tests involve all of the leading surfaces of a vehicle and result in essentially a vertically uniform residual crush profile.

When attempting to reconstruct real-world accidents the situation where bumper height mismatches occur, either through geometrical differences or braking, is often encountered. Collisions of this type, known as override/underride impacts, typically result in one of the accident vehicles experiencing a greater amount of residual crush above the bumper level than that of the bumper itself. The classical approach of utilizing stiffness values derived from full-frontal barrier test data is not intuitively applicable in this situation. In fact, considering only the bumper profile in this situation will underestimate the dissipated crush energy. A further problem arises when trying to account for the crush energy of the deformed structures above the bumper. Necessary test data is generally not available to properly characterize the crush versus displacement behavior of these structures independent of the bumper.

Statistical data indicating the real-world frequency of override/underride impact occurrence are not well documented and, hence, are not readily available. Hobbs [1] in a paper presenting a rationale for frontal offset barrier testing conducted a small sample survey of 52 fatal accident files in which frontal structures were involved. It was found that in 14 of these accidents (27%) neither of the two main longitudinal structural members (upon which the bumper is often mounted) were involved (i.e. deformed).

Much attention has been focused on the aggressivity/compatibility of light trucks and vans (LTV's) when involved in collisions with passenger cars [2], [3]. A prominent factor discussed in these studies is the weight differences between LTV's and passenger cars, though relative structural stiffnesses and geometrical differences such as ride height are also noted. Much of the focus in the agressivity/compatibility studies have been on the side impact mode. A more recent study by Barbat, et al, [4] focused on frontal impacts. In this study it was found that geometrical (bumper/rail height) differences between impacting partners plays a significant role in the measured occupant response.

OVERRIDE/UNDERRIDE IN ACCIDENT RECONSTRUCTION

A situation frequently encountered by accident reconstructionists is one where an overridden vehicle will suffer some level of bumper deformation with an even greater amount of deformation to the structures above the bumper. An example of this type of vehicle damage pattern is shown in Figure 1.

Figure 1. Example of Override/Underride Damage.

It is seen in this photograph that there is extensive damage to the structures above the front bumper level. Measurements of this vehicle indicate that there is some deformation to the bumper as well. The residual crush profiles of the bumper and the upper radiator support are presented in Figure 2. For comparison, the residual crush profile from a 35 mph NCAP frontal-barrier impact test [5] for this vehicle is also included and is presented as a straight dotted line. In a full-frontal barrier impact the resulting residual crush profile is essentially vertically uniform. That is, the leading bumper and above-bumper components are displaced rearward, relative to the vehicle, to a common vertical plane.

The measured crush profile in this example indicates that there is some level of residual bumper crush, though well below the 35 mph barrier test level. More significant residual crush exists at the upper radiator support which is generally greater than that of the 35 mph barrier test. It is clear then that applying stiffness coefficients derived from full-frontal barrier test data to the residual crush profile of the bumper of this vehicle will underestimate the actual crush energy. Other means then become necessary to accurately perform an energy analysis in the reconstruction of this type of an accident.

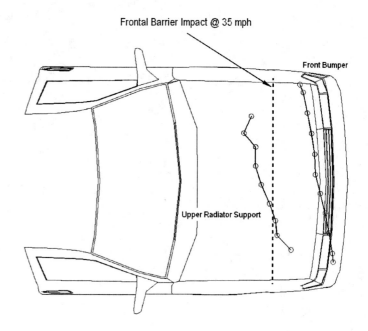

Figure 2. Override/Underride Crush Profile.

In this paper we will examine the considerations in quantifying the crush energy of override/underride crush profile patterns. We will specifically address the overridden vehicle and concentrate primarily on frontal impacts, though the discussion is also relevant for rear impacts. The extremely limited available literature and test data are reviewed. Additionally, two test programs that involved a significant range impact severities are analyzed in detail.

CRUSH ENERGY REVIEW

There are many reports and papers in the technical literature that discuss the topic of automobile crush energy. In this section we will provide a brief review of the common crush energy approaches employed in accident reconstruction. For more thorough treatments on this topic the reader is referred to references [6] – [12].

The classical approach to quantifying automobile crush energy is based on the observation that in full-frontal barrier impacts the relationship between impact velocity and the residual crush can be approximated as linear. This observation was discussed by Campbell [6] who further put forth the notion of the linear-plastic force versus residual crush relationship. These two relationships are presented in Figures 3 and 4 and are expressed as follows,

$$V_I = b_0 + b_1 c \qquad (1)$$

and

$$f = A + Bc \qquad (2)$$

Where c = residual crush

b₀ = speed at which no residual crush occurs

b₁ = slope of V_I vs. c line

A = force at which no residual crush occurs

B = slope of f vs. c line

$$E_c = \left(Ac + \frac{1}{2}Bc^2 + \frac{A^2}{2B}\right)L \quad (3)$$

This expression is then related to the initial kinetic energy of the test vehicle, through the use of equation (1). The following relationships for the constants A and B are found to be,

$$A = \frac{Wb_0b_1}{gL} \quad (4)$$

$$B = \frac{Wb_1^2}{gL} \quad (5)$$

Therefore, the crush energy can be related to the linear relationship between impact speed and residual crush through the constants b_0 and b_1. It is also noted that this approach neglects the effects of restitution as it is based only on the impact speed of the vehicle.

McHenry [7] employed this approach in the Calspan Reconstruction of Accident Speeds on the Highway (CRASH) computer program. This program, with subsequent revisions, has become a popular analysis tool even though it was initially developed as a "simple" pre-processor for arriving at approximate initial velocities as input into the Simulation Model of Automobile Collisions (SMAC) computer program [8].

Strother, et al, [9] presented an alternate approach to arriving at the crush stiffness coefficients. Observing that equation (3) is quadratic in residual crush and rearranging, the following linear expression was developed,

$$\sqrt{\frac{2E_c}{L}} = \sqrt{B}c + \frac{A}{\sqrt{B}} \quad (6)$$

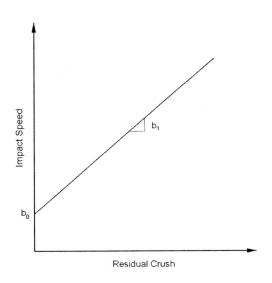

Figure 3. Impact Speed vs. Residual Crush

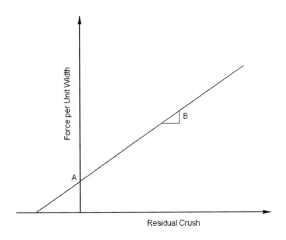

Figure 4. Force per Unit Width vs. Residual Crush

With Campbell's model, the force per unit width expression can be double integrated over residual crush and crush length (L) to arrive at the following expression for the crush energy (E_c).

If sufficient barrier test data exists for a given vehicle, the left hand side of equation (6) can be plotted against residual crush, and if the assumption of linearity is reasonable, the coefficients A and B can be found from the slope and intercept of a line fitted to the data. The left hand side of equation (6) has been called the Energy of Crush Factor (ECF) [10]. The authors of [9] presented crash test data demonstrating this approach. They also noted that a linear model does not always characterize the data over a wide range of crush levels. It was suggested that other approaches such as a constant force model or a saturation force model might be more appropriate depending on the observed vehicle crush behavior.

Fonda [11] also noted that the linear models did not necessarily characterize the crush energy versus residual crush behavior and suggested non-linear models as alternatives. He presented the use of polynomial curve fits to the impact velocity versus residual crush data while maintaining the linear force versus residual crush relationship. More recently, Woolley [12] introduced a power-law formulation for modeling the force versus crush relationship.

To develop the above non-linear models, crash test data encompassing a range of residual crush levels must be available. For many automobiles, the only crash test data available is from FMVSS 208 or NCAP testing. Because of the limited available test data, the linear approach remains the most common and most practical method of approximating the crush behavior of passenger vehicles.

An underlying assumption in the previously described crush energy approaches is that the vehicle structure is considered to be homogenous and isotropic. In the context of frontal impacts this means the entire front of the vehicle is treated as having the same force-deflection characteristics (e.g. the same stiffness) at the centerline as at locations further outboard. Furthermore, in most previous discussions the residual crush profiles have been treated as vertically uniform. Conceptual representations of damaged vehicles have been shown in plan view with the implication of vertical crush uniformity. In fact, Campbell in Reference [6] stated "This model also assumes that the damage is uniform vertically. Damage produced by underride or override collisions is not covered."

The term "residual crush" has been used quite often in discussions regarding automobile damage energy. However, rarely has it been suggested in the literature where on a damaged vehicle the residual crush profile should be measured. Tumbas and Smith [13] in one of the few papers addressing this topic discussed it in detail. They suggest that for frontal and rear vehicle damage the residual crush measurements should be made at the frame level. Measurements made at the frame level are thought to be associated with the major load bearing members of a vehicle. In many vehicles the front and rear bumpers provide a convenient measurement surface at the frame level. Tumbas and Smith also discussed the situation of vertically non-uniform crush profiles. The recommended measurement convention for underride crush profiles is that "if at any station measurement of depth of crush at other than the bumper-frame level exceeds that at the bumper-frame by 5 or more inches, the crush measurement at the frame level should be averaged with the greater measurement." This convention is suggested even if there is no residual crush at the bumper-frame level, at a given measurement station. No mention was made, however, regarding appropriate measurement surfaces at above-bumper levels.

The authors presented no data to support the above override/underride measurement protocol, rather, with the lack of research in the area of override/underride impacts, it was presented with the belief that the impact forces are related to some "average" of the two crush measurements. Additionally, it was mentioned that the protocol is invalid in the case of "severe underride". Just what constitutes a severe underride condition is left to engineering judgement. An impact in which no direct bumper contact is made with all of the vehicle crush occurring above the bumper-frame level, was considered an example of severe underride.

AVAILABLE CRASH TEST DATA

While there have been hundreds of full-frontal barrier tests conducted over the last several decades, there have been relatively few test programs in which override/underride impact conditions were studied. Furthermore, there have been even fewer programs where useful data has been gathered and presented. In this section we will review some of the limited available crash test data.

Tanner, et al, [14] presented findings related to low-speed, full-scale vehicle crash testing with override/underride conditions. No discussion of crush energy was included and sufficient vehicle residual crush measurements were not provided to independently analyze the vehicle structural behavior. Another report documented a high-speed rear impact test that was conducted for a Society of Automotive Engineers TOPTEC in October 1997 [15]. In this test a bob-tailed heavy truck (i.e., not pulling a trailer) impacted the rear of a 1984 Renault Alliance passenger car. The bumper alignment was such that an override/underride condition existed. Again, crush energy dissipation was not discussed in the test documentation, nor were crush measurements provided for independent analysis.

Goodwin, et al, [16] discussed a test program in which several vehicles of different makes were subjected to both front and rear bumper override barrier impacts. Residual crush measurements were taken after each test, however all of the tests were conducted at impacts speeds of less than 6 mph. In these tests it was observed that grill fascia and headlight assembly deformation occurred at impact velocities of approximately 3 mph.

Woolley, et al, [17] conducted an offset, override crash test on a 1992 Isuzu Rodeo using their Massive-Moving Barrier (MMB) approach. In this testing a rigid barrier was mounted to the front of a large, heavy truck. The barrier face was aligned to impact the Rodeo just above the bumper with a 20% offset overlap on the driver side. The MMB was driven into the stationary Rodeo at a speed of approximately 23 mph. The residual crush data presented by the authors seems to indicate that the crush stiffness for this impact mode is roughly 50% of the full-frontal barrier testing for this vehicle, based on a reported damage length of 12.6 inches. The stated purpose of this test was to replicate the vehicle damage for accident reconstruction purposes and to provide acceleration data for subsequent sled testing. This testing was conducted for a set of circumstances related to a specific accident and not as a part of general study of override/underride impacts.

Croteau, et al, [18], conducted crash tests where a stationary 4-door sedan was struck in the rear by a Class 8 heavy truck travelling at approximately 26 mph. There was significant override damage of the target vehicle, with some plastic deformation occurring at the bumper level as well. In determining the energy dissipated by the deformation of the target vehicle the authors suggested an approach where the total crush energy is defined as follows,

$$DE_T = f_1 DE_1 + f_2 DE_2 \quad (7)$$

Where f_1 and f_2 are arbitrary constants applied to the damage energy based on the bumper level crush (DE_1) and the damage energy based on the above-bumper level crush (DE_2) and used to solve equation (7) for a known total damage energy (DE_T). The damage energies DE_1 and DE_2 were both calculated using the A and B stiffness coefficients derived from a rigid, flat-faced moving barrier test. Two rear override crash tests were presented, one was a full-width impact and the second was an offset impact. In deriving a solution for equation (7) for both of these tests, a value of 1.0 was used for f_1. The value of f_2 for the full-width override test was determined to be 0.46 and for the offset override test f_2 was found to be 0.56. This implies that, for this vehicle in the rear impact mode, the above-bumper A and B coefficients are approximately 50 percent of those determined using bumper crush data in a rigid, flat-faced moving barrier test. The authors of this paper acknowledge that the above solution is not unique. Either f_1 or f_2 can be arbitrarily chosen and the other one can then be solved for algebraically. However, intuitively one would think that using 100 percent of the calculated bumper crush energy is a proper approach.

Presently, we are aware of only two test programs in which sufficient data is provided to gain insight into the frontal crush behavior of override/underride impacts over a wide range of crush levels. One of these programs was sponsored by NHTSA and was related to the development of heavy truck rear underride guard protection structures [19], [20]. The second test program was conducted by Synectics Road Safety Corporation as part of an SAE Accident Reconstruction Committee meeting [21].

NHTSA Testing

In the NHTSA testing a 1990 Ford Taurus 4-Door and a 1993 Honda Civic 3-Door Hatchback were subjected to repeated impacts into a rigid truck underride guard mock-up mounted to a stationary barrier. The underride mock-up was set-up just above the bumper level of the test vehicle such that there was no direct contact with the bumper in any of the tests. Crush measurements were taken after each impact at the bumper level and the underride guard mock-up level. From the test reports, it appears that the underride guard mock-up level measurements were taken at the leading edge of the engine hood.

A summary of the tests are provided in Tables 1 and 2 for the Ford Taurus and Honda Civic, respectively. Included in the tables are the impact speeds, the equivalent uniform crush at the underride guard level, the cumulative crush energy (E_c), and the cumulative ECF. The residual crush values included in Tables 1 and 2 are not based on the reported C1-C6 measurements, as these measurements were apparently not made to the same points on the vehicle from test to test. The values indicated in Tables 1 and 2 are based on the residual displacement of discrete points on the vehicle that were measured after each test. The discrete points used to generate the data in Tables 1 and 2 were 31 to 49 and 34 to 50 for the Taurus and Civic, respectively. The profiles defined by these points for each test are shown in Figures 5 and 6. The above-bumper residual crush values used in the tables and figures referred to above are based on the difference in position between the post-test damage profile and the undamaged position of the above-bumper structures.

While the direct contact damage in these tests occurred at the above-bumper level, there was induced bumper deformation in the last test (highest cumulative energy) of each series. For the Taurus there was little or no reported induced bumper deformation in the first three tests. In the Civic testing there was no reported induced bumper deformation in the first two tests. In the third test the bumper cover was displaced and the bumper measurements were not subsequently reported.

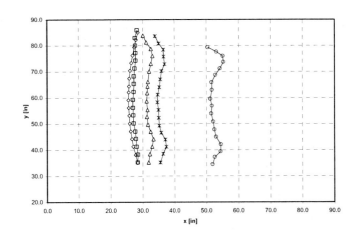

Figure 5. 1990 Ford Taurus Above-Bumper Profile

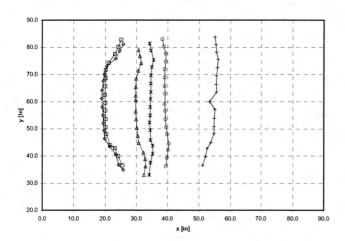

Figure 6. 1993 Honda Civic Above Bumper Profile.

Table 1. 1990 Ford Taurus Test Summary

Test No.	V_i [mph]	C_{ave} [in.]	E_c [ft-lb]	ECF [$lb^{1/2}$]
1	10.0	0.9	10918	66.7
2	15.0	5.2	35651	120.1
3	15.0	8.8	60011	156.8
4	34.7	25.4	191312	279.6

Table 2. 1993 Honda Civic Test Summary

Test No.	V_i [mph]	C_{ave} [in.]	E_c [ft-lb]	ECF [$lb^{1/2}$]
1	4.7	0.4	1724	28.0
2	14.9	10.3	18722	93.0
3	14.9	14.4	35161	128.6
4	19.9	19.5	65287	174.9
5	35.0	35.8	159974	272.1

The ECF versus residual crush plot is useful for comparing crush energy data from tests with differing test vehicle weights. As long as the crush energy and residual crush levels are known for a given test, this method can be employed for similar test types. The ECF is a function of both the crush energy and the damage length (L). Thus, there is some engineering, judgement to be made in determining the value of L in calculating the ECF. In discussing full-frontal barrier testing, Strother, et al, [22] suggest a characteristic damage length defined by the front track width of the vehicle plus six inches. Because these tests involved the full width of the vehicle at the above-bumper level, the Strother protocol was used to defined L. This resulted in damage lengths of 68 inches and 64 inches for the Taurus and Civic, respectively.

Plots of the crush energy and ECF versus residual crush for both vehicles are shown in Figures 7 through 10. Included in the plots of Figures 8 and 10 are least-squares linear regression lines for the repeated test data. Also included for comparison are data from FMVSS 208 and NCAP frontal barrier tests for both of these vehicles [23] – [27]. The same crush width values used in the above-bumper ECF calculations were used in the FMVSS and NCAP ECF calculations. In the plots for the Taurus, repeated full-frontal barrier testing for a 1986 model year vehicle are also included [28]. Aside from cosmetic differences, the 1986 model year vehicle is essentially the same as the 1990 model year. The crush values from the full-frontal barrier tests shown in these plots have not been corrected for any "air-gap" (separation between the bumper cover and the load bearing structure behind it) that may exist post-test.

From Figures 8 and 10 it is seen that the ECF versus residual crush behavior in the repeated underride guard testing for both of these vehicles can be reasonable approximated as linear. The least-squares linear regression of the repeated test data yielded the following expressions,

1990 Ford Taurus:

$$ECF = 66.5 + 7.8c \qquad [(lbf)^{1/2}] \qquad (8)$$

1993 Honda Civic:

$$ECF = 24.1 + 6.3c \qquad [(lbf)^{1/2}] \qquad (9)$$

Based on these ECF expressions, the crush stiffness coefficients for these vehicles are as follows,

1990 Ford Taurus:

$$A = 519 \qquad [lb/in]$$
$$B = 61 \qquad [lb/in^2]$$

1993 Honda Civic:

$$A = 152 \qquad [lb/in]$$
$$B = 40 \qquad [lb/in^2]$$

Figures 7 and 8 show that the full-frontal barrier test data for the Taurus is similar to the underride guard impact data, though there are some nuances to the above-bumper data that will be discussed later.

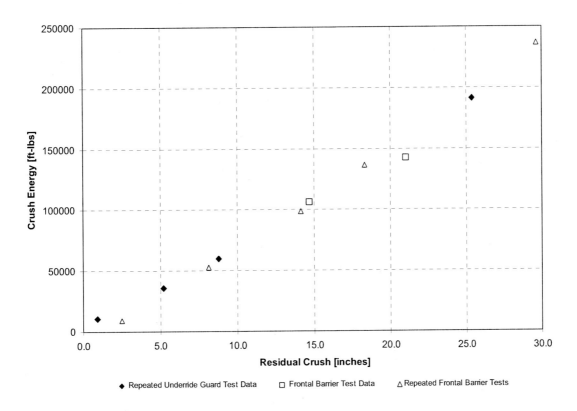

Figure 7. 1990 Ford Taurus Crush Energy vs. Residual Crush Data.

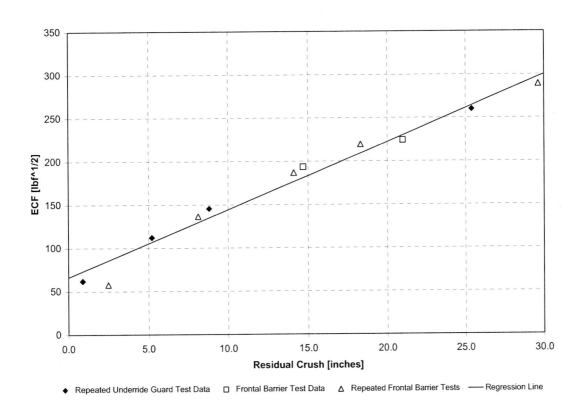

Figure 8. 1990 Ford Taurus ECF vs. Residual Crush Data.

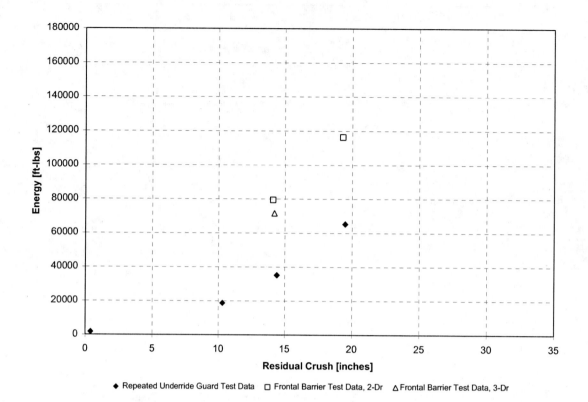

Figure 9. 1993 Honda Civic Crush Energy vs. Residual Crush Data.

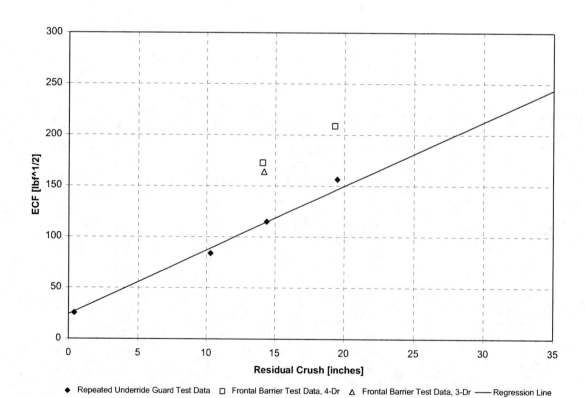

Figure 10. 1993 Honda Civic ECF vs. Residual Crush Data.

The Taurus test weight for the FMVSS 208 and NCAP barrier tests were 3700 lbs. and 3620 lbs., respectively. The test weight reported in the underride guard testing was 3331 lbs.

In the crush energy versus residual crush plot for the Honda Civic, shown in Figure 9, it can be seen that the full-frontal barrier test data is not at all consistent with the repeated underride guard data. Even when taking into account the difference in vehicle weights between the various tests, the same observation can made in the ECF plot of Figure 10. It is clear from Figures 9 and 10 that the above-bumper crush behavior of the Civic is significantly different than the full-frontal barrier test data. For this vehicle it appears that, in the full-frontal FMVSS and NCAP barrier tests, more impact energy is required to attain similar residual crush levels to those achieved in the underride guard testing.

The test weights for the Civic 2-Door FMVSS 208 and NCAP test were 2780 lbs. and 2919 lbs., respectively. The test weight for the Civic 3-Door FMVSS 208 test was 2470 lbs., and the test weight reported in the underride guard testing was 2387lbs.

Synectics Testing

In this testing, a series of nine tests were conducted as part of a mid-year meeting of the SAE Accident Reconstruction Committee. The test vehicles were subjected to impacts with a rigid truck bumper mock-up mounted on a fixed barrier. There were six tests in which the bumper mock-up was set at a height just above the test vehicle bumper. In the other tests the bumper mock-up was installed at a height in which initial contact with the test vehicle occurred at the A-pillar. In this section we will discuss the test results of the six tests which were set-up to simulate a bumper override/underride condition. The other crash tests are beyond the scope of this discussion.

In the documentation for this crash test series, the tests were identified by sequential numbering. The tests involving bumper override were identified as test numbers 1 through 5 and test number 8. In each of these tests a 1987 Plymouth Reliant was used as the test vehicle. Test 1 of this series will not be discussed as the bumper mock-up deformed significantly as a result of the impact. Without sufficient information to quantify the energy dissipated in deforming the mock-up, extracting the corresponding vehicle crush energy is not possible. Tests 2 through 5 were repeated testing of the same vehicle (test weight 2730 lbs.), while test 8 was a single impact of a previously untested vehicle (test weight 2910 lbs.). In each of these tests the bumper mock-up was arranged such that there was no direct contact of the mock-up with the test vehicle bumper. Residual crush measurements were made of the test vehicles after each test, thereby providing information for a crush energy analysis. In tests 2 through 4 and test 8, there was very little induced damage at the bumper level. In test 5 the average induced displacement at the bumper level was approximately 3 inches.

A data summary of the relevant crash tests is presented in Table 3. Similar to Tables 1 and 2, the above-bumper level residual crush data is included. As noted in the test documentation, measurements were taken at locations on the grill fascia and the upper radiator support member. In the following analysis, we use the crush measurements of the upper radiator support as it was felt that this is a more substantial structural component. In test 2 no upper radiator support measurements were reported. However, from the photographs in the published test report it appears that very little damage occurred to the cosmetic components of the grill fascia. Therefore, we assigned no residual crush to the upper radiator support.

In the available documentation for these tests the rebound velocities were not reported. Therefore, plots of the approach energy and energy of approach factor (EAF) were created. These plots are shown in Figures 11 and 12. Again, the Strother protocol was used in determining the damage width (64 inches). As with the Ford Taurus and Honda Civic data, a least-squares linear regression line is also included in Figure 12. Though there was no observable residual crush resulting from the impact in Test 2, the data was included in the linear regression analysis. From the photographs included in the test documentation it was decided that this impact was conducted at or near the residual crush threshold speed. Data from an NCAP 35 mph full frontal barrier crash test [29] was included in Figure 12 for comparison. The test weight for this vehicle was 3060 lbs. As with the Taurus and Civic full-frontal barrier test data, the residual crush values have not been corrected for air-gap.

Table 3. 1987 Plymouth Reliant Test Summary

Test No.	V_i [mph]	C_{ave} [in.]	E_A [ft-lb]	EAF [lb$^{1/2}$]
2	4.1	0.0	1534	24.0
3	10.0	2.3	10661	63.2
4	13.8	9.5	28042	102.5
5	25.0	18.6	85083	178.6
8	18.7	9.6	34019	112.9

A review of Figure 12 reveals that the EAF versus residual crush behavior is approximately linear. The least-squares linear regression of this data resulted in

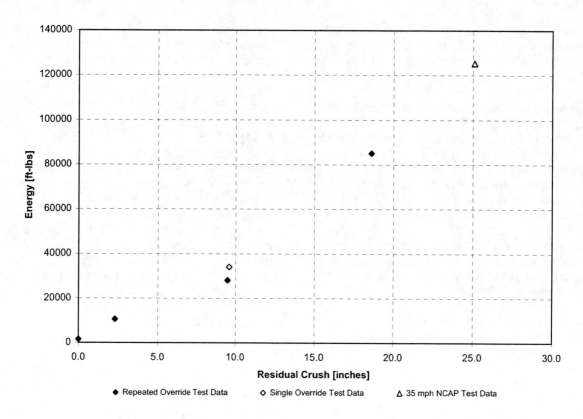

Figure 11. 1987 Plymouth Reliant Approach Energy vs. Residual Crush Data.

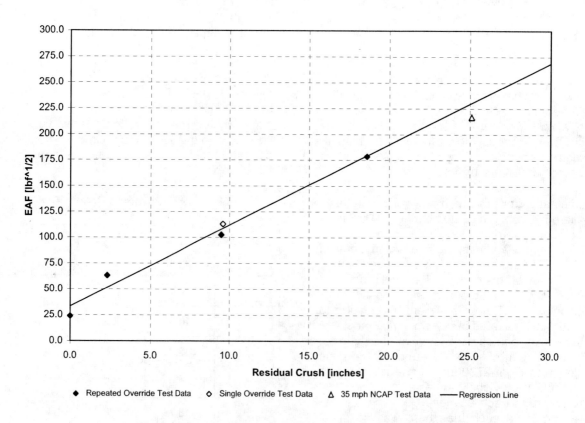

Figure 12. 1987 Plymouth Reliant EAF vs. Crush Data.

the following expression,

$$EAF = 33.5 + 7.8c \quad [(lbf)^{1/2}] \quad (10)$$

Using the EAF expression, the crush stiffness coefficients for this data are found to be,

$A = 263$ [lb/in]

$B = 61$ [lb/in^2]

From Figures 11 and 12 it is observed that the NCAP full-frontal barrier test data is quite consistent with the above-bumper behavior.

DISCUSSION

It is necessary, at this point, to discuss the fact that not all of the discrete points measured at the above-bumper level on the Ford Taurus reflect the residual crush as a result of direct contact with the underride guard mock-up. The measurements documented in the test reports and used above were measured at the leading edge of the hood. A schematic of the barrier used in these tests is shown in Figure 13. It is apparent from the photographs included in the test reports that the edge of the hood passed over the crossbeam and impacted the vertical supports. Thus, a portion of the hood directly impacted the uprights of the underride guard mockup. This can be seen in the damage profiles presented in Figure 5. Measurements at locations on the vehicle that were directly impacted by the crossbeam were not reported. The crush measurements reported along the leading edge of the hood underestimate the residual crush at the crossbeam level. A review of the reported crush measurements and the shape of the underride guard mockup suggest that the equivalent uniform crush calculations may underestimate the direct impact residual crush by approximately three inches in the most severe impact. An inspection of Figures 7 and 8 show that even if the crush values for all the above-bumper impact data points were increased by three inches, the data would still be quite similar to the full-frontal barrier test data. In the Honda Civic and Plymouth Reliant testing the reported crush measurements were in the region that was directly contacted by the underride guard mockup crossbeam.

From the data presented above it is clear that the above-bumper residual crush behavior is reasonably approximated as linear for all three vehicles. The data further indicates that the full-frontal barrier test data for the 1990 Ford Taurus and 1987 Plymouth Reliant are consistent with their corresponding above-bumper data. That is, the energy required to achieve a given level of residual crush is similar for both the full-frontal and above-bumper impact modes. This is true for the full range of crush levels shown for the Taurus and of the crush level of the single full-frontal barrier test for the Reliant. The same cannot be said about the 1993 Honda Civic test data, however. Higher levels of energy are required in the full-frontal impact mode to achieve similar residual crush as in above-bumper impacts.

We should note that in full-frontal barrier impacts, the above-bumper structures are displaced as well. However, the total above-bumper displacement from its original position on the vehicle is approximately equivalent to the residual displacement of the bumper minus the initial distance between the leading edge of the bumper and the above-bumper measurement reference. Thus, for a full-frontal barrier impact, the above-bumper residual displacement will always be less than that of the bumper.

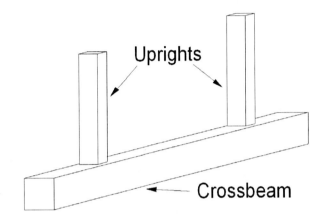

Figure 13. Underride Guard Barrier Mockup.

The above-bumper crush data from the Taurus and Reliant testing is unexpected. Intuitively, one might expect the above-bumper crush behavior of a vehicle to be less stiff than that of the full-frontal impacts with an ECF versus crush plot as in Figure 14. In this figure the parameter δ is defined as shown in Figure 15, and defines a vertically offset non-deformable barrier. A δ value of infinity represents an above-bumper-only impact, such as the previously discussed NHTSA and Synectics testing, and a value of zero represents a flat, full-frontal barrier impact where the bumper is engaged first. If we were to conduct a series of above-bumper only (δ = infinity) crash tests on a hypothetical vehicle, assuming linearity, we might define a line on the ECF vs. residual crush plot as shown in Figure 14. If δ is decreased to some value Γ for another series of tests, and the above-bumper crush data is plotted, one might expect the lower severity crush data to follow the δ = infinity line until the bumper becomes engaged. Once the bumper becomes engaged, the slope of the ECF vs. above-bumper crush line might be expected to increase, as shown.

The data for the Taurus and Reliant do not indicate this trend at all. The data seems to indicate that the δ = infinity slope and the δ = 0 slopes are approximately the same. The specific energy dissipation phenomenon that produced these results remains unclear. It may well be that in these tests, for these vehicles, the vehicle-barrier friction forces and structural failure modes, such as shearing, are significant contributions to the dissipation of the impact energy. The manner in which the engines are engaged and loaded, and other potential structural configuration considerations, for these particular vehicles may also contribute to the observed results. These mechanisms have not been fully examined at this time.

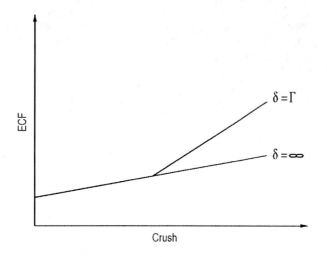

Figure 14. Above-Bumper Crush Behavior.

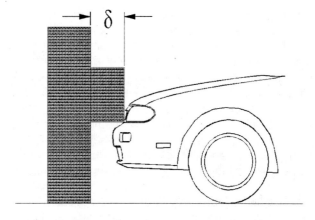

Figure 15. Vertically Offset Rigid Barrier.

In performing a crush energy analysis the engineer will usually have access to either FMVSS 208 or NCAP barrier test data only (for some vehicles data from both tests are available) to assist with the analysis. For override/underride impact energy analyses, the question arises as to if and how the barrier test data can be used to estimate the crush energy. The data presented here suggests that without above-bumper test data for other vehicles, a reconstructionist will not know, *a priori*, if full-frontal barrier test data can be reliably used to model the above-bumper behavior. For vehicles other than those used in the NHTSA and Synectics above-bumper testing, further testing is required to characterize the above-bumper crush behavior. This poses a practical problem since there are no comprehensive test programs being conducted to generate the necessary data.

Furthermore, if we were to average the bumper residual crush with the above-bumper residual crush from the above tests and use FMVSS 208 barrier stiffness data based on the commonly used b_0 value of 5 mph, the crush energy would be significantly underestimated for the Taurus and Reliant and slightly underestimated for the Civic. As mentioned previously, Tumbas and Smith suggested the use of bumper/above-bumper crush averaging except in severe override/underride impacts. This data demonstrates the potential for error when using bumper/above-bumper crush averaging in severe override/underride conditions.

Based on the data presented above, an accident reconstructionist may choose to use a range of stiffness values for the above-bumper structures to be from 50 percent to 100 percent of those found from full-frontal barrier data for the same vehicle. The crush energy may then be found using the approach described in [18] with an f_1 value of 1.0. The effects of this choice on the final analysis may be significant and engineering judgement must be used in considering the results within the context of the particular circumstances at hand.

CONCLUSIONS

The currently available override crash test data has been reviewed. For the three different passenger cars that were tested over a range of impact severity, the above-bumper ECF vs. residual crush data was reasonably approximated as linear. The NHTSA and Synectics testing consisted of impacts involving the full width of the vehicle at the underride guard mockup level. Therefore, it was decided to use a characteristic damage width defined as the track width plus six inches. For impact modes not involving the entire frontal width, other damage width definitions become necessary to compare to the ECF data presented in Figures, 8, 10, and 12. The choice of using the damage width associated with the directly impacted surfaces (direct damage) or a damage width that includes surfaces not directly impacted but have still been displaced (induced damage) requires engineering judgement by the analyst. The linear crush model was used for comparison to full-frontal test data. It is acknowledged that non-linear models could be applied as well.

Upon comparing the above-bumper-only data to the full-frontal barrier impact data (FMVSS 208 and/or NCAP) for the respective vehicles, it was found that the above-bumper data was quite similar to the full-frontal data over the available range of Ford Taurus data and the single

Plymouth Reliant data point. This was an unexpected and counter-intuitive result. The specific mechanisms behind these results remain unclear. Without full-frontal testing over a wider range of impact severities for the Reliant, it remains to be seen how favorably the data would compare to the above-bumper data. More expected results were observed for the Honda Civic where the above-bumper data was not consistent with the full-frontal data. Less energy was required in the above-bumper configuration to achieve the same level of residual crush as that of the full-frontal barrier impacts. The above-bumper data available to date do not indicate a practical correlation for an accident reconstructionist to use when confronted with quantifying crush energy in an override/underride situation with only full-frontal impact data available for a given vehicle.

The current state of override/underride impact data is very limited. The only comprehensive test programs available involved impacts into barriers simulating underride guard structures that are fabricated on the rear ends of heavy truck trailers. Thus, they involved only severe override/underride situations. The above-bumper data presented thus far is not sufficient to assist us in defining a general approach (if one is even possible) to estimate the crush energy associated with override/underride residual damage profiles. The test programs described above provide interesting and unexpected data for above-bumper only impacts. Further research of this impact mode is needed to more fully understand the force-deflection behavior and the mechanisms through which energy is dissipated. Ultimately, further research will also be needed in the area of combined crush impact modes, where both the bumper and above-bumper structures are directly involved. As mentioned previously, the combined crush situation often occurs with passenger vehicles as impacting partners and is encountered frequently in accident reconstruction analyses.

As a practical matter, when measuring above-bumper crush profiles it is recommended that the upper radiator support be used as the reference surface, if available, because it is the most significant frontal structure in the above-bumper region. If the upper radiator support is not available, then the leading edge of the engine hood may be an acceptable alternative. For instance, it was noticed in the Honda Civic testing that the engine hood latch did not appear to separate and was directly impacted by the underride guard barrier. In situations like this the upper radiator support may not be available for measurement as the hood latch may be inoperable due to the crush damage. The opposite situation occurred in the Plymouth Reliant testing where the hood latch separated and, as a result, the hood was not impacted by the underride guard barrier.

CONTACT

If you wish to contact the authors, they may be reached via e-mail at, teinc-az@qwest.net

REFERENCES

1. Hobbs, C., "The Rationale and Development of the Offset Deformable Frontal Impact Test Procedure", Society of Automotive Engineers, Paper No. 950501, 1995.

2. Gabler, H. C., and Hollowell, W. T., "The Aggressivity of Light Trucks and Vans in Traffic Crashes", Society of Automotive Engineers, Paper No. 980908, 1998.

3. Lund, A. K., O'Neill, B., Nolan, J. M., and Chapline, J. F., "Crash Compatibility Issue in Perspective," Society of Automotive Engineers, Paper No. 2000-01-1378, 2000.

4. Barbat, S., Xiawei, L., and Prisad, P., "Evaluation of Vehicle Compatibility in Various Frontal Impact Configurations," 17th International Technical Conference on the Enhanced Safety of Vehicles, Amsterdam, The Netherlands, 2001.

5. "NCAP Frontal Barrier Impact Test – 1985 Ford Tempo 4-Door Sedan", U.S. Department of Transportation, NHTSA No. CF0202, May 16, 1985.

6. Campbell, K., "Energy Basis for Collision Severity", Society of Automotive Engineers, Paper No. 740565, 1974.

7. McHenry, R., "Users Manual for the CRASH Computer Program", CALSPAN Report ZQ-5708-V-3, January 1976.

8. McHenry, R., and McHenry, B., "A Revised Damage Analysis Procedure for the CRASH Computer Program", Society of Automotive Engineers, Paper No. 861894, 1986.

9. Strother, C., Woolley, R., James, M., and Warner, C., "Crush Energy in Accident Reconstruction", Society of Automotive Engineers, Paper No. 860371, 1986.

10. Kerkhoff, J., Husher, S., Varat, M., Busenga, A., and Hamilton, K., "An Investigation into Vehicle Frontal Impact Stiffness, BEV and Repeated Testing for Reconstruction", Society of Automotive Engineers, Paper No. 930899, 1993.

11. Fonda, A., "Crush Energy Formulations and Single-Event Reconstruction", Society of Automotive Engineers, Paper No. 900099, 1990.

12. Woolley, R., "Non-Linear Damage Analysis in Accident Reconstruction", Society of Automotive Engineers, Paper No. 2001-01-0504, 2001.

13. Tumbas, N., and Smith, R., "Measurement Protocol for Quantifying Vehicle Damage From an Energy Basis Point of View", Society of Automotive Engineers, Paper No. 880072, 1988.

14. Tanner, C., Chen, H., Wiechel, J., Brown, D., and Guenther, D., "Vehicle and Occupant Responses in Heavy Truck to Car Low-Speed Rear Impacts", Society of Automotive Engineers, Paper No. 970120, 1997.

15. "High Speed Rear Impact Crash Test – Basic Test Notebook," Society of Automotive Engineers, 1997.

16. Goodwin, V., Martin, D., Sackett, R., Schaefer, G., Olson, D., and Tencer, A., "Vehicle and Occupant Response in Low Speed Car to Barrier Override Impacts", Society of Automotive Engineers, Paper No. 1999-01-0442, 1999.

17. Woolley, R., Asay, A., Jewkes, D., and Monson, C., "Crash Testing with a Massive Moving Barrier as an Accident Reconstruction Tool", Society of Automotive Engineers, Paper No. 2000-01-0604, 2000.

18. Croteau, J., Werner, S., Habberstad, J., and Golliher, J., "Determining Closing Speed in Rear Impact Collisions with Offset and Override," Society of Automotive Engineers, Paper No. 2001-01-1170, 2001.

19. "Final Report of a 1990 Ford Taurus into Heavy Truck Rigid Rear Underride Guard in Support of CRASH3 Damage Algorithm Reformation," U.S. Department of Transportation, National Highway Traffic Safety Administration, DOT HS 808 231, June 1993.

20. "Final Report of a 1993 Honda Civic DX into Heavy Truck Rigid Rear Underride Guard in Support of CRASH3 Damage Algorithm Reformation," U.S. Department of Transportation, National Highway Traffic Safety Administration, DOT HS 808 228, June 1993.

21. "Bumper Underride Crash Test Series," Synectics Road Safety Research Corporation, March 26, 1997.

22. Strother, C., Woolley, R., and James, M., "A Comparison Between NHTSA Crash Test Data and CRASH3 Frontal Stiffness Coefficients," Society of Automotive Engineers, Paper No. 900101, 1990.

23. "Vehicle Safety Compliance Testing for Occupant Crash Protection, Windshield Mounting, Windshield Zone Intrusion (Partial) and Fuel System Integrity – 1990 Ford Taurus 4-Door Sedan", U.S. Department of Transportation, National Highway Traffic Safety Administration, 208-CAL-90-04, January 1990.

24. "New Car Assessment Program Frontal Barrier Impact Test – 1990 Ford Taurus 4-Door Sedan", U.S. Department of Transportation, National Highway Traffic Safety Administration, CAL-90-N06, January 1990.

25. "Vehicle Safety Compliance Testing for Occupant Crash Protection, Windshield Mounting, Windshield Zone Intrusion (Partial) and Fuel System Integrity – 1993 Honda Civic 2-Door Coupe", U.S. Department of Transportation, National Highway Traffic Safety Administration, 208-CAL-93-10, January 1993.

26. "New Car Assessment Program Frontal Barrier Impact Test – 1993 Honda Civic 2-Door Coupe", U.S. Department of Transportation, National Highway Traffic Safety Administration, MSE-93-N09, April 1993.

27. "Vehicle Safety Compliance Testing for Occupant Crash Protection, Windshield Mounting, Windshield Zone Intrusion (Partial) and Fuel System Integrity – 1992 Honda Civic CX 2-Door Hatchback", U.S. Department of Transportation, National Highway Traffic Safety Administration, 208-CAL-92-13, May 1992.

28. "Final Report of Frontal Barrier Impacts of a 1986 Ford Taurus 4-Door Sedan in Support of Crash III Damage Algorithm Reformation", U.S. Department of Transportation, National Highway Traffic Safety Administration, DOT HS 807 350, September 1988.

29. "New Car Assessment Program Frontal Barrier Impact Test – 1985 Plymouth Reliant 4-Door Sedan", U.S. Department of Transportation, National Highway Traffic Safety Administration, CAL-85-N05, March 1985.

Reviewer Discussion

Reviewers Name: James A. Neptune

Paper Title: Crash Energy Considerations in Override/Underride Impacts

Paper Number: 2002-01-0556

The following is my review of the outstanding paper that you authored:

The analysis of the Honda Civic crash test data produced results that were expected. The above bumper-level structure was less stiff than the combined, frame-level & above bumper-level, structure. Intuitively this makes sense. The work performed on the combined structure is the sum of the frame-level and the above bumper-level deformation.

The analysis of the Ford Taurus and the Plymouth Reliant crash tests produced results that were very much unexpected. How can the stiffness of the above bumper-level structure be equivalent to the stiffness of the combined frame-level/above bumper-level structure? Does this mean that no work is performed in deforming the frame-level structure? Clearly, energy is required to deform the frame-level structure. Therefore, some previously unexplained phenomenon is occurring during the above bumper-level only deformation that does not occur in the combined deformation. This is an important revelation!

The authors are right, considerable judgment is needed when reconstructing a traffic collision where significant above bumper-level deformation has occurred. Without vehicle specific crash testing, the potential inaccuracy in such a reconstruction is significant.

Please contact me at your earliest convenience so we may discuss my review.

NEPTUNE ENGINEERING, INC.

James A. Neptune

2002-01-0557

Curb Impacts – A Continuing Study In Energy Loss and Occupant Kinematics

Steven E. Meyer, Joshua Hayden, Brian Herbst, Davis Hock and Stephen Forrest
Safety Analysis & Forensic Engineering (SAFE)

Copyright © 2002 Society of Automotive Engineers, Inc.

ABSTRACT

Accident reconstruction analysis of both pre- and post-impact vehicle trajectory wherein an involved vehicle has collided with or traversed a roadside curb often leaves the analyst with uncertainty associated with the speed loss and accelerations attributable to these impacts. A review of available published data reveals very few studies considering the energy dissipated and transferred to the vehicle's occupants. This paper quantifies the changes in vehicle velocity (delta-v) for various vehicles traversing a typical roadside curb at various approach angles and impact speeds. Vehicle accelerations are recorded in the vertical, longitudinal, and lateral directions. Resulting three-point belted driver movements are observed via an interior mounted video camera and general occupant motions are described. Curb impacts were conducted with four different passenger vehicles ranging in size from a small compact car to a large full-size sport utility vehicle. Additionally, residual vehicle damage is discussed as a function of impact speed and angle.

INTRODUCTION

Roadside curbs are a common feature of modern roadside design. Curbs serve a variety of purposes including delineation of roadway edges, controlling and providing drainage, delineation of pedestrian walkways and bike paths, easing of roadside maintenance, and redirecting vehicles in order to keep them within the roadway. Curbs are common borders to most urban intersections and are usually bordered by sidewalks, grass, plantings or fields.

Consequently, intersection collisions frequently include a post-impact traversing of a curb with the vehicle coming to rest on the elevated surface bordering the curb. Of course, such curb impacts or traversings are often also seen prior to collisions, or pre-impact, wherein vehicles travel over raised center medians or roadway-defining curbs.

Although these curb impacts may be relatively minor compared with other collisions included in the crash sequence, they must, nonetheless, be considered in a complete reconstruction of the vehicles trajectory. Additionally, there has been some debate as to what influence these impacts may have upon the vehicle's occupant's position. That is, are these impacts sufficient to move an occupant significantly out of position?

This paper reports the continuation of studies involving passenger vehicles with various weights, tire sizes, and dimensional characteristics. Instrumented passenger vehicles were driven over a typical asphaltic-concrete curb at various impact angles and speeds. Vehicle delta-V, resulting damage, and occupant motions are reported and discussed.

Initial tests were conducted with a small sub-compact passenger vehicle and a full-sized sport utility vehicle[1]. A second series of tests were conducted with a compact passenger vehicle and a small pickup truck.

METHODOLOGY

VEHICLES – A group of four late-model production passenger vehicles with varying wheel and tire sizes were tested. The tire and wheel sizes ranged from a small 13-inch (330 mm) diameter wheel to a full-size sport utility vehicle with 15-inch (381 mm) wheels. All wheel rims were standard production steel rims. Each vehicle was equipped with production 3-point restraints. Vehicle specifications are detailed in Table 1.

	Test Vehicle 1	Test Vehicle 2	Test Vehicle 3	Test Vehicle 4
Year	1981	1987	1985	1991
Make	Datsun	Jeep	Nissan	Subaru
Model	210	Grand Wagoneer	Pickup	Legacy
Weight	2162 lbs (980.7 kg)	4663 lbs (2115 kg)	3414 lbs (1552 kg)	2780 lbs (1264 kg)
Wheelbase	92" (233.7 cm)	108.5" (275.6 cm)	112" (284.5 cm)	102" (259 cm)
Tire Size	175/70R13	31X10.50R15	P175/70R15	P235/75R14
Tire Radius	11" (27.9 cm)	15" (38.1 cm)	14" (35.6 cm)	12" (30.5 cm)

Table 1: Test Vehicle Specifications

CURB – All tests were performed in a large parking lot equipped with a 6-inch (152 mm) raised asphaltic-concrete curb. The curb was a step-type curb having only one vertical side before stepping up onto an open field of dirt and grass. The curb was a roadway-defining type curb that retained the dirt field and had no gutter. The field was at the same approximate elevation as the top of the curb. (See Figure 1.)

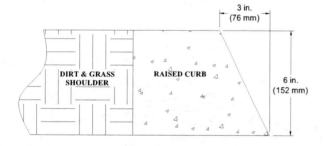

Figure 1: Test Curb

TEST PROCEDURE – For the tests conducted with Vehicles 1 and 2, a tri-axial accelerometer was mounted at the approximate center of gravity of the vehicle and was triggered by the driver via a laptop computer mounted in the right front passenger area. The driver manually engaged the instrumentation system just prior to accelerating the vehicle. For the tests conducted with Vehicles 3 and 4, in addition to the tri-axial accelerometer mounted at the approximate center of gravity of the vehicle, additional accelerometers were placed at the right front (first impacting) tire and the left rear (last impacting) tire. The two additional accelerometers were oriented in the vertical plane and used to trigger the start of data collection and to record the end of the impact with the curb. In all four vehicles the tri-axial accelerometers were oriented to record longitudinal (Gx), lateral (Gy), and vertical (Gz), vehicle accelerations.

In addition, all tests were documented photographically. An interior video camera was mounted to the front dash and recorded relative movements of the three-point belted driver. Also, an exterior real-time video camera, as well as a high-speed video camera, was used to record the vehicle kinematics and driver motions as the test vehicles traversed the curb.

A series of highway cones were utilized to define an approach lane to the curb at various angles. (See Figure 2.) The approach angles were designed such that the right front wheel made initial contact with the raised curb as the test vehicle was driven through the cones and over the curb at various speeds.

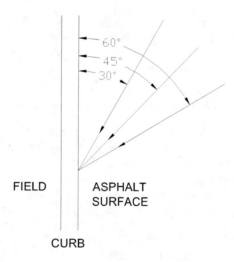

Figure 2: Vehicle Approach Angles

The drivers were fitted with a full-face style motorcycle helmet and wore the production three point seat belts provided. The drivers, three in number, ranged from 5'-10", 225 lbs, to 6'-3", 170 lbs.

Target impact speeds for each vehicle at each approach angle were 5, 10, 15, 20, 30 and 40 mph (8, 16, 24, 32, 48 and 64.4 km/hr). It was left to the driver, utilizing each vehicle's speedometer, to impact the curb as closely as possible to the intended test speed. The driver was instructed to release the accelerator just prior to curb impact and allow the vehicle to transverse the curb without additional acceleration or braking. That is, the vehicle was allowed to simply roll over the curb. Actual impact speeds were measured either by film analysis or by utilizing small "speed bumps" spaced 10-feet (3.05 m) apart just prior to curb impact. These speed bumps provided recorded accelerations, which could then be converted to vehicle speeds.

Wheel and tire damage was documented after each tests. Flat tires and damaged wheels were replaced after each test. A total of 18 tests were performed with each vehicle.

TEST RESULTS

The accelerometer data for each set of tests was CFC Class 180 filtered and analyzed to obtain acceleration plots in the vehicles' x-, y- and z-directions. Review of the Acceleration vs. Time curves shows distinct spikes corresponding to individual tire impacts as well as undercarriage impacts. Positive and negative accelerations are seen as the vehicles available suspension travel is exceeded. Accelerations in the x-direction were integrated producing translational velocity vs. time curves for each test. The time duration for the curb impact is represented in the acceleration vs. time curves by the individual spikes. Where as, in the velocity vs. time curves the impact event ends where the curve levels off prior to the relatively constant vehicle deceleration as the vehicle is braked to a stop. (See Figures 3 and 4 below for sample acceleration and velocity curves.)

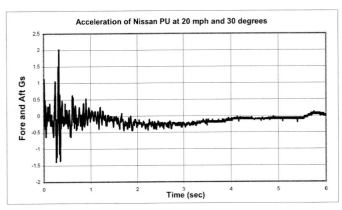

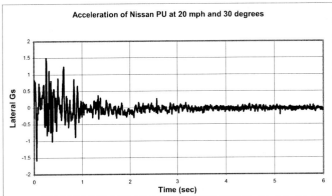

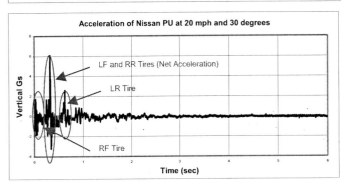

Figure 3: Acceleration Curves

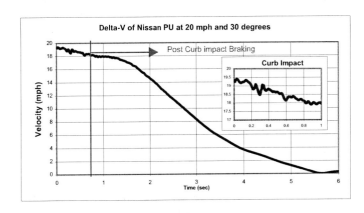

Figure 4: Velocity Curve

Additional curves were plotted to describe peak accelerations in the vertical direction and the change in vehicle translational velocities through the curb impact. The curb impact is defined as the impact event from the first, or leading, tire contact through the last, or trailing, tire contact. The reported delta-v is defined as the vehicle speed prior to impact minus the speed after all four tires have gone over the curb. In the cases where all four tires fail to transverse the curb, the delta-v is defined as the speed prior to contact plus any rebound speed. Test results are summarized in Tables 2A through 2D on the following page.

DISCUSSION

VEHICLE DYNAMICS – At the lowest impact speeds, 5 mph (8 km/hr) the 13-inch wheel Datsun failed to completely mount the 6-inch curb with all four tires. The front two wheels passed over the curb but the vehicle had slowed sufficiently such that both the rear wheels would not always climb the curb. The rear wheels, when contacting the curb, were found to either start to climb the curb before falling backward, or bounce off the curb propelling the entire vehicle in the opposite direction. Hicks, Field and Bowler reported longitudinal delta-V's for a single 13-inch tire to climb a 6-¼ inch wood curb. Their data shows that a vehicle's right front wheel experiences a change in velocity of 0.6 to 0.9 mph (0.9 to 1.5 km/h) at impact speeds of 4.4 to 7.3 mph (7 to 11.6 km/h) when rolling over a curb at 20° to the curb.[3] They quantify the change in velocity resulting from one wheel climbing the curb. The Hicks et. al. results are consistent with the low speed tests run in this study with the exception that here the change in velocity includes all four wheels and includes any rebound that may occur when the vehicle fails to completely traverse the curb.

In general, all tires were seen to more easily climb over the curb at the shallower approach angles. The larger-wheel vehicles were also found to mount and transverse the curb more easily than the smaller-wheel vehicles and found to sustain less wheel/tire damage and correspondingly lower delta-V's. However, for extremely shallow approach angles, the tires will begin to be redirected or slide along the curb as if impacting a very low guardrail instead of climbing the curb. Navin and Thomson state that, "a standard automobile with 300 mm radius tires traveling at 50 km/h would be redirected if it struck the curb at an angle of 10 to 15 degrees." They also state the maximum impact angle for a 4X4 tire to be successfully redirected when impacting a 243 mm high curb, is 8 degrees.[2]

TIRE/WHEEL DAMAGE – The smaller-diameter wheels were found to sustain significant damage at the higher test speeds. That is, the 13-inch (330 mm) wheels and 14-inch (355 mm) wheels were found to sustain significant rim flange damage beginning at speeds on the order of 30 mph. The larger-diameter 15-inch (381 mm) wheels sustained no permanent damage or even any deflation of the tires at any of the test speeds. No tires were cut or permanently damaged in the course of the testing.

As expected, the most significant damage was always seen to the leading tire and the leading rim flange. The amount of wheel damage always decreased from the first tire to impact the curb to the last tire of impact. Rim deformation was more pronounced as the impact or approach angle increased.

At 30 mph (64.4 km/hr), two to three tires for the small, 13-inch (330 mm) and 14-inch (355 mm), wheeled vehicles were found to debead at almost all angles. When the approach angle reached 60-degrees, at 40 mph, rim damage began to appear on both the inside and outside rim flanges of the leading tire (in this case the right front). (See Figure 5 for representative photograph of rim damage.) Figures 6 and 7 show the total number of rim deformations and tire rim debeadings for all tests. There was no resulting rim deformation or tire rim debeading for the Nissan Pickup or Jeep Cherokee for all test speeds.

Figure 5: Leading Tire, 60 degrees, 40 mph

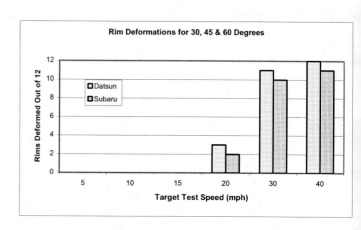

Figure 6: Rim Deformation

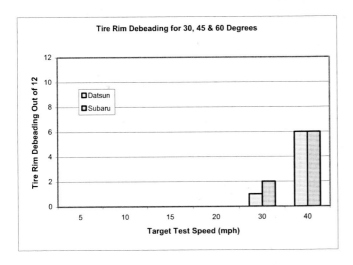

Figure 7: Tire Rim Debeading

Jeep Wagoneer - 31X10.50R15				
Actual Test Speed		Impact Angle	Peak G's Vertical	Delta-V	
mph	km/h			mph	km/h
9	14	30	0.8	3	4
15	24	30	2.5	1	2
18	30	30	4.9	2	3
24	39	30	6.0	2	3
34	55	30	6.5	3	4
38	61	30	15.8	3	4
10	16	45	1.8	1	2
13	21	45	4.6	1	2
17	27	45	5.1	1	2
20	32	45	11.1	3	4
30	48	45	10.5	3	5
36	58	45	14.0	5	7
7	11	60	0.8	1	1
12	19	60	3.9	1	1
20	32	60	4.4	2	4
20	32	60	7.1	3	5
34	55	60	11.5	3	5
40	64	60	11.8	5	8

Table 2B: Summary of Jeep Test Data

Datsun 210 - 175/70R13					
Actual Test Speed		Impact Angle	Peak G's Vertical	Delta-V	
mph	km/h			mph	km/h
5	8	30	0.47	7	11
11	17	30	2.18	4	6
16	26	30	2.82	7	12
19	30	30	4.16	4	6
27	44	30	9.63	8	12
37	60	30	8.8	3	5
5	8	45	0.88	5	8
10	16	45	2.74	1	2
16	26	45	2.05	6	9
21	33	45	6.78	6	9
27	44	45	11.87	4	6
36	57	45	10.93	5	7
5	8	60	0.96	8	12
10	16	60	3.22	1	2
17	27	60	4.76	6	9
21	33	60	8.4	6	10
27	44	60	11.21	5	8
37	60	60	12.34	6	9

NOTE: car did not climb curb with all four wheels for 5mph tests at 30 and 60 degrees

Table 2A: Summary of Datsun Test Data

Nissan PU - 235/75R15					
Actual Test Speed		Impact Angle	Peak G's Vertical	Delta-V	
mph	km/h			mph	km/h
5	9	30	0.4	2	3
9	14	30	0.9	2	3
13	21	30	2.8	1	2
19	30	30	6.1	1	2
28	46	30	11.4	2	3
40	65	30	7.4	1	2
5	8	45	0.5	2	4
10	16	45	2.3	2	3
14	22	45	2.6	2	3
20	31	45	4.0	2	3
30	48	45	5.8	2	3
38	61	45	15.1	1	2
6	9	60	0.6	2	4
11	18	60	2.6	3	4
15	24	60	3.4	3	4
21	33	60	3.7	2	4
27	44	60	7.3	2	3
38	61	60	10.9	2	4

Table 2C: Summary of Nissan PU Test Data

Subaru Legacy - 175/70R14				
Actual Test Speed		Impact Angle	Peak G's Vertical	Delta-V	
mph	km/h			mph	km/h
5	8	30	0.4	1	1
11	17	30	1.6	2	3
15	23	30	3.7	1	2
21	33	30	6.6	1	1
31	50	30	15.5	2	3
40	65	30	9.4	2	3
5	8	45	0.4	1	1
11	17	45	1.9	1	2
17	27	45	8.5	1	2
23	37	45	16.5	1	2
31	50	45	13.4	2	3
38	61	45	28.5	1	2
5	7	60	0.4	1	1
10	15	60	1.4	1	2
16	26	60	6.5	1	2
21	34	60	11.3	1	1
30	48	60	8.5	1	2
40	65	60	18.2	2	3

Table 2D: Summary of Subaru Test Data

OCCUPANT KINEMATICS – The vehicle dynamics through the curb impacts consistently resulted in significant vehicle pitch and roll. Vehicle yaw was not seen to be significant in any of the tests conducted.

The first vehicle dynamic was that of upward pitch of the front of the vehicle. This occurred as the front wheels climbed the curb. The pitch would then transverse from positive (nose-up) to negative (nose-down) as the rear of the vehicle climbed the curb. In all tests, the prepared driver, anticipating and clearly expecting the impact, with a firm grasp of the steering wheel, was able to avoid impact with any interior vehicle components. All drivers reported being able to duck and or hold onto the wheel through the pitching of the vehicle to avoid steering wheel or front header contact. However, in a situation where a vehicle occupant was unprepared for a curb impact, it is still unlikely that there would be significant contact with interior vehicle components as a result of fore and aft occupant motion. Considering the delta-Vs recorded in these tests, any fore and aft contact is expected to be minor.

Second in time to pitch, the vehicle began to roll about its longitudinal axis. The vehicle roll was found to have a more dramatic and pronounced effect on the belted driver than the pitch. Moreover, as the approach angle and the impact speed decreased the vehicle roll became more pronounced. These vehicle accelerations were found to be more difficult for the driver to control, such that a number of head contacts with upper interior surfaces lateral to his initial position were reported. Mostly the surface contacted was the near side vehicle side roof header or B-pillar. It is believed that due to the additional mass of the helmet worn by the driver, these motions may have been exaggerated. However, the general trends appear clear in that lateral occupant motions are more pronounced than fore and aft.

As for occupant motions in the vertical direction, no driver of any of the vehicles reported those as being significant. Film analysis also confirms those motions to be relatively minor (on the order of 1-2" or 3-5 cm) in these tests. The most severe vertical pulse felt by the driver was experienced at the lower speeds.[4]

CONCLUSIONS

1. In general, the change in velocity resulting from a vehicle to curb impact wherein all four tires completely traverse the curb is seen to be approximately 2 mph (3.2 km/hr) for vehicles with a tire radius of approximately two or more times the height of the curb. For vehicles with a tire radius of less than twice the height of the curb, the resulting change in velocity was seen to be approximately 5 mph (8 km/hr). These results are fairly consistent for impact speeds below 40 mph and impact angles between 30 degrees and 60 degrees.

2. Rim deformation is most likely to occur on leading wheels and begins to be seen at impact speeds between 20 and 30 mph (32.2 and 48.3 km/hr) and approach angles of 30- to 60-degrees (see Figure 6).

3. Impact speeds on the order of 40 mph (64.4 km/hr) and higher will likely result in tire debeading and deflation associated with rim damage for small-wheeled vehicles (see Figure 7).

4. Curb traversing resulting from approach angles between 30- and 60-degrees at speeds below 40 mph (64.4 km/hr) are not seen to produce significant occupant motions fore and aft or vertically, but will result in more pronounced lateral motions, including potential interior contacts.

REFERENCES

1. Hayden, J., Meyer, S., Herbst, B, Hock, D., Forrest, S., Curb Impacts – A Study in Occupant Kinematics and Energy Loss, International Conference on Accident Investigation, Reconstruction, Interpretation and the Law, 4th, Vancouver, B.C., 2001.
2. Navin FPD, Thomson R; Safety of Roadside Curbs; Society of Automotive Engineers, Inc; SAE970964; USA; 1997

3. Hicks B, Field D, Bowler J, King D; <u>Speed Change from Curb Impacts</u>; Accident Investigation Quarterly; Summer 1997; pp. 20-26; 1997

4. Ogan JS, Alcorn TL, Scott JD; <u>Acceleration Levels and Occupant Kinematics Associated with Vehicle-to-Curb Impacts</u>; Accident Reconstruction Journal; May/June 1995; pp. 54-56; 1995

Reviewer Discussion

Reviewers Name: Gustav A. Nystrom
Paper Title: Curb Impacts - A Continuing Study in Energy Loss and Occupant Kinematics
Paper Number: 2002-01-0557

SAE 2002-01-0557
Reviewer's Discussion
by Gustav A. Nystrom, Amador Newtonian Engineering
Curb Impacts - A Continuing Study in Energy Loss and Occupant Kinematics

This paper presents some curb impact measurements which complement earlier published data. Each of four vehicles impacted one curb at six speeds and three impact angles.

With regards to delta-v, the paper's definition employs a mixture of speed and velocity concepts. Surprisingly, the measured delta-v for the smallest vehicle is found to be significantly different from that of the other vehicles (regardless of speed or angle).

With regards to ability to re-direct a vehicle, the paper by Navin and Thompson is mentioned. But the authors do not discuss if the new data are consistent with the curves contained in the Navin paper, which predict when re-direction can be expected as a function of curb geometry, impact speed, and impact angle.

With regards to vehicle accelerations, wheel damage, and occupant motions, the results appear reasonable and should be quite useful.

Reviewer Discussion

by Karl F. Shuman, KEVA Engineering

SAE #2002-01-0557
"Curb Impacts – A Continuing Study in Energy Loss and Occupant Kinematics",
Steven E. Meyer, Joshua Hayden, Brian Herbst, Davis Hock and Stephen Forrest, Safety Analysis & Forensic Engineering (SAFE)

This paper has provided valuable information on the velocity change experienced by vehicles in curb strikes. Four different vehicles with different tires sizes were impacted into a six inch curb. Three different angles with a range of speeds were used to provide the test data. It is interesting that the 11" radius tire vehicle sustained larger changes in velocity, while the 12", 14" and 15" radius tire vehicles experienced fairly similar changes in velocity. This data will be very useful when reconstructing accidents involving curb strikes.

2002-01-0558

The Use of Single Moving Vehicle Testing to Duplicate the Dynamic Vehicle Response From Impacts Between Two Moving Vehicles

C. Brian Tanner, John F. Wiechel and Philip H. Cheng
S.E.A., Inc.

Dennis A. Guenther
Ohio State University

Richard Fay
Fay Engineering

Copyright © 2002 Society of Automotive Engineers, Inc.

ABSTRACT

The Federal Side Impact Test Procedure prescribed by FMVSS 214, simulates a central, orthogonal intersection collision between two moving vehicles by impacting the side of the stationary test vehicle with a moving test buck in a crabbed configuration. While the pre- and post-impact speeds of the vehicles involved in an accident can not be duplicated using this method, closing speeds, vehicle damage, vehicle speed changes and vehicle accelerations can be duplicated. These are the important parameters for the examination of vehicle restraint system performance and the prediction of occupant injury. The acceptability of this method of testing is not as obvious for the reconstruction of accidents where the impact is non-central, or the angle of impact is not orthogonal.

This paper will examine the use of crash testing with a single moving vehicle to simulate oblique or non-central collisions between two moving vehicles. First, an accident where the left front corner of a left turning passenger car struck the left side of an oncoming passenger car is described. The results of a reconstruction of the accident will also be described. Next, simulations of the accident using the EDSMAC vehicle impact model included in the HVE-2D computer program developed by EDC Corporation are presented in which both vehicles are moving, only the struck vehicle is moving, and only the striking vehicle is moving. Finally, a full scale test will be presented in which only one of the vehicles is moving and the other vehicle is stationary during the impact. The results of the test and the simulations will be compared to the actual accident, and the effects of local vehicle geometry will be discussed.

INTRODUCTION

The work described in this paper was done to verify the results of a reconstruction of an accident that occurred when the left front corner of a 1988 Dodge Daytona struck the left side of a 1991 Pontiac 6000. Both vehicles were apparently stopped at stop signs on opposite sides of an intersection and the Pontiac driver intended to go straight across the intersection while the Dodge driver intended to turn left onto the crossing street. The drivers accelerated from the stop at nearly the same time, and they either did not see each other or neither yielded to the other and the left front corner of the turning Dodge struck the driver's door of the Pontiac. Following the impact, the Pontiac spun counterclockwise and continued across the intersection coming to rest near the starting position of the Dodge, while the Dodge rotated slightly counterclockwise and was slowed, but

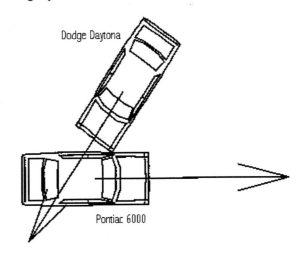

Figure 1 - The approximate impact position of the Vehicles in the accident

continued some distance before coming to rest on the cross street. Figure 1 shows the approximate orientation of the vehicles at the time of impact, based upon the reconstruction.

METHODOLOGY

ACCIDENT RECONSTRUCTION - Figures 2, 3, and 4 show the damage to the accident vehicles, and their relative rest positions following the impact at the accident scene. The information in these figures was used, along with information from the investigation of the accident to develop estimated ranges for the crush energy absorbed by each of the vehicles, and the post impact velocities of each of the vehicles. Based on the damage and scene data, a range of post impact velocities and possible impact positions and orientations for each vehicle were determined, and a principle direction of force was estimated. Equations expressing the conservation of momentum and energy were then used to find the combinations of values within these ranges that best satisfied the equations. From these methods, minimum speeds for the vehicles at impact were determined to be about 16 miles per hour for the Pontiac and about 14 miles per hour for the Dodge, with the orientations of the vehicles approximately as shown in Figure 1.

Differences in the speeds calculated in this reconstruction of the accident and in the speeds determined in a reconstruction performed by an outside party led to the consideration of performing a crash test. Because of the limited availability and the added cost of facilities capable of performing an angled, two vehicle moving test, the possibility of performing the test with one moving and one stationary vehicle was examined.

COMPUTER SIMULATIONS – To evaluate the potential of using this type of test, computer simulations were first performed using the EDSMAC portion of the HVE-2D vehicle dynamics and impact simulation software sold by Engineering Dynamics Corporation. However, before addressing the fairly complicated oblique impact that inspired this effort, it was decided to first consider a slightly less complicated orthogonal but non-central impact between the front of a 1994 Dodge Grand Caravan and the front part of the side of a 1980 Chevrolet Malibu.

Input parameters for the first simulation with both vehicles moving are shown in Figure 5, while Figures 6 and 7 show the input parameters for the Caravan moving/Chevrolet stationary and the Chevrolet moving/Caravan stationary scenarios respectively.

In the Caravan moving scenario, the x direction was taken as being parallel to the side of the Caravan and directed from its rear to front just before impact while the y direction was taken as normal to this, to the right. In the Chevrolet moving scenario, the x direction was taken as parallel to the side of the Chevrolet from the rear to the front and the y direction was taken as normal to this and to the right.

Figure 2 - The accident damage to the 1988 Dodge Daytona

Figure 3 - The accident damage to the 1991 Pontiac 6000

Figure 4 - The relative rest positions of the Dodge and the Pontiac after the impact

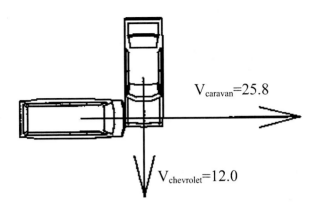

Figure 5 - Impact Configuration with both vehicles moving

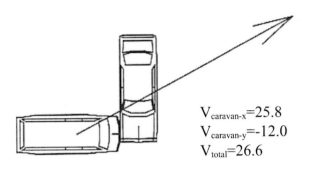

Figure 6 - Impact Configuration with only the Caravan moving

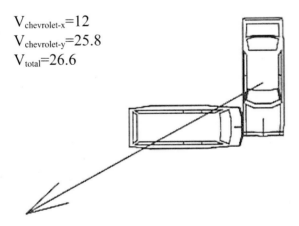

Figure 7 - Impact Configuration with only the Malibu moving

Figures 8 through 13 show the damage predicted by the simulation for both vehicles in each of the impact configurations. Examination of these figures shows that the predicted damage to the front of the Grand Caravan and the predicted damage to the front right side of the Chevrolet are very similar in each of the simulations, however there is difference among the different simulations in the predicted damage to the right rear portion of the Chevrolet and the left rear portion of the Grand Caravan. This damage is not related to the primary impact between the vehicles, but instead to the secondary impact between them.

Figures 14 through 19 show the predicted frontal, lateral and resultant accelerations for both vehicles in each simulation. Comparison of figures 14, 16 and 18, and figures 15, 17, and 19 to each other shows that the simulation predicts very similar acceleration pulses in magnitude and duration for the primary impact between the vehicles in each of the different impact configurations. There are visible differences, however, and the predicted accelerations in the secondary impact are quite different from each other.

These simulations were run using the default program settings for the integration time steps and data output intervals. The values for the integration timestep were 0.001 seconds during contact and .01 seconds during pre- and post-impact travel of the vehicles, while data was output only every 0.050 seconds. The result of this is that the program essentially averages the results at each timestep. In addition, with data only being recorded every 0.050 seconds, it is possible to have an entire 0.1 seconds impact pulse described with only a single meaningful data point. For these reasons, it was decided to re-run the simulations with the simulation variables modified to allow integration timesteps of 0.001 seconds throughout the simulation with data recorded every 0.002 seconds.

The predicted damage for each of the vehicles in the second set of simulations is shown in Figures 20 through 25. Examination of the figures shows that once again, there is excellent agreement in the damage predicted to both of the vehicles in the primary impact, regardless of the distribution of the closing velocity. Interestingly, there appears to be even greater disagreement from one scenario to the others about what happens when the secondary impact occurs. Both the extent and magnitude of the secondary impact damage varies from one scenario to the next. Again, however, the secondary damage looks significantly less severe than the primary impact damage.

This observation is amplified by examining the predicted acceleration curves for the vehicles for this set of simulations, shown in Figures 26 to 31. The figures show that while the secondary impacts occurred at slightly different times and with different severity in each of the simulations, that the secondary impact generally followed the primary impact by at least a half a second in each case and was at most about a third the magnitude of the first impact.

While the different impact scenarios with similar closing velocities will produce similar damage, acceleration and speed change for the primary impact between the vehicles, the vehicles necessarily have different post impact velocities. This follows from the fact that their speed changes during the primary impact do not change, but in the different scenarios they had different pre-impact velocities. In an accident such as the one being considered here, which involves non-central impact and relatively low impact speeds, the result is that following the primary impact the vehicle speeds are significantly reduced, and secondary impact does not occur for a significant period of time. During this time, the vehicles' tires interacting with the pavement cause relatively large changes in the vehicles' velocities before the secondary impact occurs.

In a case such as the one being considered, for the purpose of defining the interaction between vehicle occupants and the vehicle interior, the primary impact is of overriding concern. Because of the time delay between the primary and secondary impacts, any interaction between the vehicle and test subjects has generally concluded before the secondary impact occurs, and is at worse only mildly affected by the secondary impact.

In central impact between two vehicles, no matter how the vehicles hit, such as both moving, or with all the closing velocity given to either of the vehicles, there will be no secondary impact unless a light, short vehicle hits the side of a long, heavy vehicle. Thus the only serious concern regarding a problem with secondary impact for the test methodology under consideration would be in those accidents and in accidents involving non-central impact at higher speeds. However, it would seem likely that the higher closing speeds between the vehicles would result in less time passing between primary and secondary impacts, so that any differences between the different scenarios would be minimized.

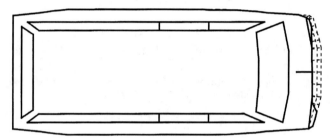

Figure 8 - Grand Caravan Damage for both vehicles moving

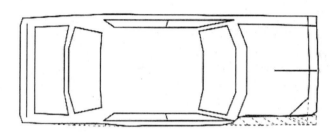

Figure 9 - Chevrolet damage for both vehicles moving

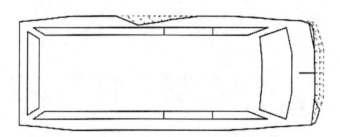

Figure 10 - Grand Caravan damage for only the Caravan moving

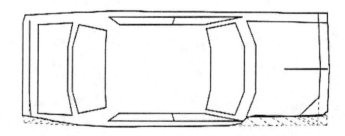

Figure 11 - Chevrolet damage for only the Caravan moving

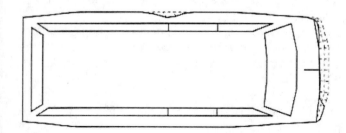

Figure 12 - Grand Caravan damage for only the Malibu moving

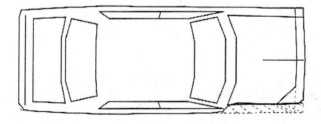

Figure 13 - Chevrolet damage for only the Malibu moving

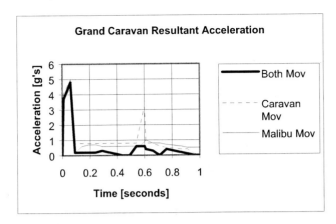

Figure 14 - Grand Caravan resultant acceleration in the simulations

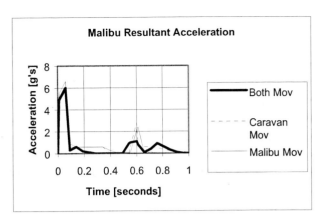

Figure 15 -- Chevrolet resultant acceleration in the simulations

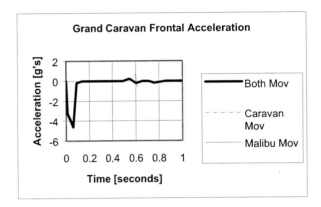

Figure 16 - Grand Caravan frontal acceleration in the simulations

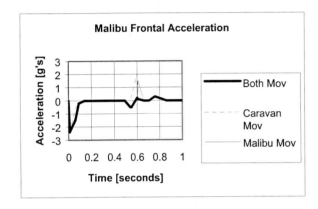

Figure 17 - Chevrolet frontal acceleration in the simulations

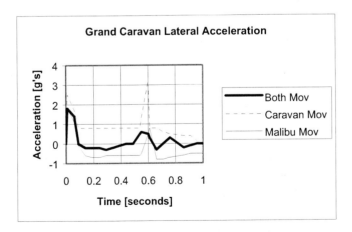

Figure 18 - Grand Caravan lateral accelerations in the simulations

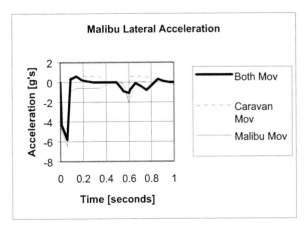

Figure 19 - Chevrolet acceleration in the third simulation

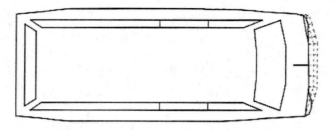

Figure 20 - Grand Caravan Damage for both vehicles moving

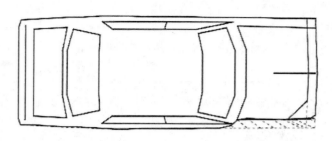

Figure 21 - Chevrolet damage for both vehicles moving

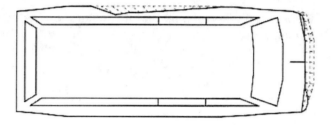

Figure 22 - Grand Caravan damage for only the Caravan moving

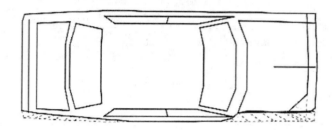

Figure 23 - Chevrolet damage for only the Caravan moving

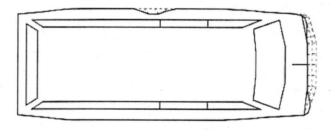

Figure 24 - Grand Caravan damage for only the Malibu moving

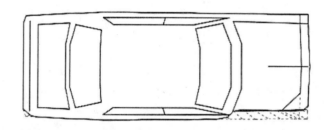

Figure 25 - Chevrolet damage for only the Malibu moving

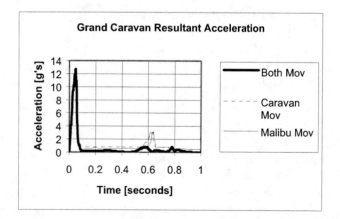

Figure 26 - Grand Caravan Resultant acceleration in the simulations

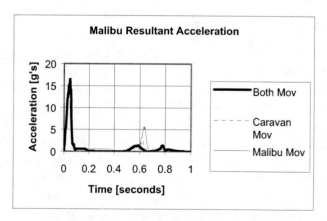

Figure 27 -- Chevrolet Resultant acceleration in the simulations

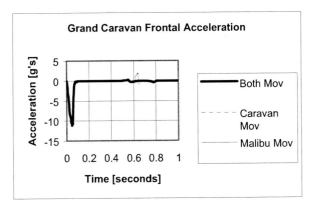

Figure 28 - Grand Caravan Frontal acceleration in the simulations

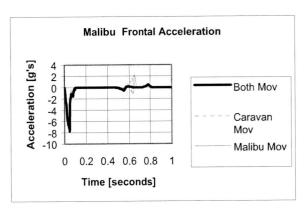

Figure 29 - Chevrolet Frontal acceleration in the simulations

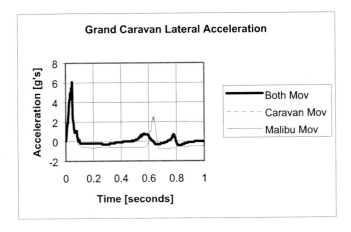

Figure 30 - Grand Caravan Lateral acceleration in the simulations

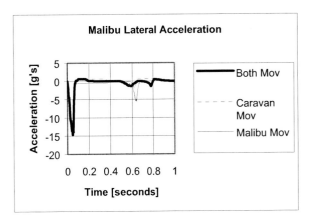

Figure 31 - Chevrolet Lateral acceleration in the simulations

This hypothesis was checked by again running the impact simulations between the Grand Caravan and Chevrolet, but now with the speed of each vehicle doubled. The damage and acceleration results are shown in Figures 32 to 43. Examination of the figures comparing the predicted damage for both vehicles in the different scenarios shows that both the damage due to primary impact and that due to secondary impact are closely matched for all scenarios. Perhaps more notably, a comparison of the acceleration pulses predicted for both vehicles shows that in all of the scenarios, not only did components of the primary impact accelerations closely agree, but the same was now true for all components of the secondary impact accelerations as well.

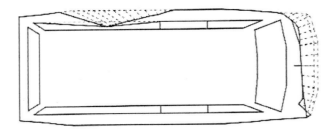

Figure 32-Grand Caravan damage for both vehicles moving

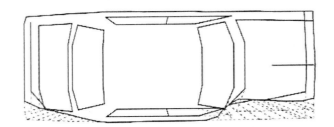

Figure 33 - Chevrolet damage for both vehicles moving

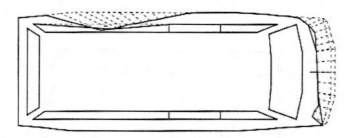

Figure 34 - Grand Caravan damage for the Caravan only moving

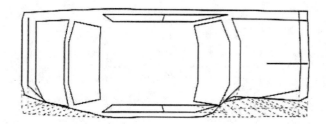

Figure 35 - Chevrolet damage for the Caravan only moving

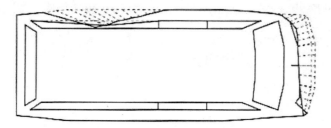

Figure 36 - Grand Caravan damage for the Chevrolet only moving

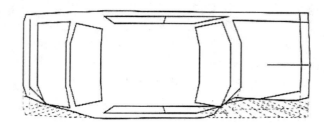

Figure 37 - Chevrolet damage for the Malibu only moving

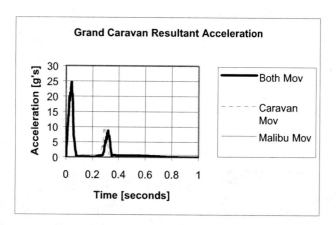

Figure 38 - Grand Caravan Resultant acceleration in the simulations

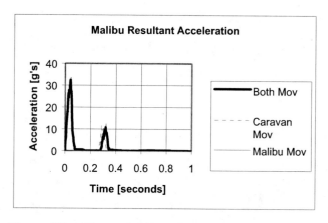

Figure 39 -- Chevrolet Resultant acceleration in the simulations

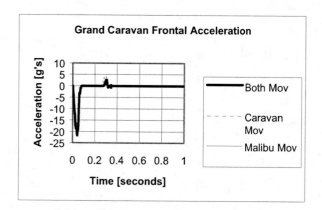

Figure 40 - Grand Caravan Frontal acceleration in the simulations

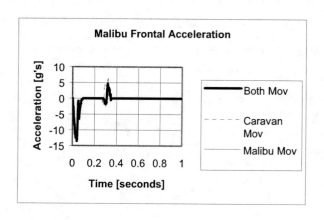

Figure 41-Chevrolet Frontal acceleration in the simulations

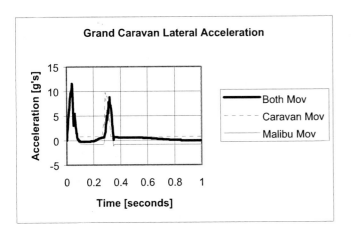

Figure 42 - Grand Caravan Lateral acceleration in the simulations

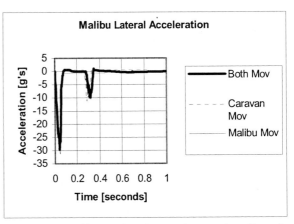

Figure 43 - Chevrolet Lateral acceleration in the simulations

The next step in the analysis of the proposed testing method is to consider its application to an oblique impact like the one shown in figure 1. Again, the impact involve the left front corner of a left turning Dodge Daytona striking the driver's side of Pontiac 6000 crossing the intersection in the opposite direction.

Before performing the full scale crash test, the different impact scenarios were again examined with the computer simulation. The simulation configurations are shown in Figures 44 to 46. In the simulation giving all of the closing velocity to the Pontiac, shown in Figure 45, the x direction was defined as parallel to the side of the Pontiac and directed from the rear to the front, with the y direction defined as perpendicular to this and positive to the right. Similarly, in the simulation giving all of the closing velocity to the Daytona, the x direction was defined as being parallel to the side of the Daytona and running from the rear toward the front, while the y direction was perpendicular to this and was positive toward the right.

In each of the simulations, the simulation variables controlling the integration timesteps were set so that a timestep of 0.001 seconds would be used throughout. In addition, the data was recorded during the simulations at every 0.002 seconds. The results of the simulations are given in Figures 47 to 58, which compare the predicted damage to the vehicles for each of the impact configurations and also compare the predicted acceleration pulses for the vehicles in all configurations. It is notable that there was no secondary impact in any of the impact scenarios. Examination of figures 47 to 52 shows that the predicted damage to the side of the Pontiac was very similar in all three cases.

There may have been a slightly greater amount of absorbed energy predicted for the Pontiac in the simulation where it was stationary and the Daytona was moving, however it was difficult to determine this from the figures. It was clear that there was slightly more energy absorbed by the Daytona in this simulation. This may have been due to slight difficulties in achieving repeatable vehicle alignment and duplicating the initial

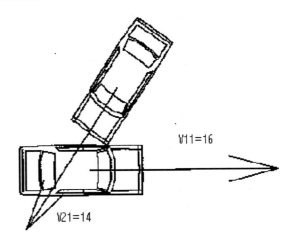

Figure 44 - Impact Configuration for both vehicles moving.

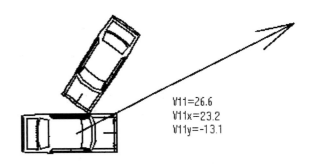

Figure 45 - Impact Configuration for only the Pontiac moving

contact point in the simulations. The acceleration data showed a similar trend to the damage, with quite close agreement between the two vehicle moving and only Pontiac moving simulation and slightly greater variation for the Daytona moving simulation. Despite this, there was generally good agreement for all three scenarios considering the complexity of the impact configuration and the limits of the contact model within the simulation package on handling sideswipe.

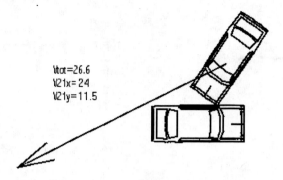

Figure 46 - Impact Configuration for only the Daytona moving

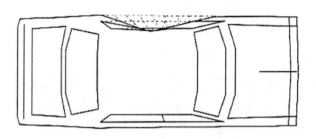

Figure 47 – Pontiac Damage for both vehicles moving

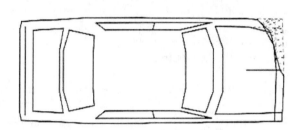

Figure 48 - Dodge damage for both vehicles moving

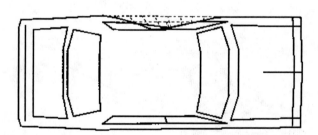

Figure 49 - Pontiac damage for only the Pontiac moving

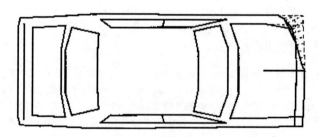

Figure 50 - Dodge damage for only the Pontiac moving

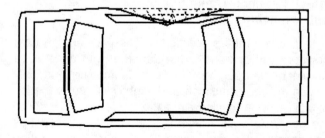

Figure 51 - Pontiac damage for only the Daytona moving

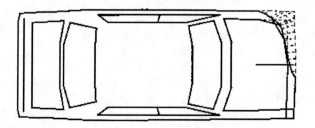

Figure 52 - Dodge damage for only the Daytona moving

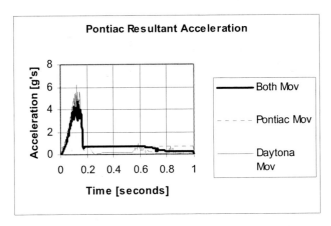

Figure 53 – Pontiac resultant acceleration in the simulations

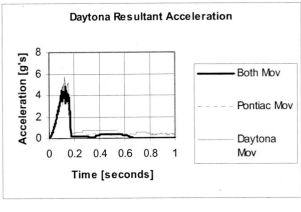

Figure 54 - Dodge Resultant acceleration in the simulations

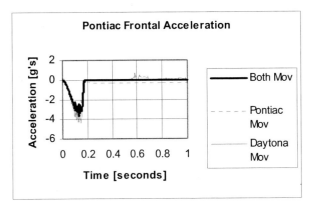

Figure 55 - Pontiac Frontal acceleration in the simulations

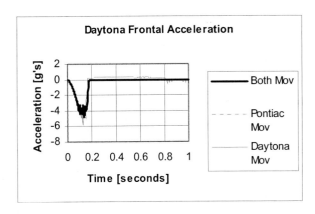

Figure 56 – Dodge Frontal acceleration in the simulations

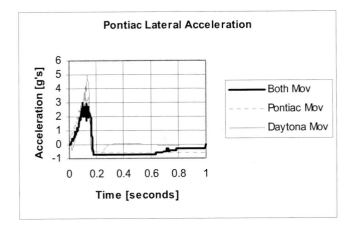

Figure 57 - Pontiac Lateral acceleration in the simulations

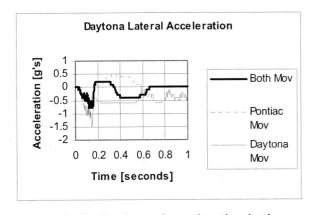

Figure 58 - Dodge Lateral acceleration in the simulations

TESTING - After completing the simulations, it was decided to conduct the test with all the closing velocity between the vehicles given to the Pontiac, and with the Dodge initially at rest. In order to achieve this and maintain the same vehicle impact orientation as in the actual accident, it was necessary to crab the wheels on the Pontiac so that it would roll at an angle of about 25 degrees with its normal longitudinal axis. This was achieved by fixing the appropriate amount of steering into the front axles, and by welding wedged shaped spacers in between the ends of the rear axle bar and the rear stub axles. This resulted in the rear tires and wheel extending significantly out beyond the body of the Pontiac. The impact configuration of both vehicles and the Pontiac are shown in Figures 59 and 60.

Figure 60 - The crabbed configuration of the Pontiac in the test

Figure 59 - The impact configuration of the 1988 Dodge Daytona and the 1991 Pontiac 6000 for the test

Figure 61 - The damage to the Daytona in the test

The damage to Dodge and the Pontiac in the test is shown in Figures 61 and 62. Comparison of these figures with Figures 1 and 2 will show that the damage in the test is very similar to what occurred in the accident. In the accident there appeared to be slightly more buckling of the Dodge front fender, while in the test the damage to the side of the Pontiac was slightly lower and possibly a little longer on the vehicle side. The probable reasons for these differences are slight differences in the heights of the vehicles and the rust on the test Pontiac's doors.

For the testing, each of the vehicles was instrumented with triaxial accelerometers near their centers of gravity. After the test, this data was integrated to get the speed changes of both the vehicles in their longitudinal (x) and lateral (y) directions. The data is shown in Figures 63 to 68.

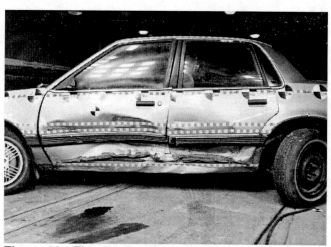

Figure 62 - The damage to the Pontiac in the test

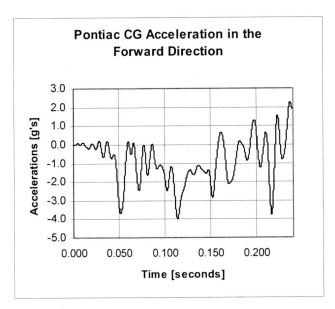

Figure 63- The Daytona Forward Acceleration

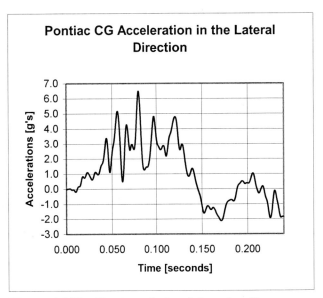

Figure 64-The Daytona Lateral Acceleration

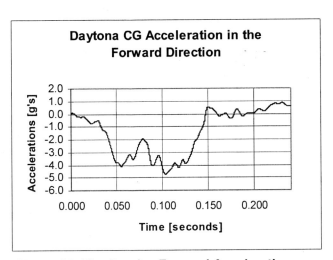

Figure 65- The Pontiac Forward Acceleration

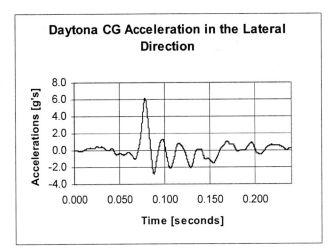

Figure 66-The Pontiac Lateral Acceleration

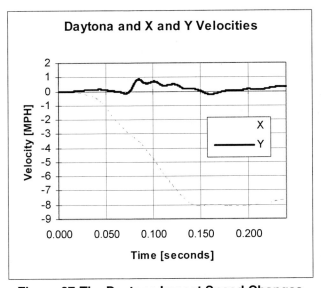

Figure 67-The Daytona Impact Speed Changes

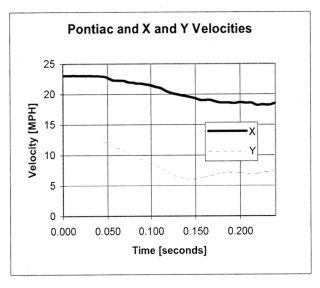

Figure 68-The Pontiac Impact Speed Changes

DISCUSSION

From the examples given, it seems that the simplified method of performing vehicle crash tests to simulate accidents in which two vehicles were moving, by using only one moving vehicle might always be permissible. However, there are instances where its use could be problematic or impractical. This would be true for example, if the objective was to demonstrate occupant kinematics and injury causation in a relatively lower impact accident in which the secondary impact was important in causing the injury. This might be the case if some body part was crushed between the two vehicles during the secondary impact, or if there was partial or full ejection of the occupant during the primary impact.

Although the method presented will almost always faithfully reproduce the accelerations, delta-V's and damage associated with a primary impact, there is often significant deviation in any secondary impact that occurs. This deviation happens because in each of the different impact scenarios, the configuration of the vehicles when the secondary impact occurs is determined by a combination of both the linear and angular motion of the vehicles between when the first impact occurs and the second impact occurs. In the proposed testing method, while both vehicles experience the same linear and angular speed changes due to the primary impact regardless of whether both vehicles are moving or only one of the vehicles are moving, in each different scenario, the relative post-impact linear velocities are different while the relative post-impact angular velocities are about the same. This is true despite the fact that the orientation of the vehicles at separation after the first impact is essentially the same after all the scenarios. The result of this is that the timing of the secondary impact and the orientation of the vehicles when the secondary impact occurs can vary from one impact scenario to the others.

As the results described earlier show, this problem is most pronounced when the initial closing velocity of the vehicles is relatively low. This is aggravated by the fact that with relatively low closing velocities, the time to secondary impact after primary impact is usually longer, and differences in vehicle motions might be accentuated by the tire forces acting on the vehicles between the impacts. Fortunately however, when occupant kinematics are the issue, the effects of the primary impacts in higher speed situations are usually over relatively quickly, and in lower speed impacts, secondary impact is often minor and occurs late enough not to effect the kinematics of the primary impact. This is not always the case, however.

For these reason, it is always recommended to perform the computer simulation of the different impact scenarios before using the test method discussed in this paper. By performing the simulations the investigator can determine any potential problems with this type of testing before expending the resources required for a full scale impact simulation. In addition, the simulation is often useful in determining which single moving vehicle configuration is likely to provide the best results. It should always be remembered, however, that there are often practical limits in how far a vehicle can be crabbed, and in some cases in the amount of weight that can be towed. These practical considerations must also be considered when using the method of testing discussed here.

Finally, a minor problem with the use of EDSMAC for performing the computer simulations needs to be addressed. While the vehicle kinematic calculations it performs are quite helpful, there are separate outputs that it generates in which the accident history, including impact and separation times and the associated speeds at those times, as well as damage based vehicle delta-V's and Principal directions of force are given. The problem with these is that in cases where there is secondary impact, the second impact is not always considered separately from the first, and the numbers for PDOF, delta-v and impact timing are not consistent. Similarly, in cases where there is sideswipe, the damage based PDOF and delta-V's often disagree with those determined from the acceleration. In such cases, the acceleration based calculations would be more accurate. The calculated delta-V's for the three simulations are given in Figures 69 through 74.

CONCLUSIONS

This paper presents a method for simulating a two vehicle moving accident by a full scale impact test involving both vehicles but requiring that only one of the vehicles be towed. The method generally requires that the closing velocity and impact positions between the vehicles be determined, and that the towed vehicle be crabbed so that this velocity and orientation can be duplicated.

The findings in the paper indicate that the simulation method generally works well, but that secondary impacts can present a problem if they occur with lower initial closing speeds or a significant period of time after the primary impact has occurred. In many cases, however, the secondary impact may not be problematic if the occupant kinematics of the primary impact are of principle concern and are completed before the secondary impact happens.

In any case, to detect such problems before performing a full scale impact, and to help determine which of the test vehicles should be towed from a mathematical standpoint, some vehicle impact kinematics computer simulation should be performed and the results carefully examined. In some cases, practical considerations and limitations will play a greater role is determine which vehicle is crabbed, and towed. The method described was finally used to successfully verify the results of the reconstruction of a real world accident for which only limited data was available.

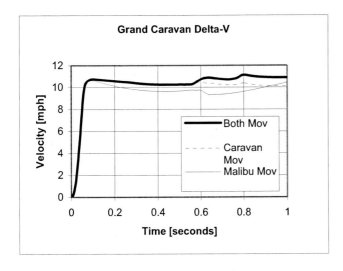

Figure 69- Simulation 1: Caravan Speed Changes

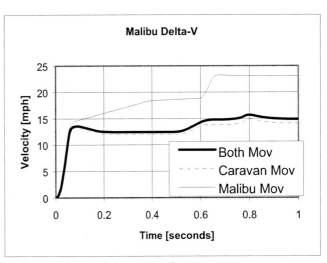

Figure 70-Simulation 1: Malibu Speed Changes

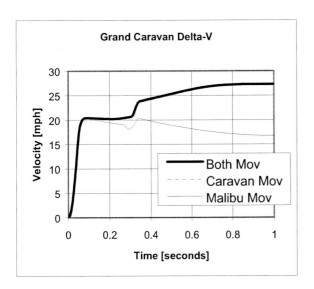

Figure 71-Simulation 2: Caravan Speed Changes

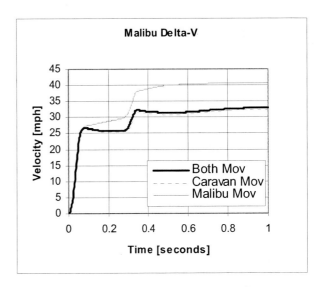

Figure 72-Simulation 2: Malibu Speed Changes

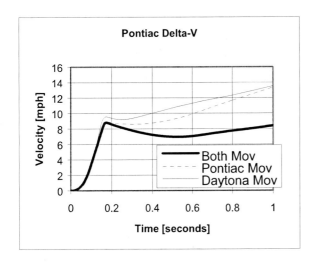

Figure 73-Simulation 3: Pontiac Speed Changes

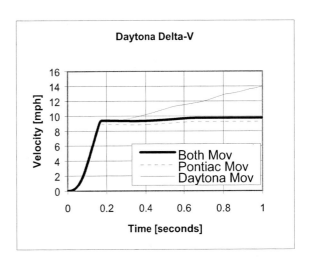

Figure 74-Simulation 3: Daytona Speed Changes

2002-01-0559

A Simulation Model for Vehicle Braking Systems Fitted with ABS

Terry D. Day and Sydney G. Roberts
Engineering Dynamics Corp.

Copyright © 2002 Society of Automotive Engineers, Inc.

ABSTRACT

Most vehicles built today are fitted with anti-lock braking systems (ABS). Accurate simulation modeling of these vehicles during braking as well as combined braking and steering maneuvers thus requires the effects of the ABS to be included. Simplified, lump parameter models are not adequate for detailed, 3-dimensional vehicle simulations that include wheel spin dynamics. This is especially true for simulating complex crash avoidance maneuvers. This paper describes a new ABS model included in the HVE simulation environment. It is a general purpose model and is available for use by any HVE-compatible vehicle simulation model. The basic operational and control characteristics for a typical ABS system are first reviewed. Then, the specific ABS model and its options as implemented in the HVE simulation environment and employed by the SIMON vehicle simulation model are described. To validate the model, pressure cycles produced by the model are compared with stated engineering requirements. In addition, pressure vs. time histories for two ABS simulations on surfaces with different frictional characteristics are compared with experimental data. Finally, the gross effects of ABS on two simulated maneuvers (straight-line braking and ISO 3888 lane-change maneuver) are presented.

IT IS AN OLD ADAGE: A locked tire can't steer. Few old adages are truer than this one, and the safety implications of a vehicle without steering are obvious.

*Numbers in brackets designate references found at the end of the paper.

Recognizing the importance of maintaining steering control, vehicle engineers have for many years designed methods to prevent tires from locking. Harned, et. al. [1] addressed the issue directly in 1969. In the mid-seventies, the US DOT enacted FMVSS 121 [2], essentially requiring some type of ABS control to achieve mandated stopping distances for heavy trucks. However, hardware technologies of that era were inadequate and unreliable. As a result, the essential elements of FMVSS 121 were revoked in the late seventies. The eighties saw substantial research into various ABS hardware technologies (e.g., [3, 4]). However, it was not until the common usage of on-board computer control systems, brought about in large part by major improvements in the speed and reliability of microprocessors and related electronic hardware, that ABS became a standard feature on most passenger cars and light trucks. The nineties saw a significant change in the focus of research from basic hardware issues to improvements and innovations in control algorithms (e.g., [5, 6]; there are countless others in the literature). By 1996, sixty-two percent of the US passenger car and light truck population incorporated ABS [7].

It is interesting to note that most researchers have concluded that the widespread usage of ABS-equipped vehicles has not brought about the expected reduction in crashes. Although there is some disagreement as to why this is the case, the most likely reason is that ABS-equipped vehicles still leave the road - albeit under the directional control of their drivers (suggesting that no technology can replace driver education and experience) [7].

Because vehicles are still crashing, safety researchers are still faced with the ongoing need to reconstruct those crashes to determine their cause. However, their reconstruction has been further complicated by the introduction of ABS-equipped vehicles in two ways. First, ABS-equipped vehicles leave little or no skidmarks during straight-line braking, and second, the directional vehicle dynamics during a combined steering and braking maneuver (such as a loss of control preceding an off-road crash) may be significantly affected by the presence of ABS. For example, a non-ABS-equipped vehicle will typically skid straight during heavy braking - regardless of the amount of steering. An ABS-equipped vehicle, on the other hand, will typically respond to the driver s steering input.

Vehicle handling simulation has been an important tool in the study of loss of control crashes. To study the loss of control of ABS-equipped vehicles thus requires that the simulation be able to model the effects of the ABS system on the resulting vehicle trajectory.

This paper describes a new ABS model implemented in the HVE [8] simulation environment. It is a general purpose model available for use by any HVE-compatible vehicle simulation model. The model is applicable to the design of ABS systems as well as to the study of loss-of-control crashes of ABS-equipped vehicles. The basic operational and control characteristics of a typical ABS system are first reviewed. Then, the specific ABS model and its options as implemented in the HVE simulation environment and used by the SIMON [9] vehicle simulation model are described. To validate the model, pressure cycles produced by the model are compared with stated engineering requirements. In addition, pressure vs. time histories for two ABS simulations on surfaces with different frictional characteristics are compared with experimental data. Finally, the gross effects of ABS on two simulated maneuvers (straight-line braking and ISO 3888 lane-change maneuver) are presented.

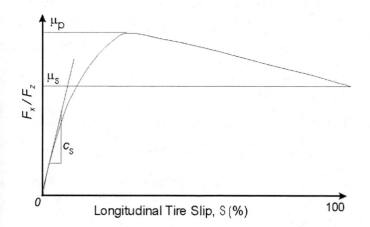

Figure 1 - Typical graph of normalized braking force vs. longitudinal tire slip (graph is also called a mu-slip curve).

OVERVIEW OF ABS

The basic concept behind ABS is quite simple and can be demonstrated by the graph of normalized braking force vs. longitudinal tire slip shown in Figure 1. This graph is traditionally called a *mu-slip curve*. It defines the relationship between longitudinal tire slip and the available longitudinal (braking) force. A key observation is that the maximum braking force occurs at μ_p (peak friction coefficient) in the vicinity of 10 to 15 percent longitudinal tire slip (this varies somewhat from tire to tire). Also, as the tire slip continues to increase to 100 percent, the available braking force falls off. The region of tire slip between μ_p and μ_s (slide friction coefficient, or 100 percent longitudinal slip, associated with locked-wheel braking) is a region of dynamic instability. As slip begins to increase beyond μ_p it quickly increases to 100 percent (i.e., the tire locks) with a commensurate reduction in available braking force.

The goal of an ABS system is simply to prevent the tire slip from increasing significantly beyond μ_p regardless of how much brake pedal effort is applied by the driver. By limiting longitudinal slip, the tire continues to roll and, therefore, maintains directional control capability (i.e., the driver can steer the vehicle). In addition, as shown in Figure 1, the available braking force is larger than for a locked tire and, therefore, braking distance can be reduced.

ABS simulation takes advantage of the simulation s wheel spin degree of freedom, wherein the braking force is calculated from first principles, rather than simply specified as a force at the tire-road interface. To *truly simulate* ABS, the algorithm must modulate the simulated brake pressure, just as it does on an actual vehicle. The procedures for accomplishing this task are described below.

ABS Methodologies

All ABS methodologies work by controlling longitudinal tire slip. This is accomplished through the use of wheel sensors that compare the tire circumferential velocity to the current *reference velocity*, V_r, normally calculated using the current spin velocities of two or more wheels (see Reference 10 for a detailed discussion of the calculation of reference velocity). On the vehicle, longitudinal tire slip cannot be measured directly. Instead, slip is calculated:

$$Slip = \frac{V_r - \dot{\Omega}_w \times R_{tire}}{V_r} \quad \text{(Eq. 1)}$$

where

V_r = Reference velocity

$\dot{\Omega}_w$ = Wheel spin velocity

R_{tire} = Tire rolling radius

State Variables

To accomplish the required control of longitudinal slip, the following state variables are monitored or calculated by the vehicle's ABS control module:

- Vehicle Velocity - Linear velocity of the vehicle sprung mass
- Wheel Spin Velocity - Angular velocity of each wheel (or axle on some systems)
- Tire Longitudinal Slip - Relative velocity between the tire and road, expressed as a fraction of vehicle velocity
- Wheel Spin Acceleration - Angular acceleration of each wheel
- Tire-Road Surface Friction - Ratio of the maximum braking force to the normal tire force
- Brake System Pressure - Pressure produced as a result of the driver's brake pedal application (input variable)
- Wheel Brake Pressure - Pressure supplied to the wheel brake assembly (output variable)

Typical Hardware

To monitor or calculate the above state variables, the typical vehicle ABS system includes the following hardware components:

- Electronic Control Unit (ECU) - This is one of the vehicle's microcomputers. It is programmed with the algorithm that reads the current state variables, determines the required pressure at each wheel and sends the appropriate signals to the brake pressure modulator (see below).
- Wheel Speed Sensor - These components directly measure the wheel spin velocity of each wheel using a wheel-mounted pulse rotor (a notched metal ring) and a fixed, magnetic sensor that measures the rotation of the pulse rotor.
- Brake Pressure Modulator - This component (or components, depending on the system) controls the wheel brake pressure according to the control conditions specified by the ECU.
- Brake Master Cylinder/Air Compressor - This component provides the fluid pressure source.
- Wheel Brake Caliper/Cylinder/Chambers - These components apply the braking force at each wheel according to the wheel brake pressure.

The basic hardware requirements are generally the same for all vehicle types, ranging from passenger cars to on-highway trucks. Reference 10 provides a detailed description of these required components.

Table 1. HVE ABS System Variables. These variables apply to the overall vehicle (compare with Wheel Variables in Table 2).

Variable	Description
Algorithm	ABS algorithm selected from a list of available algorithms
Control Method	ABS control method selected from a list of available control methods
Cycle Rate	Sets the time required for a complete ABS cycle
Threshold ABS Pressure	Minimum pressure for ABS activation
Threshold ABS Velocity	Minimum vehicle velocity for ABS activation
Friction Threshold	Tire-terrain surface friction threshold
Delay Method	Delay method selected from a list of available delay methods
Apply Delay	Time delay for controlled output pressure increase
Release Delay	Time delay for controlled output pressure release

ABS USER INTERFACE

The HVE ABS user interface allows the user to select an ABS algorithm and to enter and edit the independent parameters required by the selected ABS algorithm. The interface includes numerous options, thus, various algorithms may be supported. The interface is divided into two sections:

- System Variables - Variables that are applicable to the entire vehicle
- Wheel Variables - Variables that are applicable to (and may be specified independently for) each wheel

System Variables

The ABS system variables included in the HVE ABS model are shown in Table 1. A brief description of each variable follows:

- ABS Algorithm - This is the ABS algorithm, selected from a list of the various ABS algorithms available to the user. The algorithms currently available are *Tire Slip* and *HVE Bosch Version 1*. This list can be updated as new algorithms become available.
- Control Method - This option determines if all wheels are controlled by a single controller or if individual wheels or axles are controlled separately. Vehicle-based sampling

uses the same control cycle (see below) for all wheels; axle-based control allows different control cycles for the front and rear axles (typically only the rear axle is controlled); wheel-based control cycles allow different control cycles for each wheel.

- Cycle Rate - If the *Control Method* is vehicle-based, this parameter provides the maximum time required to perform a complete ABS cycle. It is the same for all wheels.

- Threshold ABS Pressure - This parameter provides a minimum system pressure threshold for ABS actuation. ABS is bypassed when the system pressure is below this value.

- Threshold Vehicle Velocity - This parameter provides a minimum vehicle velocity threshold for ABS actuation. ABS is bypassed when the vehicle velocity is below this value.

- Low Surface Friction Threshold - This parameter sets a threshold defining low friction surfaces. Algorithms can use this parameter to invoke friction-dependent modulation behaviors. For example, the *Delay Interval* (see below) may be reduced for low friction surfaces.

- Delay Method - This parameter determines if control pressure delays are vehicle-based, axle-based or wheel-based. Vehicle-based delay uses the same delay period for each wheel; axle-based delay allows different delay periods for each axle; wheel-based delay allows different delay periods for each wheel.

- Apply Delay - If the *Delay Method* is vehicle-based, this parameter determines the delay period for all wheels before output pressure is increased. (See below for axle-based and wheel-based delay.)

- Release Delay - If the *Delay Method* is vehicle-based, this parameter determines the delay period for all wheels before output pressure is reduced. (See below for axle-based and wheel-based delay.)

Wheel Variables

Wheel variables are those ABS parameters that are assigned independently for each wheel. The ABS wheel variables included in the HVE ABS model are shown in Table 2. A brief description of each variable follows:

- Cycle Rate - If the ABS System *Control Method* is *At Axle* or *At Wheel*, this parameter determines the maximum time required to perform a complete ABS cycle for the selected wheel.

- Threshold Wheel Velocity - This parameter specifies a minimum wheel velocity threshold for ABS actuation. ABS is bypassed for this wheel when the wheel forward velocity is below this value.

- Minimum Tire Slip - This parameter may be used in an algorithm to establish a lower threshold for tire longitudinal slip.

- Maximum Tire Slip - This parameter may be used in an algorithm to establish an upper threshold for tire longitudinal slip.

Table 2. HVE ABS Wheel Variables. These variables apply to individual wheels (compare with System Variables in Table 1).

Variable	Description
Cycle Rate	Sets the time required for a complete ABS cycle
Threshold Wheel Vel	Minimum wheel forward velocity for ABS control
Tire Min Slip	Minimum tire longitudinal slip for ABS control
Tire Max Slip	Maximum tire longitudinal slip
Wheel Min Spin Accel	Minimum wheel angular acceleration
Wheel Max Spin Accel	Maximum wheel angular acceleration
Apply Delay	Time delay for controlled output pressure increase
Primary Application Rate	Initial rate of controlled output pressure increase
Secondary Application Rate	Secondary rate of controlled output pressure increase
Release Delay	Time delay for controlled output pressure decrease
Release Rate	Rate of controlled output pressure decrease

- Minimum Wheel Spin Acceleration - This parameter may be used in an algorithm to establish a lower threshold for wheel spin acceleration.

- Maximum Wheel Spin Acceleration - This parameter may be used in an algorithm to establish an upper threshold for wheel spin acceleration.

- Apply Delay - If the *Delay Method* is *At Axle* or *At Wheel*, this parameter determines the control pressure delay period before output pressure is increased at the selected wheel.

- Primary Application Rate - This parameter determines the initial rate of output pressure increase.

- Secondary Application Rate - This parameter provides an alternative rate of output pressure increase, usually substantially lower than the *Primary Application Rate*.

- Release Delay - If the *Delay Method* is *At Axle* or *At Wheel*, this parameter determines the control pressure delay period before output pressure is reduced at the selected wheel.

- Release Rate - This parameter determines the rate of output pressure decrease.

The ABS System and Wheel Variables are provided as a palette of parameters available to the designer of an ABS system. The selection of individual parameters and their effect on the simulated characteristics of any specific ABS system are algorithm-dependent.

DESCRIPTION OF CURRENT MODELS

Two ABS algorithms are currently implemented. These are the *Tire Slip* algorithm and the *HVE Bosch Version 1* algorithm. These algorithms are described below.

Tire Slip Algorithm

This is a simple and straight-forward ABS algorithm. Its design is based on the fundamental goal of an ABS system, that is, to maintain tire slip in the vicinity of peak friction coefficient, μ_p (refer to Figure 1). It is generally applicable to any type of vehicle (passenger car, truck, etc). Tire *Minimum Slip* and *Maximum Slip* parameters are selected to be about 5 percent below and above, respectively, the tire peak friction coefficient.

Upon brake pressure application, once ABS is invoked (that is, the minimum vehicle velocity and brake pressure thresholds are exceeded), the algorithm incorporates two switches, depending on the current tire slip:

- Tire Slip ≤ Minimum Slip - Under this condition, the status of the ABS during the previous sample determines how pressure is modulated for the current sample. If the ABS modulation status was off, the output pressure is set equal to the input pressure and the ABS system parameters (delays, etc.) are reset. Otherwise, brake pressure will be controlled. One of two possibilities exists: a) Input pressure is decreasing. In this case, output pressure is set equal to input pressure and the ABS status is turned off, or b) Input pressure is constant or increasing. In this case, the output pressure is maintained at a constant for the specified *Apply Delay,* after which output pressure is increased according to the *Primary Apply Rate*.

- Minimum Slip < Tire Slip < Maximum Slip - Under this condition, pressure control will occur according to the specified *Cycle Interval*. One of two possibilities exists: a) Input pressure is decreasing. In this case, output pressure is set equal to input pressure and the ABS status is turned off, or b) Input pressure is constant or increasing. In this case, the pressure is maintained at a constant for the specified *Apply Delay,* after which output pressure is increased according to the *Secondary Apply Rate*.

- Tire Slip ≥ Maximum Slip - Under this condition, sampling will occur according to the specified *Cycle Interval*. The output pressure is maintained at a constant for the specified *Release Delay,* after which the output pressure is reduced according to the *Release Rate*.

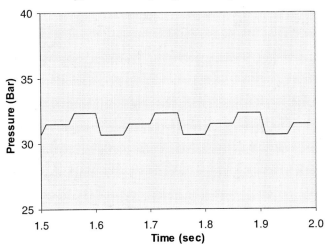

Figure 2 - Typical control cycles for Tire Slip ABS algorithm.

Figure 2 shows a typical pressure vs. time history for a few cycles of a hard brake pedal application (i.e., enough system pressure to lock the wheel). The flow chart for the *Tire Slip* algorithm is shown in Appendix II.

HVE Bosch Version 1 Algorithm

The HVE Bosch Version 1 ABS algorithm[*] is based on the information provided in reference 10. The Bosch ABS system is used on many passenger cars. The algorithm is based on wheel spin acceleration and a critical tire slip threshold.

Upon brake pressure application, once ABS is invoked (i.e., the thresholds are exceeded), the current brake pressure application is divided into eight phases (see Figure 3).

Phase 1 - Initial application. Output pressure is set equal to input pressure. This phase continues until the wheel angular acceleration (negative) drops below the *Wheel Minimum Spin Acceleration*, -α.

Phase 2 - Maintain pressure. Output pressure is set equal to previous pressure. This phase continues until the tire longitudinal slip exceeds the slip associated with the *Slip Threshold*. At this time, the current tire slip is stored and used as the slip threshold criterion in later phases. This slip corresponds to the maximum slip; the tire is beginning to lock.

---

[*]This implementation was developed by EDC based on the information provided in reference 10. Version 1 is the HVE version number, not the Bosch version number.

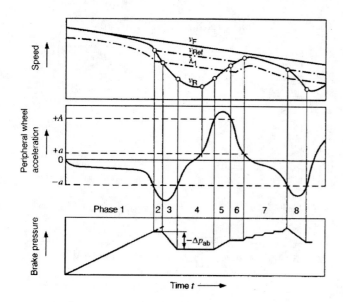

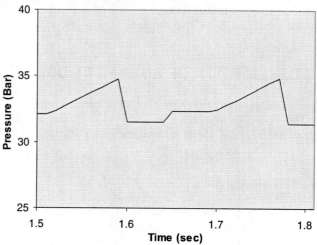

Figure 3 - Idealized control sequence for the Bosch algorithm (reprinted from reference 10). Compare the lower portion with the pressure cycles displayed in Figure 4.

Figure 4 - A single control cycle produced by the HVE Bosch Version 1 ABS algorithm. Comparison of this cycle with the idealized brake pressure control cycle shown in Figure 3 reveals the basic characteristic of the Bosch algorithm has been achieved.

Phase 3 - Reduce pressure. Output pressure is decreased according to the *Release Rate* until the wheel spin acceleration becomes positive (this is a slight modification to the sequence shown in Figure 3, in which the pressure is decreased until the spin acceleration exceeds -a).

Phase 4 - Maintain pressure. Output pressure is set equal to the previous pressure for the specified *Apply Delay*, or until the wheel spin acceleration (positive) exceeds $+A$, a multiple (normally 10x) of the *Wheel Maximum Spin Acceleration,* $+a$, (signifying the wheel spin velocity is increasing at an excessive rate).

Phase 5 - Increase pressure. Output pressure increases according to the *Primary Apply Rate*. This phase continues until the wheel spin acceleration drops and again becomes negative (this is a slight modification to the sequence shown in Figure 3, in which the pressure is increased until the spin acceleration drops below $+A$).

Phase 6 - Maintain pressure. Output pressure is set equal to previous pressure for the specified *Apply Delay*, or until wheel angular acceleration again exceeds the *Wheel Minimum Wheel Spin Acceleration* (negative).

Phase 7 - Increase pressure. Output pressure increases according to the *Secondary Apply Rate*, normally a fraction (1/10) of the *Primary Apply Rate*. This achieves greater braking performance while minimizing the potential for wheel lock-up at tire longitudinal slip in the vicinity of peak friction. This phase continues until wheel angular acceleration drops below the *Wheel Minimum Angular Acceleration* (negative), indicating wheel lock-up is eminent.

Phase 8 - Reduce pressure. At this point an individual cycle is complete, the process returns to Phase 3 and a new control cycle begins.

As stated, some minor differences exist between the HVE implementation and the Bosch description provided in Reference 10. These differences reflect some inconsistencies between the acceleration and pressure profiles shown in Figure 3. For example, unless the throttle is applied, it is physically inconsistent that the wheel acceleration would be positive, let alone increase (as shown in Phase 5), in the presence of increased brake pressure (and, therefore, brake torque).

Each of the above phases begins with a comparison between the current tire longitudinal slip and the value stored during Phase 2. If the current slip exceeds this value, the normal logic is bypassed and resumed at Phase 3. This effectively allows the algorithm to learn the wheel slip associated with wheel lock-up on the current surface. This is referred to as adaptive learning, and is a key to the success of this ABS algorithm. As the tire travels onto surfaces with differing friction characteristics, the ABS model is able to maximize its performance accordingly.

Default parameters used by the HVE Bosch Version 1 algorithm were developed through the evaluation of numerous simulation runs. See Appendix I for typical parameters applicable to a P195/75R14 passenger car tire.

Figure 4 shows a typical pressure vs. time history for a few cycles of a hard brake pedal application (i.e., enough system pressure to lock the wheel). The flow chart for the HVE Bosch Version 1 algorithm is shown in Appendix III.

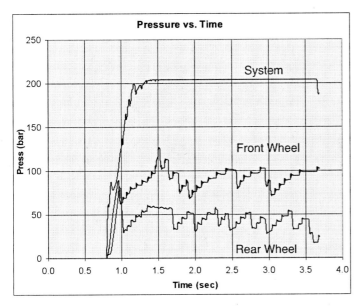

Figure 5 - Experimental braking test results for an ABS-equipped vehicle on a high-friction (asphalt) surface.

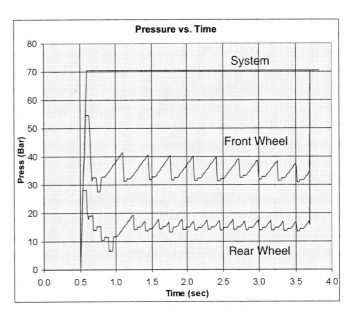

Figure 6 - Simulated braking test results for an ABS-equipped vehicle on a high-friction (asphalt) surface.

Other Algorithms

The ABS model implemented in HVE is not restrictive in terms of the algorithms it can support, other than its need to provide the parameters required by the algorithm. Endless tweaking of an algorithm is possible, resulting in different ABS system characteristics, each with its advantages and disadvantages. Thus, it is certain that new ABS algorithms will be developed and implemented in HVE over time, both to develop and to model new ABS systems.

COMPARISON WITH EXPERIMENT

Results from simulations using the HVE ABS model were compared against experimental data provided by Robert Bosch USA GmbH. These straight-line braking tests were performed on various surfaces at various speeds. The vehicle type and Bosch ABS version were not identified, although the vehicle probably used Bosch ABS 5.2. Because specific and detailed data for the vehicle, ABS system and tire-road frictional characteristics were lacking, no attempt was made to duplicate the experimental runs. Rather, the purpose of these comparisons was to isolate the general characteristic trends found in the experimental pressure vs. time histories and compare them against time histories simulated using the HVE ABS model with default parameters. The specific parameters are provided in Appendix I.

High Friction Surface

Figure 5 shows experimental test results on a high friction surface (asphalt) at an initial speed of 100 Km/h (62 mph). Master cylinder and front and rear wheel pressure histories are presented. Figure 6 shows SIMON simulation results for a Generic Class 2 Passenger Car [12] with the HVE ABS model enabled. The simulation uses the HVE Bosch Version 1 algorithm.

Comparison of Figures 5 and 6 reveals the basic characteristics are quite similar. Both show approximately 2 to 4 cycles per second for the front wheels and 4 to 6 cycles per second for the rear wheels. A similarity in the detailed pressure characteristics within each control cycle is also seen.

Comparison between experiment and simulation also shows a similar proportion of front and rear brake pressures, as well as system pressure (the actual pressures are different because the vehicle weights and brake torque ratios are different; insufficient test data were available to attempt to duplicate the Bosch tests).

Control pressure at the end of the simulation increases quickly back to system pressure as the vehicle velocity drops below the velocity threshold and comes to rest (the test vehicle's brakes are released and it does not come to rest within the time presented in Figure 5).

Low Friction Surface

Figure 7 shows experimental test results on a low friction surface (ice) at an initial speed of 50 Km/h (31 mph). Master cylinder (i.e., system) and front and rear wheel pressure histories are presented. Figure 8 shows SIMON simulation results for a Generic Class 2 Passenger Car [12] with the HVE ABS model enabled. The simulation uses the HVE Bosch Version 1 algorithm with the same default parameters as those used in the test on the high friction surface (see Appendix I).

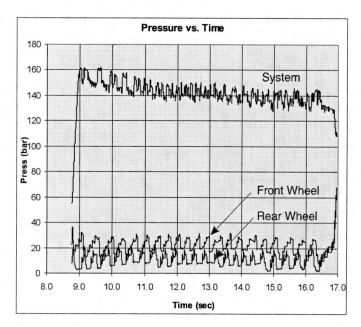

Figure 7 - Experimental braking test results for an ABS-equipped vehicle on a low-friction (ice) surface.

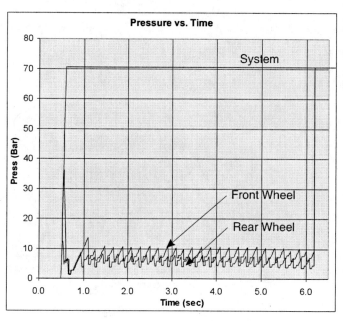

Figure 8 - Simulated braking test results for an ABS-equipped vehicle on a low-friction (ice) surface.

The source of the noise in the master cylinder pressure trace is unknown (because the master cylinder pressure is significantly greater than the controlled wheel pressures, this factor is not considered important in the test results). Again, the basic characteristics of the controlled wheel pressures are quite similar. Both show approximately 4 to 6 cycles per second for the front wheels and slightly higher cycle rates for the rear wheels. The cycle frequency is a natural consequence of the wheel spin dynamics (spin accelerations are higher on a low-friction surface; see earlier discussion of the eight phases in a Bosch cycle). Comparison between experiment and simulation again shows a similar proportion of front and rear brake pressure.

Control pressures for both the test and the simulation increase quickly to master cylinder pressure as the vehicle velocity drops below the velocity threshold comes to rest.

EXAMPLES

The effects of an ABS braking system on vehicle handling are illustrated through the use of two simulation examples. The first is a straight-line braking test with and without ABS model invoked. The second is an ISO 3888 lane-change maneuver during hard braking with and without the ABS model invoked. Both examples use default ABS parameters. The purpose of these simulations is to confirm the expected behavior of the ABS model, that is, to reduce braking distance and provide steering control during a hard braking and steering maneuver.

Straight-line Braking

A reduction in stopping distance is expected to be provided by ABS during straight-line braking. This example illustrates the reduction in stopping distance on a typical asphalt surface. The vehicle is a Generic Class 2 Passenger Car [12] fitted with Generic P195/75R14 tires. The initial velocity is 100 km/h (62 mph). A 300 N force is applied to the brake pedal. No steering is applied. The HVE Bosch Version 1 algorithm is used in this example.

Figures 9 and 10 show distance, velocity and acceleration vs. time with and without the ABS system activated. Braking begins at t = 0.5 seconds. Stopping distance is reduced from 48.7 m (160 ft.) to 47.2 m (155 ft), a 3.2 percent reduction. Braking time is reduced from 3.44 seconds to 3.33 seconds. The calculated average deceleration rate increased from 0.80 g to 0.83 g. Close inspection of the acceleration vs. time history in Figure 9 shows the modulation due to the ABS (compare with Figure 10, which is smooth due to the locked wheels).

ISO 3888 Lane-change Maneuver

The chief benefit of ABS is that the driver's ability to maintain vehicular control during a heavy braking and steering maneuver is significantly improved. To illustrate this point, a simulation of an ISO lane change maneuver is executed during a panic brake application. The vehicle is a Generic Class 2 Passenger Car. The initial velocity is 80 km/h (50 mph). The HVE Driver Model (path follower) [8] was used, and a 300 N sudden brake pedal force was applied at t= 0.5 seconds. The HVE Bosch Version 1 algorithm is used in this example.

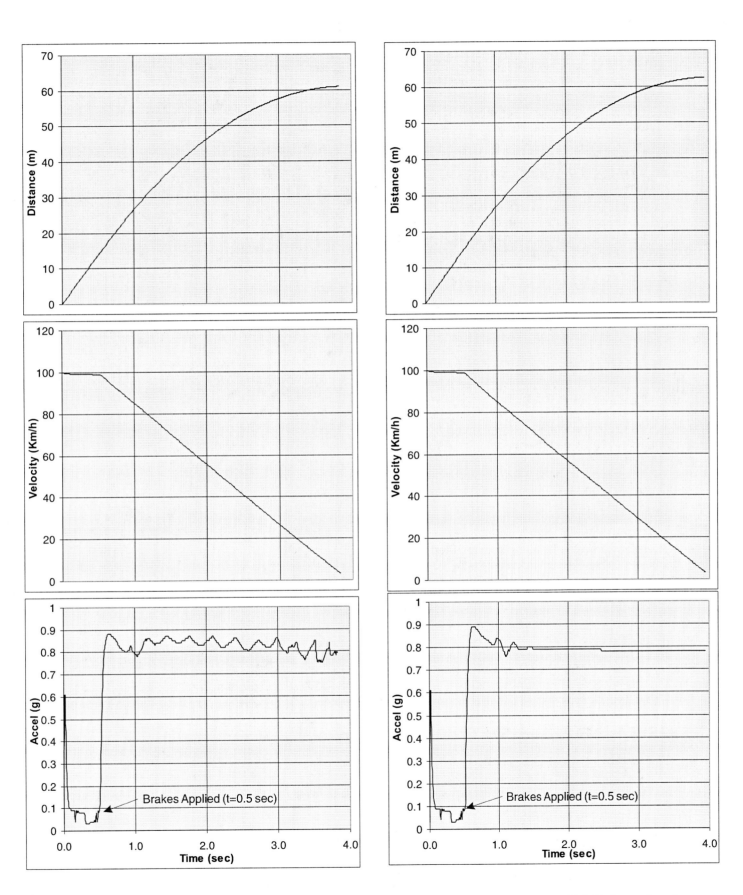

Figure 9 - Simulation results on a high-friction surface for an ABS-equipped vehicle during straight-line braking from 100 km/h.

Figure 10 - Simulation results on a high-friction surface for a non-ABS-equipped vehicle during straight-line braking from 100 km/h.

Figures 11 and 12 provide a visualization of the trajectory of each vehicle. Note that the vehicle with ABS successfully stays within the cones, while the vehicle without ABS fails to perform the lane-change, skidding almost straight ahead (because steering preceded braking in this example, there is a change in direction prior to skidding). This result is typical for non-ABS-equipped vehicles.

DATA REQUIREMENTS

From a vehicle design engineer's standpoint, the HVE ABS model provides a palette of parameters used by ABS systems. New ABS algorithms may be written and then tested directly against experiment. Parameter optimization may likewise be performed via simulation prior to reprogramming the vehicle's ABS controller (ECU) firmware.

From a crash reconstruction engineer's standpoint, the most important task of the HVE ABS model is to allow the simulation analysis of vehicular loss of control of ABS-equipped vehicles. This requires the use of parameters that maintain high levels of braking force while also preventing wheel lock-up.

For the Tire Slip algorithm, minimum and maximum tire slip values are required by the model. These values are tire-specific and can be determined by inspection of the tire's mu vs. slip curve. The default values used by the HVE ABS model are 0.05 and 0.15 for minimum and maximum slip.

For the HVE Bosch Version 1 algorithm, minimum and maximum wheel spin accelerations are required by the model. These values vary according to tires size (specifically, tire spin inertia). Default values used by the HVE ABS model were assigned according to vehicle class category [12] and were determined via simulation experiments. Default values for apply and release rates were also determined via simulation experiments.

The values chosen for the HVE Bosch comparisons presented in this paper are shown in Appendix I.

TRUCK (AIR BRAKE) SYSTEMS

Preliminary results show the HVE ABS model is applicable to air brake systems such as those used by on-highway trucks. The Tire Slip algorithm works well. However, the Bosch algorithm requires additional study to determine the minimum and maximum wheel spin accelerations required by that algorithm. The difference in required wheel spin acceleration values is attributed to the significantly larger spin inertia of a truck tire compared with a passenger car tire. It is expected that time delays will also require adjustment. Research on the parameters required for truck air brake ABS simulation is underway.

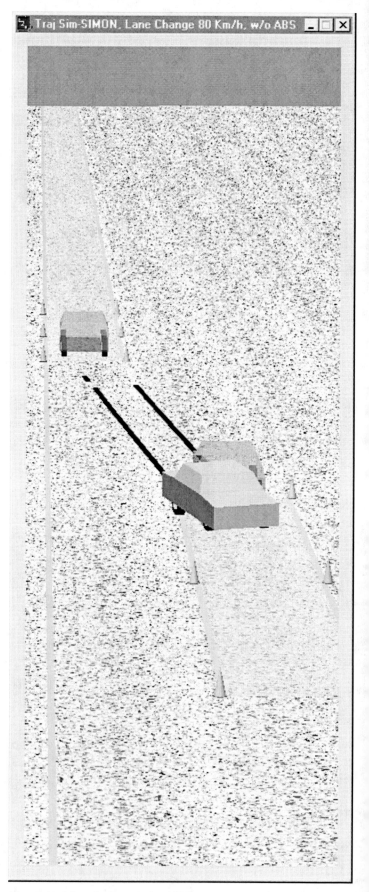

Figure 11 - Visualization of a lane-change maneuver with hard braking, non-ABS-equipped vehicle. The solid vehicle is the simulated position; the translucent vehicles are path "target" positions.

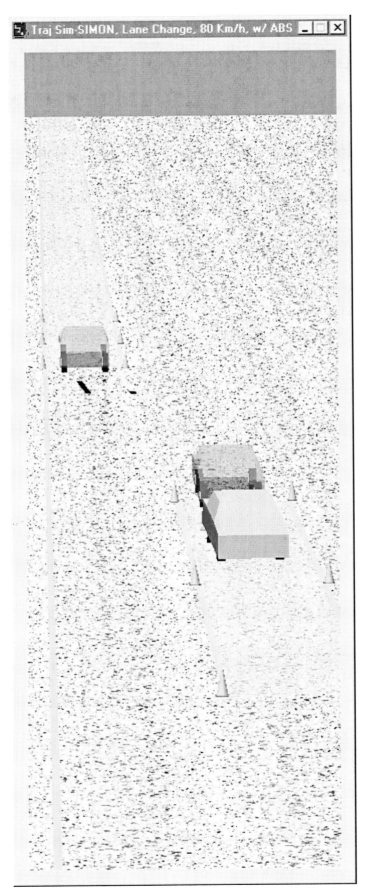

Figure 12 - Visualization of a lane-change maneuver with hard braking, ABS-equipped vehicle. The solid vehicle is the simulated position, the translucent vehicles are path "target" positions.

DISCUSSION

Validation of the HVE Bosch Version 1 algorithm is provided in the comparison between Figures 3 and 4, wherein it is shown that the Bosch control cycle has been nearly duplicated by the HVE Bosch Version 1 algorithm.

Additional comparisons between experimental maneuvers and simulations should be performed. It is important to recognize, however, that such comparisons would not serve to validate the HVE ABS model, per se, because of the difficulty in separating any differences attributable to the HVE ABS model from differences attributable to other parts of the vehicle simulation model (especially the tire model). Note, for example, that the same HVE ABS model implemented in two different vehicle simulators would undoubtedly yield slight differences in vehicle kinematics during the simulation used in the ISO 3888 lane-change example provided earlier in this paper. In addition, the vehicle and tire data requirements for such a study would be immense. It is unrealistic to believe that such data would be available to most researchers. The primary benefit from comparisons between experimental maneuvers and simulations would be an increased level of confidence in the trends predicted by simulation of ABS-equipped vehicles.

The simulations presented in this paper used a Generic Class 2 Passenger Car having a weight of 1119 kg (2469 lb), a wheelbase of 254 cm (99.9 in) and Generic P195/75R14 tires with μ_p in the range of 0.767 - 0.903 (load-dependent) and μ_s in the range of 0.684 - 0.804 (again, load-dependent) [12].

The HVE Bosch algorithm is more sophisticated than the Tire Slip algorithm, and while it has additional capabilities, such as adaptive learning, its input data requirements are also more demanding. The Tire Slip algorithm simply requires estimates for minimum and maximum slip; these values are available by inspection of the tire's mu vs. slip curve. Although the Tire Slip algorithm was not used for the examples cited in this paper, its effect on vehicle gross handling behavior is remarkably similar to the HVE Bosch Version 1 algorithm. When simulating an ABS-equipped vehicle with an unknown ABS type, the Tire Slip algorithm is recommended.

The primary purpose of ABS is to prevent the wheels from locking so directional control can be maintained. The secondary purpose of ABS is to maintain (higher) braking force associated with peak friction, rather than slide friction, in order to reduce braking time and distance.

The requirements for the vehicle simulation model include a brake system model, a robust tire model that includes μ_p and μ_s, and a spin degree of freedom for each wheel to calculate current wheel velocities and accelerations.

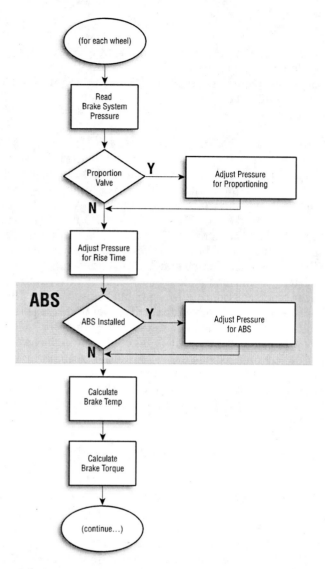

Figure 13 - Flow chart showing integration of the HVE ABS model into the simulation.

Incorporating the ABS model into an HVE-compatible simulation model is a relatively straight-forward task. The ABS model is incorporated as a C-style function call in the vehicle simulation model. The ABS function call typically is placed directly before the wheel brake temperature or brake torque function call (see Figure 13).

Implementing ABS in a simulation model is significantly easier than designing, manufacturing and installing ABS on a vehicle. There are several practical reasons for this. First, the computer simulation has no vehicle hardware issues, either mechanical or electrical. Manufacturers of brake system components will readily appreciate the difficulty in developing adequate solenoid-actuated valves, wheel speed sensors, computer controllers and electrical connectors. None of these is required for simulation of ABS on a computer.

Vehicle reference velocities are required by ABS algorithms for the computation of tire slip (a key component in all algorithms). In the vehicle, the reference velocity cannot be generated from the speedometer, since wheel lock-up at certain wheels would yield misleading information. Therefore, reference velocity is derived via comparison of multiple individual wheel velocities, typically using proprietary algorithms. In simulation, the required velocity is directly accessible since it is a dependent variable in the equations of motion.

The ABS model described in this paper is extendable to traction control systems (TCS) and yaw moment stability (YMS) as well. However, additional control algorithms are required for implementation of these models.

This paper describes two possible systems (Tire Slip and HVE Bosch Version 1). However, experience has shown that the development of ABS algorithms is a highly creative process. One could envision the possibility of endless modifications or extensions to these and other algorithms. The ABS model implemented in HVE is extendable in this regard.

SUMMARY

1. A basic overview of ABS has been provided, both in terms of the operational characteristics and the required parameters.

2. The required parameters have been defined in terms of overall vehicle system parameters and individual wheel parameters.

3. The HVE ABS user interface has also been described, and is likewise defined in terms of overall vehicle system parameters and individual wheel parameters.

4. Requirements for implementing an ABS module into a simulation were provided. These include a brake system model and spin degrees of freedom for each wheel.

5. Two ABS models have been implemented in HVE to date. These two models, the Tire Slip algorithm and the HVE Bosch Version 1 algorithm, were described and an example of a typical ABS cycle for each algorithm was presented.

6. Comparison between the Bosch control cycle and the cycle simulated using the HVE Bosch Version 1 algorithm reveals they are substantially similar.

7. Experimental results for two tests were compared with simulation results using the HVE Bosch Version 1 algorithm. The results compared favorably.

8. Two simulations illustrating the effect of ABS on vehicle handling were presented. These simulations showed that the ABS system reduced straight-line braking distance by approximately 3 percent and allowed the driver to maintain directional control during a lane-change maneuver with panic braking. Simulation of the latter maneuver without ABS resulted in a loss of control (failure to successfully execute an ISO 3888 lane-change maneuver).

9. It follows from the latter maneuver (see Summary 8, above), that successful simulation of such maneuvers requires a simulation model that has the capability to simulate ABS.

10. The HVE ABS model is a general purpose model, applicable to all vehicle types. Development of default data sets and validation for for heavy trucks is under way.

11. Since the vast majority of the US vehicle population now includes ABS, the ability to model ABS is an important advancement in vehicle simulation modeling.

ACKNOWLEDGMENT

The authors wish to extend their appreciation to Robert Bosch USA for providing the test data used in this paper.

REFERENCES

1. Harned, J.L., Johnston, L.E., Scharpf, G., Measurements of Tire Brake Force Characteristic as Related to Wheel Slip (Antilock) Control System, SAE Paper No. 690214, Society of Automotive Engineers, Warrendale, PA, 1969.

2. Federal Motor Vehicle Safety Standard 121, US Dept. of Transportation.

3. *Anti-lock Braking Systems for Passenger Cars and Light Trucks - A Review*, SAE PT-29, Society of Automotive Engineers, Warrendale, PA, 1986.

4. Suzuki, K., Kanamori, M., Development of the Quick Response Tandem Brake Booster, SAE Paper No. 971110, Society of Automotive Engineers, Warrendale, PA, 1997.

5. Khatun, P., Bingham, C.M., Mellor, P.H., Comparison of Control Methods for Electric Vehicle Antilock Braking / Traction Control Systems, SAE Paper No. 2001-01-0596, Society of Automotive Engineers, Warrendale, PA, 2001.

6. Assadian, F., Mixed H_∞ and Fuzzy Logic Controllers for the Automobile ABS, SAE Paper No. 2001-01-0594, Society of Automotive Engineers, Warrendale, PA, 2001.

7. Garrott, W.R., Mazzae, E.N., An Overview of the National Highway Traffic Safety Administration's Light Vehicle Antilock Brake Systems Research Program, SAE Paper No. 1999-01-1286, Society of Automotive Engineers, Warrendale, PA, 1999.

8. *HVE User's Manual, Version 4, Fourth Edition*, Engineering Dynamics Corporation, Beaverton, OR, February 2001.

9. *SIMON (SImulation MOdel Non-linear) User's Manual, Version 1, Second Edition*, Engineering Dynamics Corporation, Beaverton, OR, January 2001.

10. *Bosch Driving-safety systems*, 2nd Edition, Robert Bosch GmbH, ISBN 0-7680-0511-6, Society of Automotive Engineers, Warrendale, PA, 1999.

11. *HVE ABS Simulation Model - Comparison with Experimental Results*, Engineering Dynamics Corporation, Library Reference No. (TBD), Beaverton, OR 2001.

12. *HVE Generic Human/Vehicle/Tire Database, Version 4.2*, Engineering Dynamics Corporation, Beaverton, OR, December, 2001.

APPENDIX I

ABS variables and values for simulation examples using the HVE Bosch Version 1 Algorithm.

Variable	Value
Algorithm	HVE Bosch Version 1
Control Method	At Wheel
Threshold Wheel Velocity	70.4 rad/sec
Threshold ABS Pressure	70 kPa (10 psi)
Low Friction Threshold	0.35
Delay Method	At Wheel
Apply Delay	0.05 sec
Primary Application Rate	35000 kPa (5000 psi/sec)
Secondary Application Rate	3500 kPa (500 psi/sec)
Release Rate	7000 kPa/sec (10000 psi/sec)
Wheel Minimum Spin Accel	-175 rad/sec^2
Wheel Maximum Spin Accel	50 rad/sec^2
Wheel Maximum Slip	0.15

APPENDIX II

Flow chart for the HVE ABS Tire Slip Algorithm

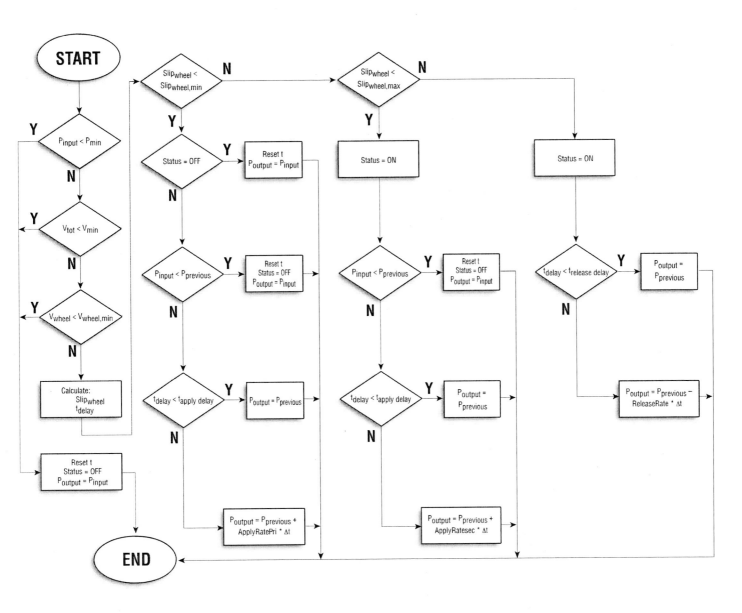

APPENDIX III - *Flow chart for the HVE Bosch Version 1 Algorithm*

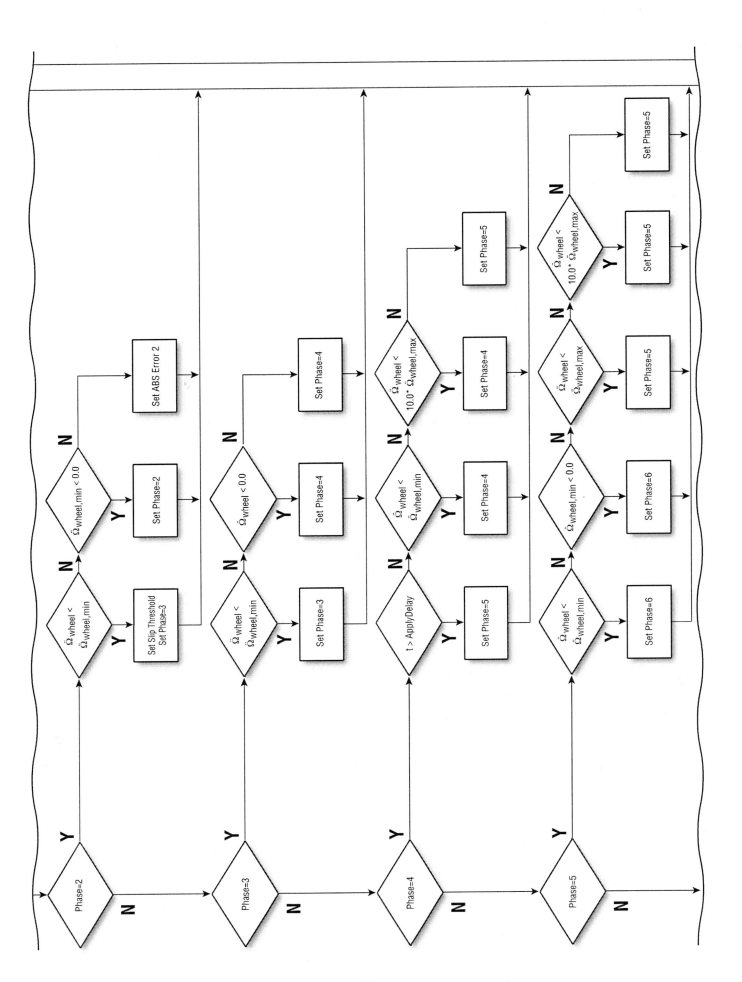

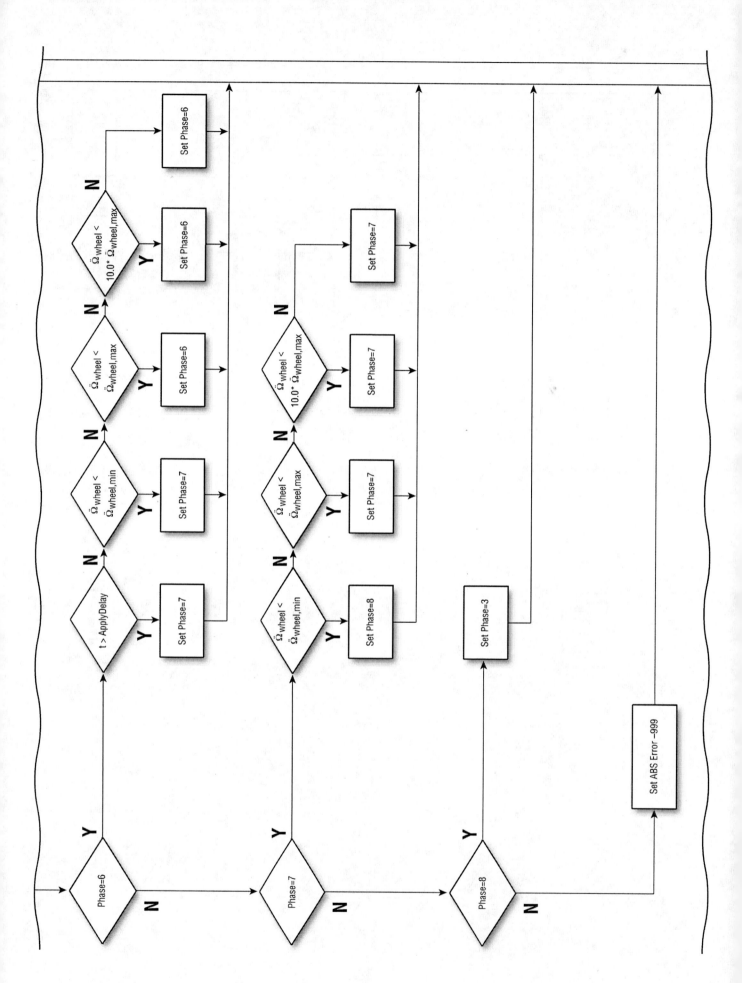

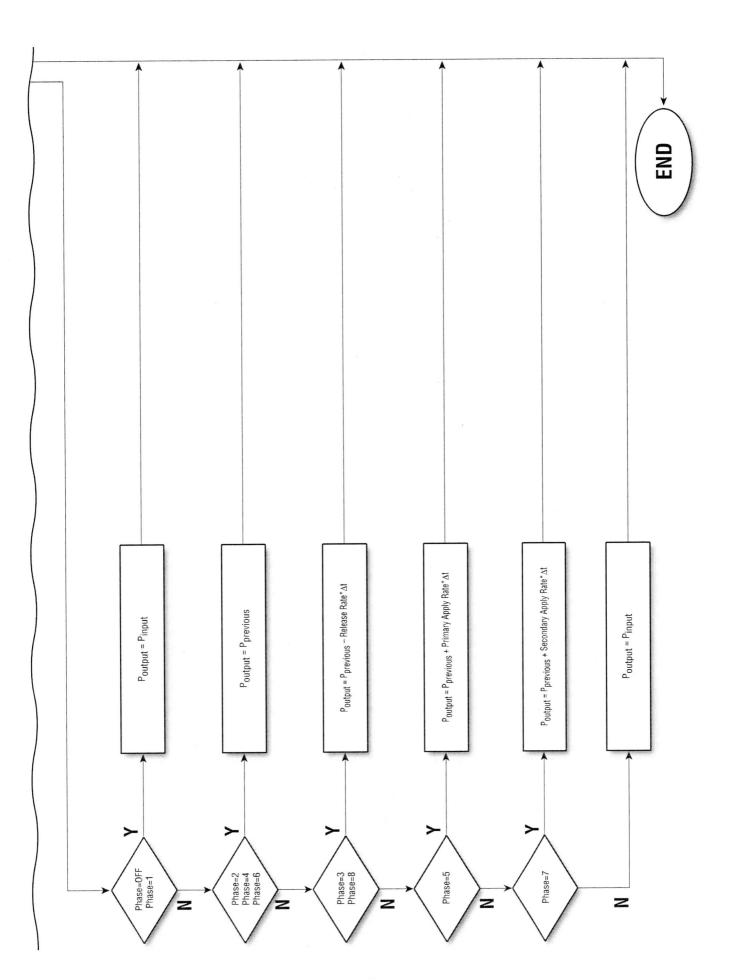

Reviewer Discussion

Reviewers Name: Kristin Bolte, Ph.D.
Paper Title: A Simulation Model for Vehicle Braking Systems Fitted with ABS
Paper Number: 2002-01-0559

Reviewer's Discussion
By Kristin Bolte, Ph.D., National Transportation Safety Board
A Simulation Model for Vehicle Braking Systems Fitted with ABS
Terry D. Day, Sydney G. Roberts, Authors

Anti-lock braking systems (ABS) are becoming much more common on passenger cars and commercial vehicles today. Collisions involving vehicles equipped with ABS are often difficult to reconstruct because the tires do not lock and leave skid marks on the roadway. Thus the vehicle's pre- and post-impact path with ABS is not always obvious due to the ability to brake and steer simultaneously. Furthermore, the ABS braking pattern can be difficult to reconstruct using traditional methods because of the increased steering ability when braking and the complex control cycles implemented within the ABS algorithm.

The modeling techniques presented by the authors provide an interesting method to simulate the trajectory and deceleration of a vehicle after the ABS has been activated. The program also enables the user to develop new algorithms as braking algorithms change and advance.

Future work should concentrate on comparing the simulation results to data available from engine control modules or other such systems that record the vehicle's deceleration or braking over time.

Reviewer Discussion

Reviewers Name: Rolly Kinney, P.E.
Paper Title: A Simulation Model for Vehicle Braking Systems Fitted with ABS
Paper Number: 2002-01-0559

An ABS simulation scheme is described and shown to approximate the general characteristics and vehicle stability observed in actual driving with ABS control. That is a necessary condition if simulation is used to study vehicle maneuvers which approach or might have approached the limiting traction condition. However, in the present state of development, it is unclear how precisely an ABS simulation can be shown to represent an actual vehicle maneuver but as ABS simulation matures, a sufficient basis may evolve to show adequate fidelity. In the present state, ABS simulation may prove helpful in accident analysis because the considerable uncertainty associated with the actual driver control demands may be greater than the uncertainty involved with the ABS functional model.

This ABS model implementation is more informed than an actual ABS controller because the simulation uses the simulation speed derived from the physics of the simulation for control reference while the controller has only a reference computed from actual wheel spin rates. The inclusion of an option to compute the reference speed from simulated wheel spin rates would permit the study of how the reference computation scheme may or may not degrade performance compared to a fully informed controller.

2002-01-0560

A Demographic Analysis and Reconstruction of Selected Cases from the Pedestrian Crash Data Study

Jason A. Stammen
Vehicle Research and Test Center, National Highway Traffic Safety Administration

Sung-won (Brian) Ko and Dennis A. Guenther
The Ohio State Univ.

Gary Heydinger
FTI/SEA, Inc.

ABSTRACT

This study involves two areas of research. The first is the finalization of the Pedestrian Crash Data Study (PCDS) in order to provide detailed information regarding the vehicle/pedestrian accident environment and how it has changed from the interim PCDS information. The pedestrian kinematics, injury contact sources, and injuries were analyzed relative to vehicle geometry.

The second area presented is full-scale attempts at reconstruction of two selected PCDS cases using the Polar II pedestrian dummy to determine if the pre-crash motion of the pedestrian and vehicle could somehow be linked to the injuries and vehicle damage documented in the case.

INTRODUCTION

In the mid-1970's, pedestrian fatalities in the U.S. reached nearly 8,000 per year [1]. Since then, automakers have incorporated more streamlined, less aggressive front ends into their new vehicle designs. Since then, pedestrian fatalities in the U.S. have steadily declined to less than 5,000 per year [1].

The National Highway Traffic Safety Administration (NHTSA) implemented the Pedestrian Crash Data Study (PCDS) to provide data on vehicle/pedestrian accidents (All acronyms used in this paper are described in the Abbreviations section at the end of this paper). The interim PCDS data was obtained through the National Automotive Sampling System (NASS) database. The interim PCDS analysis, consisting of 292 cases, was conducted by NHTSA from 1994 through 1996 [1]. The PCDS was actually conducted to update the Pedestrian Injury Causation Study (PICS), a similar study conducted in the late 1970's and early 1980's [2]. By 1998, the PCDS was updated to include data from 521 cases acquired at six sites across the United States [3,4]. The first part of this study describes the evaluation and comparison of the results from the interim PCDS dataset consisting of 292 cases and the final dataset of 521 cases (including the 292 from the original PCDS). The pedestrian kinematics, injury contact sources, and injuries documented in the updated PCDS were analyzed relative to vehicle geometric properties.

The second part of this study is full-scale sled testing that attempted to reconstruct two selected PCDS cases using the Polar II pedestrian dummy. Cases were selected based on information such as dummy size, vehicle availability, and conditions relevant to trends present in the PCDS. The objective was to determine if the pre-crash motion of the pedestrian and vehicle could somehow be linked to the injuries and vehicle damage documented in the case.

UPDATED PCDS ANALYSIS

BACKGROUND

The PCDS case information included the events of the crash, the vehicle and pedestrian interaction, and the resultant injuries. The PCDS gathered information from investigation teams, police reports, medical records, and interviews with the pedestrian, driver, and any witnesses to the accident. After all of the necessary information was gathered, the final case report was reviewed and recorded in the PCDS database [5].

Analysis of the PCDS database focused on several aspects of pedestrian and vehicle information, including proposals by the International Harmonization Research Activities (IHRA) pedestrian safety working group [1]. Pedestrian injury data was correlated to age, impact speed, vehicle contact regions and parts (injury sources), and other aspects of the pedestrian/vehicle collision environment. A thorough comparison was made between the interim PCDS (292 cases) and final PCDS (521 cases) results to evaluate how changes in vehicle geometry have affected the severity and location of injuries by using Statistical Analysis Software (SAS).

METHODS

DATA PROCESSING

The PCDS data was obtained from the National Automotive Sampling System (NASS) PCDS database and the data was then manipulated with Statistical Analysis Software (SAS) [4,5]. Numeric results were input into spreadsheets to produce graphs describing the relationship between pedestrian injuries and vehicle parameters.

DEFINITIONS

Bumper height, hood height, bumper lead, hood length, lead angle, and wrap-around distance (WAD) are defined in Figure 1:

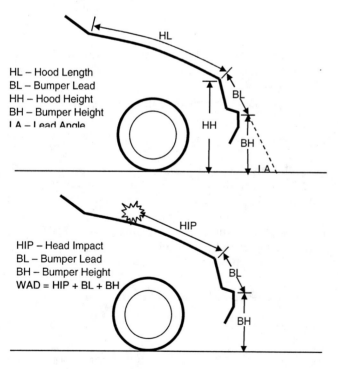

Figure 1: Definitions

VEHICLE AND DRIVER ANALYSIS

Vehicle parameters such as those described in the previous section, as well as vehicle make and category, were calculated and the results of the interim and final PCDS datasets were compared. In addition, collision situation parameters such as vehicle maneuvers prior to impact, number of travel lanes, alcohol involvement and impact speed were compared.

PEDESTRIAN ANALYSIS

Pedestrian characteristics such as gender, height, weight, age, maximum Abbreviated Injury Scale (AIS) injury (Table 1), and Injury Severity Score (ISS) were tallied for the interim and final PCDS and compared. As with the vehicle, collision environment information such as pre-impact motion of the pedestrian, pedestrian avoidance actions, and relative pedestrian orientation to the vehicle was also examined.

AIS CODE	DESCRIPTION
AIS = 0	No Injury
AIS = 1	Minor Injury
AIS = 2	Moderate Injury
AIS = 3	Serious Injury
AIS = 4	Severe Injury
AIS = 5	Critical Injury
AIS = 6	Maximum Injury
AIS = 7	Injured, Severity Unknown
AIS = 9	Unknown if Injured

Table 1: AIS Code Description

INTERACTION OF PEDESTRIAN AND VEHICLE

The results of the vehicle and pedestrian analyses were compiled to get a more complete picture of the collision environment. Pedestrian motion in response to the impact, WAD versus reported impact velocity, injury severity/type versus vehicle contact source, and injury severity versus impact velocity were all evaluated for the final PCDS and compared with the interim PCDS.

FATALITY ANALYSIS

The influence of vehicle body type, impact velocity, pedestrian orientation relative to the vehicle, and pre-impact pedestrian motion on the occurrence and nature of pedestrian fatalities was examined and compared with the interim PCDS data.

STATISTICS

Student T-tests (significance level of 0.05) were done to compare mean values in the interim and final PCDS databases.

RESULTS & DISCUSSION

VEHICLE ANALYSIS

Table 2 presents vehicle specifications from the interim and final PCDS analysis. The average model year of the vehicle involved did not significantly change (p>0.05), which explains why none of the hood and bumper measurements were significantly different than the interim PCDS analysis of 292 cases. Even though the increase was not statistically significant, the hood height increased by 5.9 percent, indicating the presence of more sport utility vehicles as well as minivans in the final database. Similarly, the lead angle was expected to be higher because of the presence of more high profile vehicles, but it did not increase significantly (p>0.05) since the interim PCDS data.

	Interim PCDS (n=292)		Final PCDS (n=521)		Variatio
	Mean	Std Dev	Mean	Std Dev	
Model Year	1992.8	2.1	1993.1	2.4	0.3 years
Bumper Height	44.1 cm	22.2	44.6 cm	22.1	1.1% ↑
Hood Height	64.4 cm	34.0	68.2 cm	51.1	5.9% ↑
Bumper Lead	7.8 cm	7.0	8.0 cm	7.4	2.6% ↑
Hood Length	103.5 cm	17.6	103.2 cm	19.6	0.3% ↓
Lead Angle	66.9 deg	19.3	67.2 deg	17.6	0.4% ↑

Table 2: Average Vehicle Characteristics

The most common vehicle maneuvers prior to colliding with pedestrians are driving straight and turning left [3,4]. The frequency of accidents is much higher for vehicles turning left than those turning right. This perhaps is due to the driver's side A-pillar, which can more easily impede the driver's frontal view when making a left turn as opposed to a right turn. It may also be due to the considerably longer time it takes to complete a left turn maneuver than a right turn. The pedestrian hazard may not materialize until after the driver has made the decision to initiate the left turn. The driver having made the decision that it is safe to go may no longer be alerted to pedestrian hazards. Conversely, right turns are much shorter from the time the decision is made until the turn is negotiated, and the vehicle is not in the intersection as long as it would be in a left turn situation.

There were significant decreases in the vehicle speed categories of 9-16 km/hr and 25-32 km/hr from the interim PCDS data (p<0.05) (Figure 2), but the percentage of unknown vehicle speeds has increased, perhaps hiding instances in these speed categories.

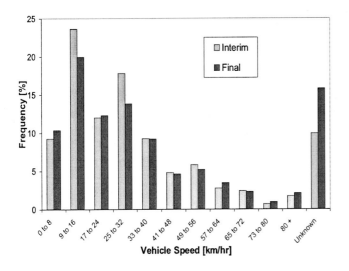

Figure 2: Vehicle Speed Distribution

The interim PCDS analysis contained an even distribution between male and female pedestrians. However, the percentage of male involvement (51%) in the final PCDS analysis was slightly higher than the female percentage (49%) [3,4]. The average height (161 cm) and weight (63 kg) of pedestrians involved in the interim PCDS analysis and the final study were nearly identical [3,4]. The final PCDS dataset contained a significantly higher number of pedestrians in the 11-15 year old age group (p<0.05), jumping from 2% to 11% of the cases, as shown in Figure 3. It is unknown exactly why this abrupt increase has occurred, but it may be due in part to the paucity of the data in this age group. Additionally, this trend could perhaps reflect an increase in vehicle per person ratio, as individuals of legal driving age are more likely to be inside the vehicle than outside, thus reflecting the percentage decrease in crash involvement for ages 21 to 60. In the span of only a couple of years, however, it seems that this reason is unlikely.

The distributions of pedestrian pre-impact motion in the interim and final PCDS data were not significantly different from each other [3,4], with walking as the most common activity. Over 70% of pedestrians either stopped moving or did not react prior to getting hit as documented in witness and participant statements included in the case information [3,4]. Since the majority of pedestrians did not react much to the oncoming vehicle, testing using a stationary standing pedestrian dummy represents the typical situation in pedestrian collisions.

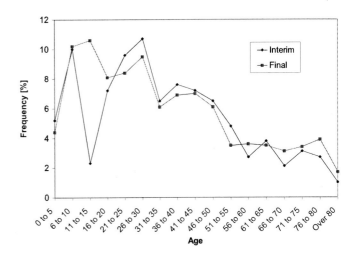

Figure 3: Pedestrian Age Distribution

Figure 4 indicates a large increase in the percentage of pedestrians carried by the vehicle. This number increased significantly from 32.5% to 44.8% (p<0.05) [3]. This change is reflected in the noticeable decrease in pedestrians knocked to the pavement (38.6% to 27.7%).

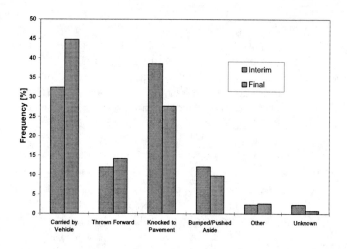

Figure 4: Post-Impact Pedestrian Motion [3,4]

The maximum AIS injuries suffered are shown in Figure 5. In the interim PCDS analysis with 292 cases, over 50% were AIS 1 and 11% were AIS 4 or higher [1]. While the frequency of MAIS 1-2 injuries has decreased by 14% ($p>0.05$), there has been an 8% increase in MAIS 4 and higher level of injuries ($p>0.05$). The reason for this increase in injury severity is unclear, but it may have to do with the popularity of sport utility vehicles (SUVs) in the last few years, which are more likely to cause broken ribs and fractured upper legs and hips because of their geometry. These injuries are AIS 3 and higher in many cases. The frequency of upper and lower extremity injuries has in fact increased, especially in the AIS 2 and higher severity ranges [1,3,4].

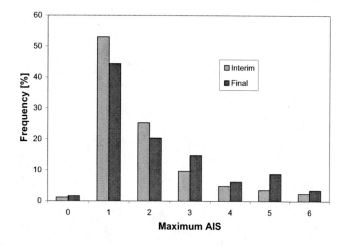

Figure 5: Maximum AIS Injury Suffered by Pedestrian

PEDESTRIAN-VEHICLE INTERACTION

The average 15 year old in the U.S. is 166 cm tall according to CDC growth charts, regardless of gender [6]. The wrap around distance (WAD) is influenced by vehicle impact speed, age, and stature [3]. In Figure 6, the average WAD for adults regardless of using age (>15 years) or stature (>166 cm) as the independent variable is 197 cm. The average WAD for children when using stature as the independent variable is 171 cm, which is larger than when using age (162 cm). This difference in WAD for adults and children reflects the need for concentrating on different areas of the vehicle, depending on whether a child or adult-sized test device is being used. For example, the windshield and A-pillars are important vehicle structures for adults, but not necessarily for children. The bumper and front end of hood are more prevalent injury-causing structures for ages 0-15.

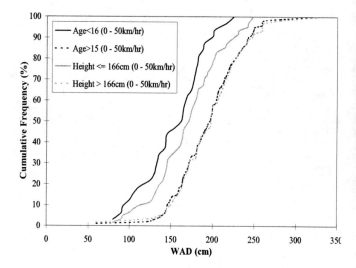

Figure 6: Child and Adult WAD by Age and Stature [7]

There were noticeable changes in the frequency of AIS 2 or greater injuries caused by the bumper and hood areas from the interim PCDS. The frequency of injuries caused by contact with the bumper decreased from 25.5% to 21.5% ($p<0.05$), while injury frequency due to hood surface contact increased by 4.5% ($p<0.05$). In the category of "environment" (road surface), the percentage was 20% of all AIS injuries but only 7% of cases with AIS 2 or greater [3]. This trend may have been caused by the lower ride height, cab-forward design, and the lower lead angle of more recent model vehicles resulting in less energy absorption by the leg and pedestrians being carried by the vehicle more frequently, as shown in Figure 4.

As in the interim PCDS, the most frequent injury regions were the head and lower leg (combined 50% of all injuries), most of which were caused by contact with the hood surface, windshield, and bumper areas of the vehicle [1,3,4].

The impact velocity curve of AIS 5-6 injuries for the final PCDS data showed lower percentages of these injuries for the same velocity than in the interim PCDS (Figure 7). As the average vehicle year had not changed significantly from the interim study, it seems that this change is mostly due to sparseness of data for high severity injuries. In fact, there were only 122 AIS 5-6 injuries out of 4,184 total injuries (3%) [3,4].

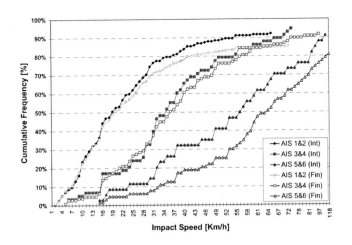

Figure 7: PCDS Impact Speed versus Injury AIS

FATALITY ANALYSIS

The percentage of SUVs involved in fatal cases increased significantly from 11% of the interim cases to 35% of the finalized 521 cases [3], while the percentages of automobiles (sedan and coupe), minivans, and pick-up trucks decreased significantly from the interim PCDS analysis (p<0.05) (Figure 8). Even though this change can perhaps be explained again by paucity of data (only 63 fatalities in the final PCDS), this significant increase illustrates the popularity of SUVs in recent years, as well as emphasizing the need to address pedestrian safety concerns with these types of vehicles.

There was a slight increase in the average impact speed of fatalities in the final PCDS analysis, but no significant changes occurred (p>0.05).

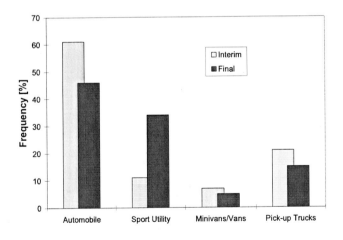

Figure 8: Vehicle Body Type Involvement (Fatalities)

There was a 38% decrease in the percentage of left-side pedestrian impacts in fatal cases (15 out of 28 fatal cases in the interim database but only 21 out of 63 fatal cases in the final database) [3,4], while the increases in the percentage of right side (28%), facing toward the vehicle (46%), and unknown orientations (39%) were also significant (p<0.05) (Figure 9). It is difficult to assess whether there really was a substantial decrease in left side impacts because of the eight additional unknown cases. Even with these cases added, there is a significant decrease in fatalities resulting from left side impacts. It is unclear why this trend has changed so much in the past few years.

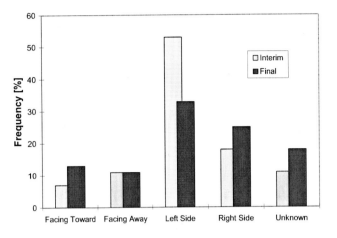

Figure 9: Fatally-Injured Pedestrian Body Orientation Relative to the Vehicle

The percentage of pedestrians walking increased from 39% to 51%, and there was a decrease in those running or jogging prior to being fatally injured (p<0.05) (Figure 10) [3,4]. It is unclear why this has changed so much since the interim PCDS, but it seems that the faster the pedestrian is moving laterally across the vehicle front, the less chance there is of receiving a direct fatal injury from one of the vehicle components. As discussed in Part II of this study, an increase in the height of the pedestrian center of gravity and lateral rate of motion relative to the vehicle front is more conducive to the pedestrian sliding across instead of impacting rigid front-end vehicle components.

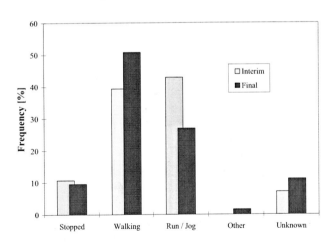

Figure 10: Pedestrian Motion (Fatalities Only)

CONCLUSIONS

The first part of this study accomplished two major objectives. First, it has shown how vehicles, pedestrians, and their interactions have changed from the interim database (292 cases) to the final database (521 cases). Secondly, the current state of pedestrian collisions has been broken down and quantified. Due to the large number of pedestrian cases and situations, test procedures evaluating the aggressiveness of vehicle front ends should consider the common conditions present in these PCDS accidents. Using computer simulations, test conditions such as head impact speed, angle, and location can be determined based on the vehicle shape in question [7]. The following conditions sum up the present state of pedestrian/vehicle interactions and differences from the interim PCDS [3,4]:

- The average injuring vehicle front had a bumper height of 44.6 cm, hood height of 68.2 cm, bumper lead of 8 cm, hood length of 103.2 cm, and lead angle of 67.2 degrees.

- The most common vehicle motion prior to colliding with pedestrians is driving straight ahead on a two-lane road (one lane going each direction).

- The highest proportion of pedestrian impacts occurred between 9-16 km/hr (6-10 mph), and roughly two-thirds of all accidents occur at or below 40 km/hr (25 mph).

- Males (51%) were involved in slightly more pedestrian accidents than were females (49%).

- The average injured pedestrian regardless of gender was 34.3 years old, with a height of 161 cm and weight of 63.6 kg.

- Most pedestrian stances were walking with their left side facing the front of the vehicle and did not react prior to getting hit.

- New front-end designs are "carrying" instead of "running over" pedestrians, as shown by an increase in AIS 2 and greater upper leg and pelvis fractures.

- For Wrap Around Distance (WAD), stature is a better parameter than age for distinguishing children and adults. The average WAD for adults regardless of using age or stature as the independent variable is 197 cm. The average WAD for children is 171 cm when using stature.

- The most frequent AIS 2 and greater injury regions were head (32%) and lower leg (19%).

- Maximum AIS (MAIS) 1-2 injuries per case have decreased by 14%, while MAIS 3 and greater injuries have increased by that same percentage of total cases.

- The percentage of fatalities has increased from 9.5% to 12% of all cases, likely due to the increase in popularity of sport utility vehicles, which have aggressive front-end profiles toward pedestrians.

These characteristics were taken into account when selecting representative cases from the PCDS for reconstruction in full-scale dummy tests, as described in Part II of this study. These tests were done to demonstrate the experimental application of the collision information contained in the PCDS database.

CASE RECONSTRUCTIONS

BACKGROUND

In addition to updating the PCDS database and analyzing the current state of pedestrian/vehicle crashes in the United States, there is a need to apply this information to the improvement of vehicles. Full-scale sled testing provides a realistic view of the entire collision environment, and it can lend information toward the development of more realistic component-based test procedures [7].

Two cases involving low and high profile vehicle fronts were tested by using a 50th percentile size pedestrian dummy (Polar II, Honda R&D) [8]. A series of experimental collisions were conducted at varying impact speeds, angles, and pedestrian stances with vehicles representative of the various front-end profiles present on United States highways. The purpose was to study how these vehicle profiles, speeds, and pedestrian position parameters affect the resultant kinematics of the adult through the use of a post-mortem human subject (PMHS)-validated pedestrian dummy.

METHODS

POLAR II DUMMY

The Polar II dummy is a second-generation 50th percentile adult male pedestrian dummy (Figure 11). Table 3 shows the size of the dummy, which is identical to the size of a 50th percentile American male [8].

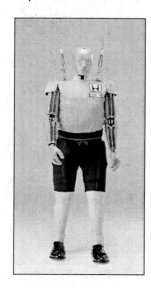

Figure 11: Polar II Pedestrian Dummy [8]

	50th Percentile American Male
Weight	165.3 lb / 75 kg
Stature	69 in / 175 cm
Sitting Height	34.8 in / 88.4 cm

Table 3: Polar II Dummy Stature [8]

The dummy was designed using THOR, the NHTSA Advanced Frontal Dummy, as a base model for its initial design [8]. The dummy was specified to be durable enough to handle up to a 50-km/hr impact [8], and it possessed the following instrumentation scheme:

Location	Type	No. of Channels
Head C. G.	Accelerometer	3
Upper Neck	Load Cell	3
Lower Neck	Load Cell	3
Neck Spring	Load Cell	2
O.C.	Rotary Pot	1*
Chest C. G.	Accelerometer	3
Lateral Ribcage	Rotary Pot	3*
Lateral Abdomen	String Pot	1*
Pelvis	Accelerometer	3
Femur	Load Cell	4
Femur	Lin. Accelerometer	1
Femur	Ang. Accelerometer	1
Upper Tibia	Load Cell	4
Upper Tibia	Lin. Accelerometer	1
Upper Tibia	Ang. Accelerometer	1
Lower Tibia	Load Cell	3

*Not connected to on-board DAS
Total No. of Channels Available for Current Testing = 37
No. of Channels Connected to On-Board DAS = 32

Table 4: POLAR II Instrumentation [8]

PCDS CASE SELECTION

Test vehicle selections were made based on the available vehicles on-site at the Vehicle Research & Test Center (VRTC) facility. To observe the extremes of pedestrian collisions, two types of vehicle were considered for full-scale sled tests: low profile automobiles and high profile vehicles (SUV, trucks, minivans).

The following criteria were used in selecting the cases:

- Adequacy of case documentation (photographs of vehicle damage, injury descriptions, diagrams, etc)
- Approximate pedestrian weight and stature of 50th percentile male
- Impact velocity in the range of 30-50 km/hr

After applying the above criteria, the two cases selected included a low profile vehicle, the Honda Civic coupe, and the Chevrolet Silverado, a pick-up truck with a high profile.

LOW PROFILE CASE INFORMATION

Table 6 contains the Honda Civic case information:

PSU:	82
Case No.:	651P
Accident Year:	1996
Description of Pedestrian:	
1. Age:	34
2. Gender:	Male
3. Pedestrian's Height:	178 cm (5 ft 10 in)
4. Pedestrian's Weight:	75 kg (165 lbs.)
Description of Vehicle:	
1. Class of Vehicle:	Compact
2. Year / Make / Model:	1992 / Honda / Civic (2 Dr coupe)
Description of Injury patterns and Vehicle damages:	
1. AIS:	1
2. Injury Source:	Windshield
3. Speed Limit:	48 kph (30 mph)
4. Impact Speed:	46 kph (29 mph), [Accuracy range of impact speed ; 2 - 8 kph]
5. Damage Plane:	Front (dents, smears and scuffs as well as smashed holed windshield)
6. Impact Angle:	17-20 degree
Descriptions of Accident:	
Vehicle one was eastbound in lane one of a 5 lane, two way street, the only lane to travel straight through an intersection. A pedestrian was running in the crosswalk which angle southeasterly across the street. The front of vehicle one impacted the right and backside of the pedestrian. The pedestrian wrapped and struck the windshield and flipped to the right side as vehicle one continued and then stopped in the middle of the intersection. It happened in clear daylight.	

Table 6: Honda Civic Case Information [9,10]

Figure 12 shows that the longitudinal travel distance of the pedestrian from leg-to-bumper contact to head impact across the vehicle front was 164 cm, and the lateral distance between these two points was 55 cm. The resulting angle is 18.5 degrees [9], and this angle was used as the rotation of the vehicle on the sled buck since the motion of the sled was linear [9]. A tolerance of three degrees rotation was allowed in each direction in case the angle had to be adjusted to facilitate a change in the resulting pedestrian path and vehicle damage. The sled buck was thus fabricated to allow an impact angle between the pedestrian and Honda Civic to be 15.5 to 21.5 degrees [9].

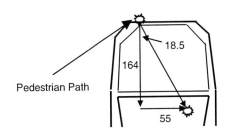

Figure 12: Rotation of Vehicle on Sled

Figure 13 outlines the injuries suffered by the pedestrian in the case:

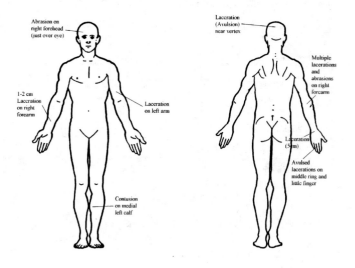

Figure 13: Soft Tissue Injuries [10]

The vehicle contact point descriptions are detailed in Figure 14 and Table 7. These points were marked on the test vehicles so the proximity of test vehicle damage marks could be compared with the case damage [9].

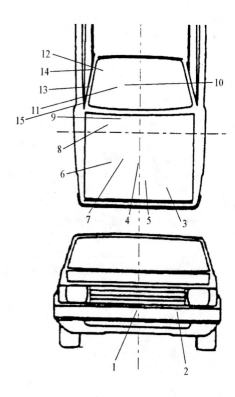

Figure 14: Vehicle Exterior Damage Patterns [10]

Label	Component Contacted	Longit. Location (X)	Lateral Location (Y)	Crush [cm]	Suspected Body Region	Supporting Physical Evidence
1	License Plate	94	0 to -10	0	Left leg	Scuffed at top bumper
2	Bumper	94	-30	0	Right shoe	Scuffed
3	Hood	76	-20	0	Right shoe	Rubber transfer
4	Hood	48	2	0	Right shoe	Rubber transfer
5	Hood	73	-5	0	Right shoe	Curve rubber transfer
6	Hood	30	35	0 <= 1	Left knee	Small dent
7	Hood	12	22	0 <= 1	Hip region	Wide area of scratch
8	Hood	-6	42	0 <= 1	Hip region	Streak & minor dents
9	Hood	-22	38	0	Hip / Legs	Curved scratch marks
10	Windshield	-70	28	5 - 10	Right arm	Skin / Tiny hair
11	Windshield	-71	58	0	Right arm	Blood
12	Windshield	-103	55	3 - 5	Head	Hair / Skin blood
13	A-pillar	-77	70	0	Right arm	Lateral scratch streaks
14	A-pillar	-116	58	0	Head	Lateral scratch streaks
15	Side mirror	-79	88	0	Arm	Scuffed

Table 7: Vehicle Contact Locations and Descriptions [10]

HIGH PROFILE CASE INFORMATION

As Table 8 describes below, the pedestrian's weight and height varied slightly from the dummy size, but not significantly. Additionally, unlike the Honda Civic case, the high profile vehicle case with the Chevrolet Silverado reported a straight-on impact angle [9].

PSU :	90
Case No. :	628P
Accident Year:	1998
Description of Pedestrian:	
1. Age :	77
2. Gender :	Female
3. Pedestrian's Height :	169 cm (5 ft 6.5 in)
4. Pedestrian's Weight :	71 kg (155 lbs)
Description of Vehicle :	
1. Class of Vehicle :	Large Pick-up
2. Year / Make / Model :	1991 / Chevrolet / Silverado (1500 series, regular 2WD)
Description of Injury patterns and Vehicle damages :	
1. AIS :	4
2. Injury Source :	Hood edge (straight path accident)
3. Speed Limit :	64 kph (40 mph)
4. Impact Speed :	50 kph (31 mph), [Accuracy range of impact speed ; < 2 kph]
5. Damage Plane :	Front (dents, smudges and scratches on hood surface as well as cracked front hood)
6. Impact Angle :	90 degree
Descriptions of Accident:	
Vehicle one was eastbound in lane 2. The pedestrian was running across the road in a southerly direction carrying a bag filled with empty aluminum cans. The vehicle struck the pedestrian, who then rotated partly onto the hood and rolled off as the vehicle applied brakes. According to a witness, the pedestrian ran into the path of vehicle one and the vehicle did not have enough time to stop (brake was applied). It happened in clear daylight (11:05 AM)	

Table 8: Chevrolet Silverado Case Information [11]

Figures 15-17 describe the large number of injuries suffered by the pedestrian in the case. There were numerous injuries both externally and internally, with the most severe (highest AIS) injury being several broken ribs and a bruised lung [11].

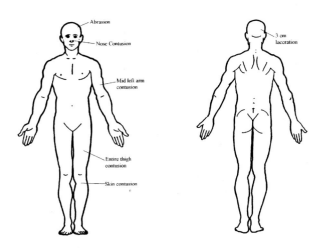

Figure 15: Soft Tissue Injuries [11]

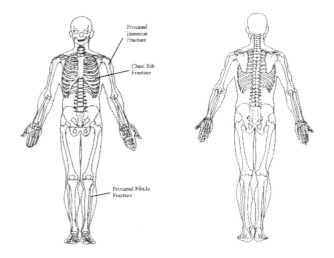

Figure 16: Skeletal Injuries [11]

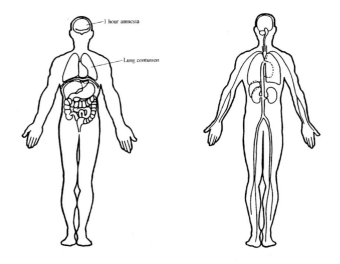

Figure 17: Internal Injuries [11]

Using the same method as for the Honda Civic case, the vehicle contact points were marked on the test vehicle according to the post-accident damage on the vehicle to provide an accurate comparison with data from the sled tests (Figure 18 and Table 9).

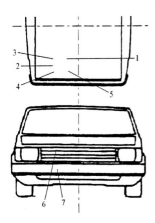

Figure 18: Vehicle Exterior Damage Patterns [11]

Label	Component Contacted	Longitudinal Location (X)	Lateral Location (Y)	Crush [cm]	Suspected Body Region	Supporting Physical Evidence
1	Hood	21	29	2	Shoulder	Dent
2	Hood	49	55	2	Shoulder	Dent
3	Hood surface	20	55	2	Chest	Dent
4	Hood surface	70	51	2	Chest	Dent
5	Hood surface	73	24	2	Chest	Dent
6	Grill	95	36	0	Hip	Cracked
7	Bumper	137	36	0	Left leg	Scuff

Table 9: Vehicle Contact Descriptions [11]

1991 vs. 1999 SILVERADO DESIGN COMPARISON

The actual year of the vehicle involved in the Silverado collision was a 1991 model. The actual testing vehicle available at VRTC was a 1999 Chevrolet Silverado. Therefore, exterior body frame measurements and the location of under-hood components were compared using photographs and measurements to check the feasibility of performing the tests with the 1999 Chevrolet Silverado instead of the 1991 model. There were slight differences between the hood angle and bumper height of the two models, but it was concluded that the differences were not large enough to refrain from using the 1999 model in the test series.

FULL-SCALE DUMMY HYGE SLED TESTS

The HYGE sled test facility at Transportation Research Center, Inc. (TRC) in East Liberty, OH was used to conduct the tests. Five iterative tests were performed with each vehicle using a 10-g sled pulse to achieve the required impact velocity [9]. The number of five tests was chosen for two reasons. First, the dummy was only available for a short amount of time. Secondly, it was estimated that five iterative attempts to get a close approximation of the accident would be sufficient [9]. Dummy position parameters were changed between tests based on the results of the previous iteration to adjust the path and damage to the vehicle and pedestrian.

Since the pedestrian dummy was only instrumented on the left side, the low profile case had to be reconstructed as a mirror image because the pedestrian was hit on the right side [9]. Since the exterior body of the vehicle was symmetric laterally, this was not seen as detrimental to the resulting kinematics.

In preparation for the sled attachment, both of the test vehicles were cut behind the B-pillar and the interior compartment was emptied to reduce the weight of the vehicle (the under hood components were kept intact). The vehicle was rigidly attached to a buck via a circular steel interface plate. Because the vehicle was offset from the track, as shown in Figure 19 [9], a ballast equal to the vehicle weight was attached on the opposite side of the track to balance the buck-vehicle assembly and reduce torsion effects. A net was attached to the buck over the sled to prevent the dummy from getting caught in the sled.

Figure 19: Pedestrian Case Reconstruction Test Setup

The damage locations from the actual cases were marked on the test vehicles prior to the tests in order to facilitate comparison of the damage patterns.

Figure 20: Measurement Reference Lines for Civic (above) and Silverado (below) [9]

Prior to each test, detailed measurements of the dummy stance were made to ensure that the dummy positions were the same for each test, except for changes made between tests to get a more accurate reconstruction of the case. Different colors of chalk were applied to the left side of the dummy to distinguish which area of the dummy contacted which portion of the vehicle in post-test analysis. These body areas included the head, arm, lumbar/pelvis, upper leg, and lower leg. Photographic targets were placed on important body landmarks for trajectory analysis (Figure 21).

The dummy was held in a standing position by a magnetic holding device that was triggered to release just prior to impact of the dummy by the vehicle so that the entire weight of the dummy was on the ground and downward acceleration of the dummy was minimal [9].

Figure 21: Target and Dummy Stance [9]

RESULTS & DISCUSSION

HONDA CIVIC (LOW PROFILE)

In the Honda Civic tests, the third test out of five was most similar to the actual collision in terms of injury measurement levels, vehicle damage patterns, and the trajectory of the dummy. In Figure 22, the dotted white line indicates the region of vehicle damage, while the chalk on the hood and cracks on the windshield show that the test damage falls within this region. The head injury criteria (HIC) value was representative of an AIS 1 injury (688), which was the level suffered by the pedestrian in the case, and the WAD was 239 cm (9 cm short of the case WAD of 248 cm) [9].

Figure 22a: Civic Test

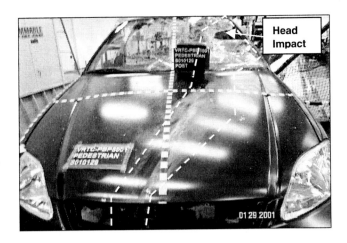

Figure 22b: Pedestrian Path of Civic Case vs. Test 3 [9]

It was found that the dummy stance was a major factor influencing damage patterns to the vehicle as well as dummy injury severity measurements (Figure 23 and Table 10). The dummy struck-side leg position controlled the rotation of the dummy following initial impact. Since the back of the pedestrian's head was injured in this case, it was determined that the dummy's left leg had to be in front of the right to get the proper rotation. This was accomplished by rotating the dummy so that it was facing slightly away from the vehicle, which could have been the case given the relative angles of travel of the vehicle and pedestrian [9].

Figure 23: Dummy Stances; Legs Even (Left) and Rotated Away/Increase in Height (Right)

Test	Stance	Elevation (inches)	Head Injury*	Match Path?	WAD (cm)
1	Legs Even	10.6**	AIS 5-6	No	182
2	Legs Even	17.0	AIS 3-4	No	222
3	Rotated Away	16.0	AIS 1-2	Yes	239
4	Rotated Away	17.0	AIS 3-4	No	216
5	Rotated Away	16.3	AIS 5-6	No	221

*Estimated AIS 1-2 corresponds to HIC of 0-800, AIS 3-4 is HIC of 800-1500, and AIS 5-6 is HIC above 1500 [9]
**Calculated ground level in the case

Table 10: Civic Test Results [9]

The elevation of the dummy off the ground was another major factor in dictating the WAD (impact point of the head) [9]. The initial test was performed with a dummy elevation of 10.6 inches off the ground (the dummy had to be raised because the vehicle was raised). The height was calculated by the matching the pedestrian height and bumper height of the Honda Civic. In this initial test, the dummy slammed down upon the hood and made two deep dents, which was very different from the actual Civic case. The actual case had many scuffmarks instead of dents, indicating that the pedestrian slid over the surface of hood. The WAD was much too low as well, so the elevation of the dummy was raised about 6 inches to simulate jogging, which includes a portion of the gait cycle when both feet are off the ground. This resulted in WAD measurements very similar to that of the actual case [9]. An increase in relative height of the pedestrian could have also occurred due to pre-impact braking, which was not noted in the case information [9].

It was indicated in Part I of this study that the impact velocity influenced the WAD. Although the impact speeds for all five tests were the same, the WAD varied between 182 and 239 cm. This shows that WAD depends not only on the impact speed of actual cases, but can also be affected by the pedestrian stance, elevation off the ground, and pre-impact motion [9].

CHEVROLET SILVERADO (HIGH PROFILE)

The documented impact velocity in this case was about 50 km/hr, but the dummy manufacturer had concerns about the durability of the dummy in high speed, high vehicle profile impacts [9]. Because this prototype dummy was needed for other testing immediately following this test series, lower impact speeds of 20 and 25 km/hr were applied in this test series. Using estimated linear relationships of trajectories and injuries with impact velocity, results were extrapolated to 50 km/hr to get a comparison with the case. Because of this indirect comparison, more emphasis was put on how changes in stance and velocity affected the kinematics of the dummy [9] than injury replication. The third test was closest to the case both in terms of damage patterns on the vehicle and the path of the dummy (Figure 24).

Figure 24a: Silverado Test

Figure 24b: Damage of Silverado Case vs. Test 3 [9]

For these tests, three dummy stances were used (Figure 25), incorporating combinations of two different leg positions (legs even, right leg in front of right) and two arm positions (bent in front at elbow, wrists tied in front). The "right leg in front" configuration modeled a walking pedestrian. The reason for tying the wrists together in one arm configuration is to negate the effect the arms would have on the kinematics. The effect of these stances and changing velocity on WAD and estimated injury levels (Tables 11 and 12) was evaluated. The elevation of the dummy was consistent from test to test because the tests were done at lower velocities than the case warranted.

Figure 25: Silverado Test Dummy Stances for Tests 1 (Left), 2 and 4 (Middle), and 3 and 5 (Right)

#	Legs	Arms	Velocity (km/hr)	WAD (cm)
1	Legs Even	Bent 90 Deg at Elbow	20	145
2	Right Leg in Front	Bent 90 Deg at Elbow	20	142
3	Right Leg in Front	Wrists tied in front of body	20	139
4	Right Leg in Front	Bent 90 Deg at Elbow	25	149
5	Right Leg in Front	Wrists tied in front of body	25	147

Table 11: Silverado Test Results [9]

Case Injury	Dummy Measurement Extrapolated to 50 km/hr[1]	Estimated Injury
Proximal Tibia Fracture	2700 N Tibia Shear, 490 N-m Tibia Bending Moment	AIS 2-3 Heavy Ligament Damage or Fracture[2]
5 Fractured Ribs	180 g Chest Acceleration	AIS 3-5 Injuries to the Chest Area[3]
Bruised Lung	180 g Chest Acceleration	AIS 4 Internal Organ Contusion or Rupture[3]
AIS 2 Head Contusion	4500 HIC	Fatal (AIS 6) Blow to the Head[4]

[1] Using linear fitting of 20 and 25 km/hr peak data [9]
[2] The bending mode shear damage threshold is 1600 N, and the shear moment threshold is 350-400 N-m according to Kajzer [12]
[3] The chest injury threshold using acceleration is estimated to be approximately 60 g [13]
[4] Anything over 1000 HIC is generally considered to be a serious head injury [9]

Table 12: Estimated Dummy Measured AIS·2 Injuries

The test results for the head were quite different from the actual case results. The pedestrian received an AIS 2 contusion while extrapolated dummy results indicated a HIC of 4500. While this discrepancy may be partially do to the assumption of linearity between velocity and HIC, the dummy's upper body may be too stiff causing the dummy to rotate and the head to strike the hood of the Silverado before the shoulder and torso can absorb the majority of energy [9]. Another theory is that pre-impact braking of the vehicle could have increased the amount of time between initial contact with the body and head impact, allowing the upper body to decrease kinetic energy of the head.

There was good agreement between the case and estimated test injury levels for the tibia and chest, indicating that the dummy responded to the high profile impact in a human-like manner. A linear relationship between injury levels and impact velocity provided an estimation of these injuries. Regardless of whether a linear or non-linear approximation was used, the measurements at low velocity were close enough to published injury thresholds to assume injury at 50 km/hr. Figure 26 gives an example of how the tests at two velocities were used to estimate the measurements at the case impact velocity of roughly 50 km/hr.

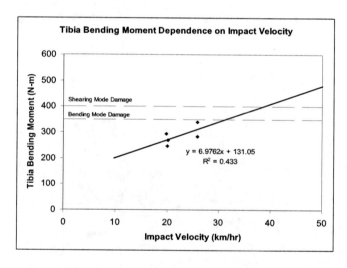

Figure 26: Extrapolation of test results to 50-km/hr

As was expected with the minimal change in velocity and constant pedestrian height relative to the vehicle, the WAD was very consistent. The WAD of the Chevrolet Silverado remained between 139 and 149 cm [9]. There was a slight increase in WAD with increasing velocity, and there was also a slight decrease in WAD when the wrists were tied together for both velocities, although the small number of samples prevents a statistical conclusion. It is not clear why this change occurred, but a possible reason for this small decrease in WAD may be that free arms would reduce the effective mass and allow the dummy to travel a bit further [9]. The wrists separated almost immediately from the force of impact, however, and it is doubtful that in that short of time the arms could make a substantial difference in WAD.

CONCLUSIONS

It was very difficult to reconstruct the two pedestrian cases for two main reasons. First, the case information is only an estimation of the impact situation. A large amount of uncertainty is contained in the collision documentation, and approximation is required to evaluate the accuracy of the reconstruction in terms of pedestrian motion and injuries. Secondly, while the Polar II dummy trajectories have been validated with human cadaver corridors in the Civic case, uncertainties in its durability in high front-end profile vehicles such as the Silverado prevented an exact replication of the impact speed. Even though these difficulties existed, there were some good results that reflected and supported information contained in the PCDS database.

The relative position of the right and left legs and orientation of the pedestrian relative to the vehicle dictates the rotation of the pedestrian and location of body contact with the vehicle. The relative positions of initial contact (usually bumper) and pedestrian center of gravity dictate the wrap-around-distance (WAD), and subsequent vehicle structures that are impacted. Lower bumper height by design or from pre-impact braking, as well as pedestrian activity such as jogging (when the pedestrian is elevated off the ground) are conducive to sliding onto and being carried by the vehicle. In this situation, impact energy is dissipated due to sliding, and by the time the head reaches the windshield, the relative velocity of the pedestrian to the vehicle is low, thereby decreasing injury severity. The occurrence of this situation has increased in the finalized PCDS (Figure 4), which shows that a combination of lower bumpers and possibly more responsive braking systems have been developed in newer automobiles.

WAD increased slightly with increasing velocity in high front-end profile vehicle tests and the occurrence of an AIS 2-3 level leg injury was caused by the high profile impact. This situation reflected the trend in the final PCDS database toward more severe leg injuries. There was also a hint of arm position affecting WAD (increasing WAD when the wrists were tied), but the small sample size prevented any firm conclusion.

The relationships of pre-impact conditions with resulting impact parameters such as pedestrian reaction, motion, and injuries are vital because they can be applied directly to the design of less aggressive vehicle fronts. From an efficiency standpoint, use of a mathematical model would be ideal for reconstructing a large number of accidents. However, full-scale sled tests give the most realistic view for determining these relationships, and the method presented here allows the application of many pedestrian stances and a large number of vehicle shapes and sizes.

ACKNOWLEDGEMENTS

The authors would like to thank Honda R&D for allowing us to use a prototype version of the Polar II dummy in our tests, JARI for inviting us to Japan to view sled tests with the dummy, and GESAC for supplying dummy replacement parts in a timely manner.

ABBREVIATIONS

PCDS	Pedestrian Crash Data Study
NHTSA	National Highway Traffic Safety Administration
NASS	National Automotive Sampling System
PICS	Pedestrian Injury Causation Study
IHRA	International Harmonization Research Activities
SAS	Statistical Analysis Software
WAD	Wrap Around Distance
AIS	Abbreviated Injury Scale
ISS	Injury Severity Score
MAIS	Maximum AIS
SUV	Sport Utility Vehicle
CDC	Centers for Disease Control
PMHS	Post Mortem Human Subject
VRTC	Vehicle Research and Test Center
HIC	Head Injury Criteria

REFERENCES

1. Jarrett, K.L. (1999), "Pedestrian Injury – A Comparison and Analysis of the Pedestrian Crash Data Study Field Collision Data," Wright State University.

2. Toth, G.R., Parada, L.O., Garrett, J.W. (1982), "Documentation for the Data File of the Pedestrian Injury Causation Study," National Center for Statistics and Analysis, National Highway Traffic Safety Administration.

3. Ko, S.B. "Demographic Analysis and Reconstruction of Selected Cases from the Pedestrian Crash Data Study," Master's Thesis, The Ohio State University (2001).

4. Chidester, A.B., Isenberg, R.A. (2001), "Final Report – The Pedestrian Crash Data Study," Paper No. 248, Seventeenth International Technical Conference on the Enhanced Safety of Vehicles.

5. National Automotive Sampling System, (1996), "Pedestrian Crash Data Study / 1996 Data Collection, Coding, and Editing Manual," U.S. DOT, NHTSA, National Center for Statistics and Analysis.

6. Centers for Disease Control and Prevention (CDC) Growth Charts, June 8, 2000.

7. Stammen, J.A., Saul, R.A., Ko, S.B. (2001), "Pedestrian Head Impact Testing and PCDS Reconstructions," Paper No. 326, Seventeenth International Technical Conference on the Enhanced Safety of Vehicles.

8. Akiyama, A., Okamoto, M., Rangarajan, N. (2001), "Development and Application of the New Pedestrian Dummy," Paper No. 463, Seventeenth International Technical Conference on the Enhanced Safety of Vehicles.

9. Stammen, J.A., Ko, S.B. (2001), "Assessment of Polar II Pedestrian Dummy for Use in Full-Scale Case Reconstructions," NHTSA Technical Report DOT HS 809 372.

10. National Highway Traffic Safety Administration (1996), "Honda Civic / PSU number: 82-651p-1996", National Accident Sampling System.

11. National Highway Traffic Safety Administration (1998), "Chevrolet Silverado / PSU number: 90-628p-1998", National Accident Sampling System.

12. Kajzer, J., Schroeder, G., Ishikawa, H., Matsui, Y., Bosch, U. "Shearing and Bending Effects at the Knee Joint at High Speed Lateral Loading," SAE 973326, Forty-First Stapp Car Crash Conference Proceedings: 151-165.

13. McIntosh, A.S., Kallieris, D., Mattern, R. Miltner, E. "Head and Neck Injury Resulting from Low Velocity Direct Impact," SAE 933112, Thirty-Seventh Stapp Car Crash Conference Proceedings: 43-57.

Reviewer's Discussion
By Amrit Toor
INTECH Engineering Ltd.

SAE 2002-01-0560
"A Demographic Analysis of Reconstruction of Selected Cases from Pedestrian Crash Data Study"
Jason Stamen, Sung Wun Ko, Dennis Gunther, Gary Headinger

The first half of this paper is dedicated to updating an earlier pedestrian collision data study (PCDS). This part of the study compares the original study's 292 cases to the updated data set of 521 cases. One of the findings of the study is that the new vehicles are now "carrying" instead of "running over" the pedestrian. The authors also suggest that there is a relationship between WAD (wrap around distance) and vehicle speed.

The authors staged vehicle/pedestrian collisions to duplicate two real world cases from the PCDS. A total of 10 vehicle/pedestrian collisions were conducted (5 for each case). A 1992 Honda Civic and a 1999 Chevrolet Silverado were the vehicles used in these tests. In both cases a Honda Polar II Dummy was used. In the tests involving the Silverado, the vehicle speed was approximately halved (compared to the real case) to avoid damaging the dummy. The results for the Silverado case were extrapolated to the real world collision speed.

The real world case information available to the authors was very limited and somewhat uncertain. The authors attempted to address these deficiencies by varying the test configurations: the dummies were lifted and oriented with respect to the vehicle.

Furthermore, the differences between a real pedestrian and the Polar II Dummy resulted in inconsistent injuries sustained by the real world pedestrian and the predicted injuries in the staged collision. The largest difference occuring in the "Silverado" staged collisions.

The first part of the study indicated that there was a relationship between WAD and vehicle speed. The vehicle speed was kept constant in each set of staged collisions. However, there was quite a variance in the WAD. This suggests the dummy geometry and orientations (test variables) may also be influencing the WAD.

On the whole, this is a good study. The first part of the study provides a reference for vehicle designers to develop pedestrian friendly vehicles. The second part of the study illustrates the difficulty in reconstructing real world pedestrian collisions.

2002-01-1305

Re-Analysis of the RICSAC Car Crash Accelerometer Data

Raymond M. Brach
University of Notre Dame

Russell A. Smith
U. S. Naval Academy

Copyright © 2002 Society of Automotive Engineers

Abstract

Data from the RICSAC1 car crashes have been presented and analyzed in the original reports and in technical papers by others. Some issues dealt primarily with respect to the transformation of data from the accelerometer locations to the centers of gravity. Accelerometers were attached to the moving vehicles and so most results have been presented in moving coordinates. It appears that an additional step to transform the velocity changes and final velocities into inertial coordinates remains to be done. The initial and final inertial velocities and vehicle physical properties are used to compare the experimental data to the law of conservation of momentum. Some of the collisions lead to a loss in total system momentum, as expected. Some show a gain in system momentum which is not physically possible. An analysis of variance of this momentum data shows that the loss or gain of momentum is not systematically related to the type of collision. Consequently, it can be concluded that the variations in the change of system momentum are random and due to factors that were not under control in the experimental collisions. The data indicate that the observed changes in system momentum of the 4 RICSAC impact configurations are not significantly different. Finally, the experimentally measured linear and angular velocity changes, ΔV and $\Delta \Omega$, of the vehicles and total energy loss of each collision are compared to theoretical values calculated using planar impact mechanics. These comparisons show that the theoretical energy losses tend to be lower than the experimental values whereas some of the theoretical ΔV and $\Delta \Omega$ values are lower and some are higher, with no discernable systematic trend.

IMPACT PHASE OF RICSAC[1] CRASHES

Jones and Baum (1978) report on a group of staged and instrumented vehicle crashes organized and carried out to provide research data. The project was named "Research Input for Computer Simulation of Automobile Collisions" and has come to be referred by the acronym "RICSAC". The data from recording transducers on-board the vehicles were subject to random experimental error and they were subject to systematic error as well. Because accelerometers were fixed to the vehicles and were located away from the centers of gravity some of the reported data were incomplete and require additional analysis. Such analyses have been presented in the past (see Brach, 1987, Cliff 1996, McHenry and McHenry, 1997 and Woolley, 1987). This problem is reviewed here but extended to include the effect of the changing orientation of the vehicle fixed accelerometers relative to a fixed inertial reference frame. Unexamined issues remain such as the estimation of experimental variation for the tests and the conformance of the test data to fundamental laws of physics, such as the conservation of momentum. These issues are covered here for the impact phase only of the RICSAC tests; that is, pre-impact and post-impact motions are not analyzed. Experimental impact results are compared to computer solutions using

---

[1] *Research Input for Computer Simulation of Automobile Collisions*, Jones & Baum (1978)

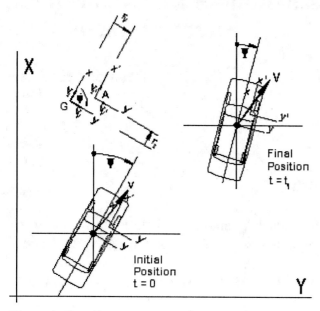

Figure 1. Coordinates and coordinate system

methods of planar impact mechanics (Brach, 1991, Woolley, 1987). Bundorf (1996) laid out the kinematics and the process for transforming accelerations from measurement location to the center of gravity. This process is reviewed and used here. Bundorf presented numerical results for multiple accelerometer locations of only 1 vehicle of a single RICSAC collision. The approach is applied here to the firewall accelerometer locations of 11 RICSAC collisions. (Of 12 tests, RICSAC 2 experienced an instrumentation failure for the firewall accelerometers and is excluded from consideration.) The test acceleration signals are in a translating and rotating coordinate system. These signals must be transformed as functions of time in order to determine the final velocities of each vehicle in a single, fixed inertial system. The results permit the calculation of the experimental final velocity, ΔV and $\Delta \Omega$ values for each vehicle and kinetic energy losses for each crash. Only final velocity values expressed in an inertial coordinate system can be used to calculate ΔV's and energy loss.

Kinematic transformation

Figure 1 shows a vehicle in arbitrary initial and final positions and orientations. Two SAE coordinate systems are attached to and move with each vehicle. One, (x,y), is located at point G, the center of gravity, and the other, (x',y'), is at point A, the location of an accelerometer. The motion considered here spans the beginning to the end of the contact duration of two colliding vehicles, that is, the duration from initiation of contact at time, $\tau = 0$ to the time, $\tau = \tau_1$, of separation (loss of contact)[2]. Symbols for velocities and angles displayed in capitals, or upper case, signify final values at separation, τ_1; lower case symbols correspond to initial conditions at $\tau = 0$ and, in some cases, to quantities that vary with time. A primary objective is to determine the velocity components of the center of gravity in an earth-fixed inertial coordinate system, (X,Y), from the accelerations measured in a coordinate system translating and rotating with the vehicle at Point A. According to Bundorf, the x component of the accelerations at point A can be related to accelerations at point G using:

$$a_{x'}(\tau) = a_x(\tau) - r_y \ddot{\psi}(\tau) - r_x \dot{\psi}^2(\tau) \quad (1)$$

Similarly, the y component can be found from:

$$a_{y'}(\tau) = a_y(\tau) + r_x \ddot{\psi}(\tau) - r_y \dot{\psi}^2(\tau) \quad (2)$$

These accelerations are referenced and measured in a moving coordinate system. A coordinate transformation is necessary to express them in inertial coordinates. That is,

$$a_X(\tau) = a_x(\tau) \cos\psi(\tau) - a_y \sin\psi(\tau) \quad (3)$$

and

$$a_Y(\tau) = a_x(\tau) \sin\psi(\tau) + a_y \cos\psi(\tau) \quad (4)$$

where $a_x(\tau)$ and $a_y(\tau)$ are obtained from Eq 1 and 2 and where $a_{x'}(\tau)$ and $a_{y'}(\tau)$ are the experimentally measured signals. Substituting Eqs 1 & 2 into Eqs 3 & 4, and integrating them to obtain velocity, one obtains Eqs 5 and 6 below. The process includes algebraic manipulation and integration by parts where the limits of integration are $\dot{\psi} = 0$ at $\tau = 0$ and $\dot{\psi} = \omega_{sep}$ at $\tau = \tau_1$ where ω_{sep} is the angular velocity at separation.

---

[2] Some collisions had a second contact duration. These are not considered here since the primary objective is data analysis and collision modeling, not reconstruction.

Table 1. Comparison of ΔV's Obtained by Integrating Acceleration Traces (local, moving vehicle coordinates)

RICSAC TEST & VEHICLE	Brach & Smith Digititized (Untransformed)		Jones & Baum Digititized (Untransformed)	
	ΔV_x mph (kph)	ΔV_y mph (kph)	ΔV_x mph (kph)	ΔV_y mph (kph)
1-V1	-10.3 (-16.6)	5.8 (9.3)	-10.6 (-17.1)	6.0 (9.7)
1-V2	-11.7 (-18.8)	-9.1 (-14.6)	-12.1 (-19.5)	-9.8 (-15.8)
3-V1	-9.4 (-15.1)	-0.1 (-0.2)	-9.5 (-15.3)	-0.4 (-0.6)
3-V2	15.0 (24.1)	-0.3 (-0.5)	15.8 (25.4)	-0.2 (-0.3)
4-V1	-18.6 (-29.9)	0.2 (0.3)	-18.7 (-30.1)	0.4 (0.6)
4-V2	21.7 (34.9)	2.5 (4.0)	22.2 (35.7)	2.8 (4.5)
5-V1	-15.7 (-25.3)	-0.3 (-0.5)	-16.3 (-26.2)	-0.2 (-.03)
5-V2	24.8 (39.9)	1.4 (2.3)	25.0 (40.2)	1.8 (2.9)
6-V1	-8.5 (-13.7)	2.9 (4.7)	-8.5 (-13.7)	3.0 (4.8)
6-V2	-11.4 (-18.3)	-3.2 (-5.2)	-11.5 (-18.5)	-3.2 (-5.2)
7-V1	-11.4 (-18.3)	3.0 (4.8)	-11.5 (-18.5)	3.5 (5.6)
7-V2	-14.0 (-22.5)	-8.4 (-13.5)	-14.1 (-22.7)	-8.5 (-13.7)
8-V1	-13.0 (-20.9)	9.1 (14.6)	-12.7 (-20.4)	8.6 (13.8)
8-V2	-6.9 (-11.1)	-8.0 (-12.9)	-7.2 (-11.6)	-8.0 (-12.9)
9-V1	-17.9 (-28.8)	11.5 (18.5)	-17.7 (-28.5)	12.0 (19.4)
9-V2	-5.3 (-8.5)	-7.8 (-12.6)	-5.0 (-8.0)	-7.4 (-11.9)
10-V1	-28.1 (-45.2)	21.4 (34.4)	-27.3 (-43.9)	22.0 (35.4)
10-V2	-9.1 (-14.6)	-11.3 (-18.2)	-8.8 (-14.2)	-11.0 (-17.7)
11-V1	-24.7 (-39.7)	0.1 (0.2)	-24.0 (-38.6)	0.8 (1.3)
11-V2	-16.1 (-25.9)	2.2 (3.5)	-15.6 (-25.1)	2.0 (3.2)
12-V1	-40.9 (-65.8)	2.6 (4.2)	-40.0 (-64.4)	2.2 (3.5)
12-V2	-24.7 (-39.7)	3.9 (6.3)	-26.0 (-41.8)	4.8 (7.7)

ms. For a 200 ms impact period this resulted in 40 sampling intervals and 41 base points for numerical integration which was done by a Simpson's Rule algorithm. (In two instances, vehicle 2 in RICSAC 7 & 8, a sampling interval of 2.5 ms was found necessary.) The sampling interval used by Jones and Baum is not indicated in their report. A comparison of results for velocity component change in local vehicle coordinates obtained with the sampling interval used herein and that used by Jones and Baum by integrating these acceleration traces is presented in the left-most columns of Table 1. These data are presented as comparisons of the integration process and do not include effects of rotation in transformations of Eqs 5 and 6. Although different sampling intervals were used by Jones and Baum than sampling intervals used herein, the comparison between velocity-change components is favorable. The differences are less than 5 percent except for one instance where the component was less than 5 mph and in that case the difference was 0.9 mph.

$$v_X(\tau) - v_X = \int_0^{\tau_1} a_{x'}(\tau)\cos\psi(\tau)d\tau +$$
$$- \int_0^{\tau_1} a_{y'}(\tau)\sin\psi(\tau)d\tau +$$
$$+ r_y \cos\psi_{sep}\omega_{sep} + r_x \sin\psi_{sep}\omega_{sep} \quad (5)$$

$$v_Y(\tau) - v_Y = \int_0^{\tau_1} a_{x'}(\tau)\sin\psi(\tau)d\tau +$$
$$+ \int_0^{\tau_1} a_{y'}(\tau)\cos\psi(\tau)d\tau +$$
$$+ r_y \sin\psi_{sep}\omega_{sep} - r_x \cos\psi_{sep}\omega_{sep} \quad (6)$$

These equations were evaluated numerically using digitized data obtained by sampling the analog traces for acceleration and heading angle presented in the data of Jones and Baum (1978). For acceleration data, the traces were sampled every 5

To complete the transformations of Eqs 5 and 6 it was necessary to have heading angle, $\psi(\tau)$, in addition to acceleration. The digitized data for heading angle differed from that for acceleration in that the analog data traces of heading angle extended over the entire time period of impact and subsequent vehicle motion to final rest. As a consequence the time period of impact was greatly compressed on the analog traces and only 10-18 samples of heading angle were obtainable during the impact period with satisfactory fidelity. To obtain digitized data at the interval rate consistent with the acceleration data cubic-spline interpolation

Table 2-a. Comparison Between ΔV's in Inertial Coordinates (U.S. units)
(Coordinate System based on Initial Orientation of Vehicle 1)

RICSAC TEST	Jones & Baum Untransformed			Brach & Smith Transformed by Eq 5 & 6			McHenry & McHenry Transformed			Separation Conditions	
	ΔV_x mph	ΔV_y mph	ΔV mph	ΔV_x mph	ΔV_y mph	ΔV mph	ΔV_x mph	ΔV_y mph	ΔV mph	Ψ_{sep} degrees	ω_{sep} deg/sec
1-V1	-10.6	6.0	12.2	-10.9	4.5	11.8	-11.3	4.8	12.3	15	90
1-V2	14.5	-5.6	15.5	13.6	-5.8	14.8	16.0	-5.3	16.9	0	0
3-V1	-9.5	-0.4	9.5	-9.4	-0.4	9.4	-9.5	-0.7	9.5	1	15
3-V2	15.6	2.5	15.8	14.9	2.2	15.4	15.6	2.5	15.8	0	0
4-V1	-18.7	0.4	18.7	-18.6	-0.3	18.6	-18.7	-0.3	18.7	1*	37
4-V2	21.5	6.2	22.4	20.9	5.4	21.6	22.2	0.6	22.2	3*	30
5-V1	-16.3	-0.2	16.3	-15.8	-0.3	15.8	-16.2	-0.1	16.2	5	12
5-V2	24.3	6.1	25.1	24.4	4.6	24.9	25.5	1.5	25.5	3*	70
6-V1	-8.5	3.0	9.0	-8.9	2.4	9.2	-8.8	2.4	9.1	5	30
6-V2	8.5	-8.4	12.0	14.0	-4.1	14.6	13.7	-4.2	14.3	20	180
7-V1	-11.5	3.5	12.0	-11.7	2.4	12.0	-11.7	2.5	12.0	12	30
7-V2	14.4	-8.0	16.5	19.5	-2.6	19.7	19.5	-3.5	19.8	22	192
8-V1	-12.7	8.6	15.3	-13.9	7.1	15.6	-13.4	6.6	14.9	15	114
8-V2	8.0	-7.2	10.8	8.0	-7.1	10.7	8.2	-7.4	11.0	0	18
9-V1	-17.7	12.0	21.4	-17.9	7.8	19.5	-18.0	8.5	19.9	27	180
9-V2	8.9	-0.6	8.9	6.9	-5.0	8.6	6.9	-4.4	8.2	-10	-45
10-V1	-27.3	22.0	35.1	-25.6	13.0	28.7	-28.8	18.1	34.0	55	300
10-V2	11.0	-8.8	14.1	9.8	-8.7	13.1	9.9	-7.7	12.5	12	-72
11-V1	-24.0	0.8	24.0	-25.0	0.4	25.0	-24.4	1.6	24.5	-5	-30
11-V2	15.7	0.5	15.7	16.1	0.7	16.1	15.7	0.5	15.7	0	0
12-V1	-40.0	2.2	40.1	-41.8	3.8	42.0	-40.8	-0.8	40.8	-10	-90
12-V2	26.4	-0.7	26.4	25.2	-0.7	25.2	26.7	-1.5	26.7	-2	-60

* data lost; value calculated by integrating angular velocity over period of impact

was used. In three instances, both vehicles from RICSAC 4 and vehicle 2 from RICSAC 5, analog traces for heading angle were lost due to instrumentation problems; however, angular velocity data was obtained. For these cases, the angular velocity data was integrated to obtain heading angle. In this manner digitized data for acceleration and heading angle could be used to evaluate the integrals in the transformation equations, Eqs 5 and 6. Final results are presented in Table 2-a and 2-b. Here the ΔV results are reported in an inertial reference frame based on the inital heading of vehicle 1.

Discussion of Transformed RICSAC data

Table 2 contains summarized ΔV data from all collisions, results of transformations as discussed above and comparisons with transformed results of McHenry and McHenry (1997), as well as the original RICSAC results from Jones and Baum (1978). **In this table the ΔV results for vehicles 1 and 2 have been expressed in inertial coordinates determined by the initial heading of vehicle 1.** Also presented in this table is the vehicle rotation and angular velocity at the moment of separation as reported by Jones and Baum. The latter data help to understand the occurrence of differences between Jones and Baum and the results transformed by Eqs 5 and 6. When rotation is present the ΔV values in inertial coordinates can differ significantly from their corresponding values in the untransformed moving coordinate system. This is notably so for vehicle 2 in RICSAC 6 and 7,

Table 2-b. Comparison Between ΔV's in Inertial Coordinates (SI.units)
(Coordinate System based on Initial Orientation of Vehicle 1)

RICSAC TEST	Jones & Baum Untransformed			Brach & Smith Transformed by Eq 5 & 6			McHenry & McHenry Transformed			Separation Conditions	
	ΔV_x kph	ΔV_y kph	ΔV kph	ΔV_x kph	ΔV_y kph	ΔV kph	ΔV_x kph	ΔV_y kph	ΔV kph	Ψ_{sep} deg	ω_{sep} deg/sec
1-V1	-17.1	9.7	19.6	-17.5	7.2	19.0	-18.2	7.7	19.8	15	90
1-V2	23.3	-9.0	24.9	21.9	-9.3	23.8	25.7	-8.5	27.2	0	0
3-V1	-15.3	-0.6	15.3	-15.3	-0.6	15.1	-15.3	-1.1	15.3	1	15
3-V2	25.1	4.0	25.4	24.0	3.5	24.8	25.1	3.5	25.4	0	0
4-V1	-30.1	0.6	30.1	-29.9	-0..5	29.9	-30.1	-0.5	30.1	1*	37
4-V2	34.6	10.0	36.0	33.6	8.7	34.8	35.7	1.0	35.7	3*	30
5-V1	-26.2	-0.3	26.2	-25.4	-0.5	25.4	-26.1	-0.2	26.1	5	12
5-V2	39.1	9.8	40.4	39.3	7.4	40.1	41.0	2.4	41.0	3*	70
6-V1	-13.7	4.8	14.5	-14.3	3.9	14.8	-14.2	3.9	14.6	5	30
6-V2	13.7	-13.5	19.3	22.5	-6.6	23.5	22.0	-6.8	23.0	20	180
7-V1	-18.5	5.6	19.3	-18.8	3.9	19.3	-18.8	4.0	19.3	12	30
7-V2	23.2	-12.9	26.6	31.4	-4.2	31.7	31.4	-5.6	31.9	22	192
8-V1	-20.4	13.8	24.6	-22.4	11.4	25.1	-21.6	10.6	24.0	15	114
8-V2	12.9	-11.6	17.4	12.9	-11.4	17.2	13.2	-11.9	17.7	0	18
9-V1	-28.5	19.3	34.4	-28.8	12.6	31.4	-29.0	13.7	32.0	27	180
9-V2	14.3	-1.0	14.3	11.1	-8.0	13.8	11.1	-7.1	13.2	-10	-45
10-V1	-43.9	35.4	56.5	-41.2	20.9	46.2	-46.3	29.1	54.7	55	300
10-V2	17.7	-14.2	22.7	15.8	-14.0	21.1	15.9	-12.4	20.1	12	-72
11-V1	-38.6	1.3	38.6	-40.2	0.6	40.2	-39.3	2.6	39.4	-5	-30
11-V2	25.3	0.8	25.3	25.9	1.1	25.9	25.3	0.8	25.3	0	0
12-V1	-64.4	3.5	64.5	-67.3	6.1	67.6	-65.7	-1.3	65.7	-10	-90
12-V2	42.5	-1.1	42.5	40.6	-1.1	40.6	43.0	-2.4	43.0	-2	-60

* data lost; value calculated by integrating angular velocity over period of impact

and vehicle 1 in RICSAC 9 and 10. Each of these cases involve side impacts and angular velocity at separation of 180 degrees per second or larger. Vehicles 2 in RICSAC 6 and 7 were struck in the side (60 degree orientation) and at a location that created forces passing well behind the center-of-mass of the vehicle. RICSAC 1 was also a 60-degree side impact; however, the location of the impact on the side of vehicle 2 was sufficiently far forward that the forces of impact did not create exceptionally large rotation. Vehicles 1 in RICSAC 9 and 10 were striking vehicles in a 90-degree side impact and the mass of the striking vehicle was approximately one half that of the struck vehicle. In both cases the forces acting on the front of the striking vehicle produced large moments about the center of mass and concomitant large rotational velocity. In contrast, RICSAC 8 was a similar 90-degree impact; however, there was no remarkable mass imbalance. All other cases were frontal or rear-end impacts with relatively small rotation or angular velocity.

The second comparison from Table 2 to be discussed is one between the transformed data from Eqs. 5 and 6 and that from McHenry and McHenry. They created a transformation of the data that required assumptions about the form of the angular acceleration and the velocity during the impact time period. As described above for Eqs. 5 and 6, it is possible to complete the transformation without such assumptions. The results of McHenry and McHenry agree closely with the transformed data of Eqs 5 and 6 with one significant exception.

Table 3. Comparison Between Transformed ΔV & ΔV Calculated by Planar Impact Equations
All results in inertial coordinates based on initial heading of Vehicle 1

RICSAC TEST & VEHICLE	Brach & Smith Transformed Eq 5 & 6 ΔV mph (kph)	Woolley(IMPAC) ΔV mph (kph)	Brach (PIM) ΔV mph (kph)	Separation Conditions Ψ_{sep} degrees	ω_{sep} deg/sec
1-V1	11.8 (19.0)	11.4 (18.3)	10.3 (16.6)	15	90
1-V2	14.8 (23.8)	17.0 (27.4)	15.4 (24.8)	0	0
3-V1	9.4 (15.1)	8.0 (12.9)	7.8 (12.6)	1	15
3-V2	15.4 (24.8)	12.7 (20.4)	12.4 (20.0)	0	0
4-V1	18.6 (29.9)	14.8 (23.8)	14.4 (23.2)	1*	37
4-V2	21.6 (34.8)	23.0 (37.0)	22.5 (36.2)	3*	30
5-V1	15.8 (25.4)	13.7 (22.0)	13.4 (21.6)	5	12
5-V2	24.9 (40.1)	24.9 (40.1)	24.4 (39.3)	3*	70
6-V1	9.2 (14.8)	11.2 (14.8)	10.3 (16.6)	5	30
6-V2	14.6 (23.5)	18.2 (29.3)	16.9 (27.2)	20	180
7-V1	12.0 (19.3)	15.0 (24.1)	15.1 (24.3)	12	30
7-V2	19.7 (31.7)	24.8 (39.9)	21.4 (34.4)	22	192
8-V1	15.6 (25.1)	11.3 (18.2)	11.7 (18.8)	15	114
8-V2	10.7 (17.2)	10.8 (17.4)	11.1 (17.9)	0	18
9-V1	19.5 (31.4)	14.6 (23.5)	15.3 (24.6)	27	180
9-V2	8.6 (13.8)	6.7 (10.8)	7.1 (11.4)	-10	-45
10-V1	28.7 (46.2)	31.1 (50.0)	24.7 (39.7)	55	300
10-V2	13.1 (21.1)	15.2 (24.5)	12.0 (19.3)	12	-72
11-V1	25.0 (40.2)	24.6 (39.6)	24.4 (39.3)	-5	-30
11-V2	16.1 (25.9)	15.5 (24.9)	15.3 (24.6)	0	0
12-V1	42.0 (67.6)	36.8 (59.2)	36.2 (58.3)	-10	-90
12-V2	25.2 (40.6)	25.6 (41.2)	25.1 (40.4)	-2	-60

* data lost; value calculated by integrating over angular velocity

Their transformation did not produce significant difference from Jones and Baum's untransformed results for vehicle 1 in RICSAC 10, a surprising outcome given the large rotation and angular velocity that this vehicle experienced.

The purpose of the RICSAC tests was to establish a data base of staged collisions data that was useful in evaluating computer models of impacts and accident reconstructions. Two examples of computer models are those of Brach (1983-a) and Woolley (1985). Comparisons are made in Table 3 for the ΔV predicted by these models and the transformed results obtained by Eqs. 5 and 6. There is no association or trend in these data with the occurrence of rotation and angular velocity as indicated by the separation conditions. There are differences between the models and the transformed RICSAC data; however, such differences are to be expected. The RICSAC data is subject to random experimental error (discussed more fully below) and the models themselves include simplifying assumptions.

Analysis of system linear momentum changes

All experimentally measured data, including those from the RICSAC tests, are subject to statistical variations or "measurement error". The RICSAC experiments were the first relatively comprehensive and complete set of realistic vehicle collisions and so significant experimental variations may exist. This section examines experimental variations. One of the intentions of the RICSAC experiments is to provide data for comparisons and evaluations of theoretical models of collision dynamics. One of the principles cited and often used in the analysis of vehicle collisions is that of conservation of linear (and angular) momentum; see for example, Brach (1991). In an ideal collision of vehicles with no significant external forces the total system momentum is conserved during the duration of contact. On the other hand, contact interaction between the vehicle and roadway is present in actual collisions and so external forces, though usually small, are not zero. The impulses of these forces exist and so the change of momentum

Table 4.
System Momentum Changes

RICSAC#	Linear Momentum			Angular Momentum (about Point C)	
	$\Delta q_t/q$	$\Delta q_n/q$	IMPAC*	Case I†	Case 2
1	-0.055	0.096	-0.055	-0.421	-0.521
6	-0.018	0.009	+0.036	-0.232	-0.175
7	0.023	0.024	-0.013	-0.238	-0.178
8	-0.135	0.124	+0.070	-0.460	-0.520
9	-0.080	0.025	+0.031	+0.122	-0.043
10	-0.088	0.038	+0.051	+0.035	-0.074
11	-0.073	-0.033	+0.125	-0.514	-0.533
12	0.234	-0.334	-0.175	-0.103	+1.005
3	-0.001	0.085	-0.018	-0.630	-0.366
4	-0.134	0.097	+0.113	-0.388	-0.348
5	-0.060	0.072	+0.065	-0.464	-0.380

* Woolley (1987)
† For Case I, impact center C is visually chosen; for Case 2 it is taken as the centroid of the CRASH3 residual crush region

never is exactly zero. Changes in linear momentum in the RICSAC tests now are analyzed.

Let q_i be the initial linear momentum vector at $\tau = 0$ of vehicle i and Q_i be the corresponding final linear momentum vector [3] at $\tau = \tau_f$. The initial momentum of the system of vehicles is $q = q_1 + q_2$ and the final system momentum is $Q = Q_1 + Q_2$. The change in momentum is $\Delta q = Q - q$. For comparative purposes, components, Δq_t and Δq_n of Δq along the direction, t, of the initial system momentum vector and a direction, n, perpendicular to the initial system momentum are computed for each test. These components then are normalized by the magnitude of the initial momentum to provide the nondimensional and normalized quantities, $\Delta q_t / |q|$ and $\Delta q_n / |q|$. By normalizing the initial system momentum, the experiments within each of the 4 collision configurations is placed on a common basis and permits a comparison of the experimental variations.

For an idealized collision of two bodies with no external forces such as road interaction aerodynamic forces, etc., momentum will be conserved exactly. In a real system such as the RICSAC tests, changes in momentum are expected. But what sort of changes are realistic? The results from the tests can help answer this question. It is helpful also to bring in concepts related to work and energy. The change in kinetic energy of the system is equal to the combined work of the intervehicular contact (crushing and sliding) forces and the work of the forces external to the vehicles. Unfortunately, the separate work of these forces cannot be deduced from the data.

Table 4 lists the components of the relative change of the system linear momentum, $\Delta q_t /|q|$ and $\Delta q_n /|q|$, expressed in inertial coordinates for the 11 tests. Figure 2 is a plot of the normalized momentum components. Two tests, numbers 7 and 12, show increases. One of these positive changes

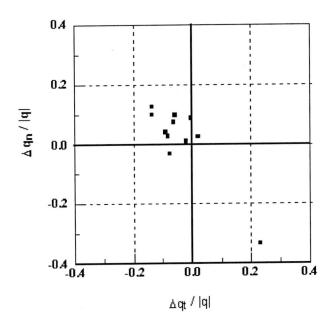

Figure 2. Changes in normalized system linear momentum, RICSAC test results.

---

[3] Quantities in bold type are two-dimensional vectors.

Table 5, PLANAR IMPACT MECHANICS SOLUTION, RICSAC 9
(common velocity conditions)

	Veh 1	Veh 2			
mass, m	70.1 lb-s^2/ft (1023 kg)	152.2 lb-s^2/ft (2221 kg)			
inertia, I	976 ft-lb-s^2 (1323 kg-m^2)	3953 ft-lb-s^2 (5360 kg-m^2)			
distance, d	4.8 ft (1.5 m)	5.6 ft (1.7 m)			
angle, ϕ	6°	-29.7°	coeff rest'n	tan'l impulse ratio	
angle, θ	0°	90°	e = 0	μ_0 = 100%	

initial velocity	Veh 1	Veh 2	final velocity	Veh 1	Veh 2
v_x	-31.09 ft/s (-9.5 m/s)	0	V_x	-12.1 ft/s (3.7 m/s)	-8.7 ft/s (2.7 m/s)
v_y	0	31.09 ft/s (9.5 m/s)	V_y	12.0 ft/s (3.7 m/s)	25.5 ft/s (7.8 m/s)
ω	0	0	Ω	-197.1 °/s (-3.4 rad/s)	59.9 °/s (1.1 rad/s)
v	31.09 ft/s (9.5 m/s)	31.09 ft/s (9.5 m/s)	V	17.1 ft/s (5.2 m/s)	27.0 ft/s (8.2 m/s)
v_n	-31.09 ft/s (-9.5 m/s)	0	V_n	-12.1 ft/s (3.7 m/s)	-8.7 ft/s (-2.7 m/s)
v_t	0	31.09 ft/s (9.5 m/s)	V_t	12.0 ft/s (3.7 m/s)	25.5 ft/s (7.8 m/s)
v_{cn}	-31.09 ft/s (-9.5 m/s)	0	V_{cn}	-13.8 ft/s -4.2 m/s)	-13.8 ft/s (4.2 m/s)
v_{ct}	0	31.09 ft/s (9.5 m/s)	V_{ct}	28.5 ft/s (8.7 m/s)	28.5 ft/s (8.7 m/s)
			ΔV	22.5 ft/s (6.8 m/s)	10.4 ft/s (3.2 ms/)

Impulses	P_x	P_y	System Kinetic Energy		
	1331 lb-s (5.92 kN-s)	843 lb-s (3.75 kN-s)	Initial	107436 ft-lb (145.7 kJ)	
	P_n	P_t	Final	73640 ft-lb (99.8 kJ)	
	1331 lb-s (5.92 kN-s)	843 lb-s (3.75 kN-s)	Loss	33796 ft-lb (45.8 kJ)	31.46%

is small but Test 12 stands out with an increase approximately 23%. Such a change is physically impossible yet there is no obvious indication from the data and information in the RICSAC test reports to indicate a reason. Crashes with decreases in momentum have reductions that range from near 0 % to -13.5%. Wooley (1987) also analyzes the change in momentum. His values also are presented in Table 4, showing differences with the values from this paper.

To examine a possibility that the momentum changes are affected by the collision configuration, an analysis of variance was carried out of the momentum change data in Table 4. At the 5% level of significance, the data indicate that the observed changes in momentum of the 4 impact configurations do not differ significantly. In other words, the changes are not systematic and associated with the collision configuration but rather are random in nature. Since the percentage change in system momentum of RICSAC 12 appears to be an outlier, the analysis of variance was repeated without RICSAC 11 and 12. The conclusion is unchanged since the variations within the categories of collisions remain relatively large compared to the variations between categories. One of the results of the analysis of variance is a measure of the level of the variance in the measurements of the system momentum, that is, the sum of squares of deviations. From the analysis omitting tests 11 and 12, the sum of squares is SS = 0.0626. So the data indicate that a standard deviation of relative system momentum change, $(Q - q)/q$, for these collisions of approximately $\sqrt{SS}$ = 0.25. There seems to be no pattern of the momentum changes related to the test configurations and so the variations from test to test are controlled by factors other than the type of collision or are due to random, experimental error.

Analysis of system angular momentum changes

Ideally, when bodies collide in the absence of external forces and moments, both linear and angular momentum are conserved. In reality the presence of external forces such as tire-ground frictional forces will cause impulse couples and cause changes to angular momentum. The change in system angular momentum for each of the RICSAC tests is examined. In order to compute and compare the change in angular momentum of the system of 2 vehicles from the beginning to end of contact, a common point must be chosen about which angular momentum is computed. Two different points are used in this analysis to allow a

Table 6. Comparison of Energy Loss in RICSAC Tests

Test No.	RICSAC Test Results			Planar Impact Mechanics		
	Energy Loss,%	ΔV_1 mph (kph)	ΔV_2 mph (kph)	Energy Loss, %	ΔV_1 mph (kph)	ΔV_2 mph (kph)
01	66%	11.8 (19.0)	14.8 (23.8)	52%	10.3 (16.6)	15.4 (24.8)
06	52%	9.2 (14.8)	14.6 (23.5)	48%	10.3 (16.6)	16.9 (27.2)
07	48%	12.0 (19.3)	19.7 (31.7)	49%	15.1 (24.3)	21.4 (34.4)
08	54%	15.6 (25.1)	10.7 (17.2)	36%	11.7 (18.8)	11.1 (17.9)
09	42%	19.5 (31.4)	8.6 (13.8)	32%	15.3 (24.6)	7.1 (11.4)
10	44%	28.7 (46.2)	13.1 (21.1)	34%	24.7 (39.7)	12.0 (19.3)
11	95%	25.0 (40.2)	16.1 (25.9)	91%	24.4 (39.3)	15.3 (24.6)
12	89%	42.0 (67.6)	25.2 (40.6)	93%	36.2 (58.3)	25.1 (40.4)
03	37%	9.4 (15.1)	15.4 (24.8)	36%	7.8 (12.6)	12.4 (20.0)
04	53%	18.6 (29.9)	21.6 (34.8)	36%	14.4 (23.2)	22.5 (36.2)
05	42%	15.8 (25.4)	24.9 (40.1)	32%	13.4 (21.6)	24.4 (39.3)

done in Brach (1987) based on a visual appraisal of the crush surfaces. Another approach is to use the centroid of the damage profile corresponding to the CRASH3 protocol. The initial and final moments of momentum were computed for each test about these two points and the fraction changes were calculated relative to the initial values. The values of the changes in system angular momentum are given in Table 4. An analysis of variance was run for each of the 2 cases (choice of the impact centers) to determine if a significant relationship exists between the fraction angular momentum change and the RICSAC collision configuration. In both cases the relationship is statistically insignificant. That is, the change in angular momentum is not significantly and systematically related to the type of collision configuration. It is worth noting that Collision 12 appears to show anomalous results with a 100 % gain in angular momentum when using Point C at the crush centroid.

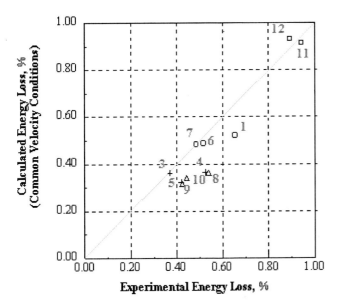

Figure 3. Comparison of experimental energy loss and energy loss computed by planar impact mechanics

comparison. In each collision analysis and for each vehicle, a distance d and angle ϕ are chosen from the center of gravity to a common point of application of the intervehicular impulse. This was

IMPACT PHASE OF RICSAC CRASHES

Planar Impact Mechanics and Common Velocity Conditions

The equations of impulse and momentum have been derived and presented for planar collisions and applied to vehicle collisions. Some examples are Brach (1983, 1991), Woolley (1987), Ishikawa (1994) and Stefan (1996). Such algebraic solutions[4] do not account for external impulses and thus satisfy the conditions of conservation of both linear and angular momentum. The equations are algebraic

(Brach, 1991) with 8 unknowns (six final velocity components for the 2 vehicles and 2 impulse components) and 8 equations. Six of the 8 equations come from Newton's laws of impulse and momentum. Two of the 8 equations arise from definitions of coefficients, a coefficient of restitution, e, and a tangential coefficient, μ. When properly formulated, these equations can be solved in closed form to provide a general solution for the planar impact problem. Table 5 is an example solution for RICSAC collision 9 under common velocity conditions $e = 0$ and $\mu = \mu_0$ (μ_0 is the ratio of the tangential impulse to the normal impulse that causes relative tangential motion to end before separation). A question that arises when using planar impact mechanics is: what values of coefficients are appropriate? A general review of the question of coefficient of restitution is provided by Monson and Germane (1999). The RICSAC collisions were used (Brach 1983b, 1991) with the method of least squares to fit the theoretical solutions to the experimental data and obtain

Table 7. Comparison of angular velocity at end of impact (deg/sec)

	$\dot{\psi}_1$		$\dot{\psi}_2$	
	Jones & Baum	PIM*	Jones & Baum	PIM*
60-deg Side Impact				
1	90	141	0	-28
6	30	129	180	88
7	30	178	192	165
90-deg Side Impact				
8	114	121	18	5
9	180	197	-45	-60
10	300	318	-77	-80
Offset-Frontal Impact				
11	-30	-77	0	6
12	-90	-121	-60	14
Rear-End Impact				
3	15	15	0	37
4	37	36	30	62
5	12	35	70	104

*Planar Impact Mechanics (Brach, 1991)

coefficient values. Specifically, the values of the coefficients e and μ were found that make the sum of squares of differences between the final velocities of the planar impacts mechanics solution and the experimental final velocities a minimum. Still retaining the method of least squares, a different criterion has been used for this study, however. Planar impact mechanics solutions again were fit using the method of least squares to each of the RICSAC test crashes. Rather than matching final velocities, solutions were sought that minimized a weighted combination of ΔV and energy loss differences. Details are not presented here, but in all cases the best fit was for the common velocity conditions. Consequently these conditions are used in the following for all comparisons.

Comparison of experimental results with solutions using Planar Impact Mechanics

Solutions for all of the RICSAC collisions have been obtained (such as the example in Table 5) and comparisons can be made. Table 6 contains ΔV values and energy loss values in such a comparison. Figures 3 and 4 show the data plotted for visual comparisons. Table 6 also contains values obtained by McHenry and McHenry (1997). Figure 6 shows that the theoretically computed energy loss values generally are lower than the experimental RICSAC values. This seems to make sense since the planar impact solutions account only for the crush and sliding energy losses at the contact interface and do not include energy dissipated during the contact duration by vehicle-ground interaction. Figure 4 shows graphically that the computed and experimental ΔV values conform reasonably well but with scatter

---

[4] A rigorous solution of the planar impact collision problem that takes into account external frictional impulses is not possible using the rigid body algebraic approach because the direction of sliding along the ground determines the external frictional impulse yet depends on the unknown, final velocity components.

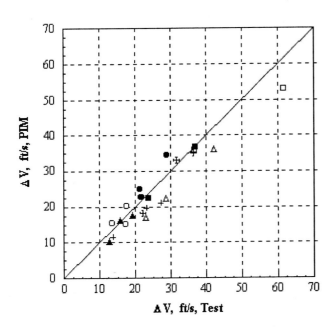

Figure 4. Comparison of experimental velocity changes, ΔV, and changes calculated by planar impact mechanics.

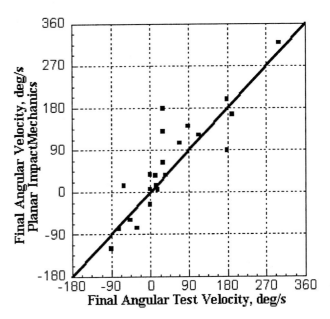

Figure 5. Comparison of experimental angular velocity changes and changes predicted by planar impact mechanics.

both above and below the 45, equality line.

The angular velocities of the vehicles in each impact are listed in Table 7, and plotted in Figure 5. Experimental values from the RICSAC tests and those calculated using planar impact mechanics are included. The comparison is favorable in most cases, although the 60-degree oblique side impact cases exhibit notable differences for the striking vehicle. In these cases the angular velocity of the striking vehicle was much larger for the theoretical PIM results than was observed in the experimental RICSAC results. This circumstance may be a consequence of the difficulty in estimating the effective point-of-impulse in the theoretical model. It also may be the effect of wheel-ground contact impulses at large distances from the centers of gravity. The standard deviation of the difference between the experimentally measured final angular velocities and corresponding PIM values is 51 deg/sec. This provides an estimate of the uncertainty when using planar impact mechanics for accident reconstruction.

CONCLUSIONS

1. The systematic error in the ΔV's derived from RICSAC test data over the time interval of contact and associated with the rotation of the transducer coordinates and the location of the accelerometers away from the vehicle center-of-mass has been eliminated using the equations of mechanics and the raw RICSAC data. No assumptions are needed or were used as to shape or form of selected kinematic responses such as angular velocity and angular acceleration versus time.

2. When corrected for this systematic error, RICSAC data in some instances does exhibit a significant increase in system momentum over the time interval of the impact. Analysis of variance indicates that this increase is statistically significant, is not related to the collision configuration and is within bounds of random experimental error. The instances of system momentum increase appear to be attributable to random experimental error that is inevitable in any experiment. This random error is not a cause for invalidating RICSAC data.

3. RICSAC data when corrected for any systematic error is useful and valid for its intended purpose of comparing experimental data with predictions from accident reconstruction models. Tests with increases of system momentum should be noted in any comparisons.

4. System kinetic energy changes have been calculated with the corrected RICSAC data necessarily stated in an inertial coordinate system. The system kinetic energy changes manifested in the RICSAC data typically exceed those predicted using planar impact models, an outcome that is likely due to the experimental error and a consequence of the simplifying assumptions in those models.

RECOMMENDATIONS

It is recommended that staged collisions involving vehicle yaw give attention to recording yaw and yaw-rate over the time interval of the impact impulse. Yaw angle data is necessary for correcting accelerometer data when it is located at the vehicle center-of-mass, and both yaw and yaw-rate data are necessary to correct acceleration data obtain at locations other than the center-of-mass.

REFERENCES

Brach, R. M., 1983-a, "Analysis of Planar Vehicle Collisions using Equations of Impulse and Momentum", *J Accd Anal & Prev*, Vol 15, No 2.

Brach, R. M., 1983-b, "Identification of Vehicle and Collision Impact Parameters from Crash Tests", Paper 83-DET-13, ASME Design Technical Conference, Dearborn, MI, 1983.

Brach, R. M., 1987, "Energy Loss in Vehicle Collisions", Paper 871993, SAE, Warrendale, PA, 15096.

Brach, R M., 1991, *Mechanical Impact Dynamics*, John Wiley, New York

Bundorf, R. T., 1996, "Analysis and Calculation of Delta-V from Crash Test data", Paper 960899, SAE, Warrendale, PA, 15096.

Cliff, W. E. and D.T. Montgomery, 1996, "Validation of PC-Crash - A Momentum-Based Accident Reconstruction Program", Paper 960885, SAE, Warrendale, PA, 15096.

Ishikawa, H., 1994, "Impact Center and Restitution Coefficients for Accident Reconstruction", Paper 940564, SAE, Warrendale, PA, 15096.

Jones, Ian S. and A. S. Baum, 1978, "Research Input for Computer Simulation of Automobile Collisions", Vol. IV: *Staged Collision Reconstructions*. DOT HS 805 040, December.

McHenry, B. G. and R. R. McHenry, 1997, "RICSAC-97 - A Revaluation of the Reference Set of Full Scale Crash Tests", Paper 970961, SAE, Warrendale, PA, 15096.

Monson, K. L. and G. J. Germane, 1999, "Determination and Mechanisms of Motor Vehicle Structural Restitution from Crash Test Data", Paper 1999-01-0097, SAE, Warrendale, PA 15096.

Steffan, H. and A. Moser, 1996, The Collision and Trajectory Models of PC-CRASH", Paper 960886, SAE, Warrendale, PA, 15096.

Woolley, R. L., 1987, The "IMPAC" Program for Collision Analysis", Paper 870046, SAE, Warrendale, PA, 15096.